Building with Concrete, Brick and Stone

Fine Homebuilding®

BUILDER'S LIBRARY

Building with Concrete, Brick and Stone

The Taunton Press

Cover photo by Lefty Kreh

First printing: December 1988
International Standard Book Number: 0-942391-13-6
Library of Congress Catalog Card Number: 88-50564
Printed in the United States of America

A FINE HOMEBUILDING Book

FINE HOMEBUILDING is a trademark of The Taunton Press, Inc.,
registered in the U.S. Patent and Trademark Office.

The Taunton Press, Inc.
63 South Main Street
Box 355
Newtown, Connecticut 06470

CONTENTS

INTRODUCTION

Building with concrete, brick and stone is specialized work that's generally left to tradesmen. But given the right instruction, considerable patience and a willingness to do repetitive work, non-professionals can achieve surprisingly good results. For this book we've selected 43 articles about masonry building techniques from the back issues of FINE HOMEBUILDING magazine.* Many of the articles cover the fundamentals of building foundations. Others are in-depth treatments of tightly focused subjects, such as building fireplaces and chimneys, veneering a block wall with stone and laying a brick floor.

Whether you want to tackle a masonry project yourself, or contract with a tradesman to do the work, you can benefit from the information in this book.

*The six volumes in the Builder's Library are from FINE HOMEBUILDING magazine numbers 1 through 46, 1981 through early 1988. A footnote with each article tells when it was originally published. Product availability, suppliers' addresses, and prices may have changed since then.

The other five titles in the Builder's Library are *Tools for Building; Frame Carpentry; Building Floors, Walls and Stairs; Building Doors, Windows and Skylights; Building Baths and Kitchens.*

Concrete

Understanding the characteristics of this material can take some of the anxious moments out of your pour and ensure a finished product of high quality

by Trey Loy

Concrete is a remarkable material that can be cast into almost any shape. It will sustain and transmit tremendous loads, and once hardened, it is practically indestructible. Yet few builders feel as affectionate about concrete as did Slim Gaillard and Lee Ricks in their scat tune from the 1940s.

> *Cement mixer, put-ti, put-ti,*
> *Cement mixer, put-ti, put-ti,*
> *Puddle-o-votty, puddle-o-goody,*
> *Puddle-o-scooty, puddle-o-vett.*
> *Who wants a bucket of cement?*
>
> *First you get some gravel,*
> *Pour it in the vout.*
> *To mix a mess of mortar*
> *You add cement and grout.*
> *See the mellow rooney come out.*
>
> *Slurp, slurp, slurp.*
>
> *Cement mixer, put-ti, put-ti,*
> *Cement mixer, put-ti, put-ti,*
> *I can never get enough*
> *of that wonderful stuff.*

Cement mixers have been largely replaced by huge batch plants that measure out hundreds of cubic yards of ready-mix a day to waiting transit mixers. The leisurely pace of pouring concrete is also a thing of the past. The distant rumblings of an approaching concrete truck can strike fear into the heart of a carpenter still bracing the forms. At nearly two tons a cubic yard, concrete has to be poured immediately, with no time to ponder problems or locate tools. For many people, a pour is considered successful when forms don't break; and a sigh of relief can be heard when a slab is smooth and unblemished by cracks the next day.

Contrary to its reputation, concrete reacts predictably, and the builder can regulate many of the variables that affect its working properties while plastic, and its strength when hardened. Some practical knowledge of the kinds of cements and aggregates and the correct proportions of each, admixtures, slump, and the effects of weather during pouring and curing can make the difference between feeling confident about a pour, and feeling you are constantly dodging disaster.

Concrete is a mixture of water, portland cement, and fine and coarse aggregates. The active ingredients are water and cement. They combine in a chemical reaction called hydra-

The advent of ready-mix has brought a change to the quality, quantity and pace of concrete pours. The physical properties of concrete in its plastic and hardened states are well understood by researchers, engineers and batch-plant operators, but this information seldom trickles down to the builder who is actually working with the material. Knowing how mixes are designed, and the different kinds of cements, aggregates and admixtures that are used to alter the concrete can give the builder the ability to predict how it is going to react and why. For more on hand- and machine-mixing concrete on site, see pp. 14-15.

tion. Although concrete that is beginning to set appears to be drying out, or dehydrating, about half the water is actually incorporated in the hydration process and becomes a permanent part of the bonding paste. This is why concrete needs to be kept moist during the first few days of curing. In fact, concrete will harden quite effectively under water.

Aggregates—The major function of aggregates is to make concrete more economical. While the cement paste binds the aggregates together in a solid mass, the aggregates keep the concrete from shrinking and cracking as hydration and evaporation take place during setting and hardening. Neat, or pure, cement is not nearly as strong as concrete correctly proportioned with aggregate.

Fine aggregate can be either sand or rock screenings. Fine-aggregate particles range in size from very fine sand to ¼ in. Coarse aggregate is either gravel or crushed stone ¼ in. to 1½ in. in diameter. The aggregate mix should be proportioned so that the smaller particles fill in the voids between the larger ones. For thick foundations and footings, gravel or rock with a diameter of 1½ inches is used. For ordinary walls, the largest pieces should be not more than one-fifth the thickness of the finished wall section. And for slabs, the maximum thickness of the rock aggregate should not be more than one-third that of the slab.

The amounts and sizes of the sand and gravel are also adjusted for the strength and workability of the mix. The plasticity of concrete in its wet state, as well as its ultimate density, depend in part on the aggregates meshing. How well the mud moves down the chute, how easily it fills the forms, and how well it finishes are measures of its workability.

Cement—An English stonemason, Joseph Aspdin, patented the process for manufacturing portland cement in 1824. He used this name because it produced a hardened concrete with a color that reminded him of the natural grey stone on the Isle of Portland. Portland cement was first produced in North America in 1872. Interestingly, the use of steel reinforcing bar was introduced a few years later, around 1880. And the first patent for prestressed concrete using steel-wire rope was issued in 1888.

Today, portland cement is produced in huge rotary kilns at temperatures of nearly

2,700°F. At this temperature, lime, silica, alumina and iron oxides, which are derived from limestone, oyster shells, marl, shale, iron ore and clay, undergo a kind of molecular reformation called calcination. After cooling, the resulting greenish-black clinkers are pulverized with small amounts of gypsum to control the set time of the cement.

The American Society for Testing and Materials (ASTM) recognizes five types of portland cement. Each is intended for a specific purpose, although they all achieve about the same strength after curing for three months.

Type I. This is the most common type of general-purpose cement, and is used when a specific type isn't called out. Most residential construction uses Type I.

Type II. A moderately sulfate-resistant cement, it is sometimes specified for walkways where de-icing chemicals will be heavily used. It sets more slowly than Type I, an advantage during the summer, when getting a finish on concrete can be a real race. Also, because it generates less heat during curing than Type I, it is better suited for mass pours, where heat radiating from hundreds of yards of curing concrete can cause problems.

Type III. This type is called high-early-strength cement because it achieves most of its strength within the first week of curing. This is useful if the concrete has to be put under full load within a few weeks of pouring, or if forms have to be stripped early. It is not widely stocked by concrete companies, but adding an extra bag of Type I or Type II cement per cubic yard of concrete and mixing at high speeds will produce similar results.

Type IV. A slow-curing variety that generates very little heat by hydration, it is used exclusively in mass concrete, such as dams.

Type V. This cement will withstand severe sulfate action that occurs in heavily alkaline soil or groundwater. Concrete can deteriorate because of physical and chemical reactions between sulfates and compounds formed by hydrated portland cement. Type V gains strength much more slowly than Type I.

The Canadian Standards Association (CSA) has three categories—normal, high-early-strength, and sulfate-resisting. These correspond to ASTM Types I, III, and V.

There are a number of ways that these cement types can be altered to meet special conditions. Portland cement is normally grey. White portland cement, light in color because it's made with a minimum of iron and magnesium oxides, can be tinted by adding pigments or used as is. Blast-furnace slag and pozzolan are two materials that are ground up and blended with portland cement to bring down its cost. When pozzolan is added to Type I it is designated IP; for slag, the abbreviation is IS. Another application of portland cement is

Jitterbugging a slab settles the large aggregate just below the surface to allow a smooth, troweled finish. Most problems in residential concrete are surface faults—crazing, dusting, scaling, honeycombing and shrinkage cracking—rather than strength failures.

From *Fine Homebuilding* magazine (February 1983) 13:28-33

lightweight concrete, which uses artificial aggregates and gas-forming admixtures to make it lighter. Types I, II, and III are available with air-entrainers (discussed under admixtures) interground, and designated with an A after the type number.

Portland cement is usually packaged in paper bags. Each bag weighs 94 lb. and contains 1 cu. ft. of cement. A common unit of measure in the past was the barrel, which contained the equivalent of four bags.

The cement content of a mix has a lot to do with its strength. One method of ordering ready-mix concrete is to specify how much cement should be used for each cubic yard of concrete. Producers of ready-mix prefer that you give cement content by weight (such as 470 lb. per cu. yd.), but ordering a certain number of bags, or sacks, of cement per cubic yard (such as a five-bag mix) is still very common. A four-bag mix is the minimum for most residential uses; five-bag is better. You should ask for six-bag or seven-bag if you use smaller aggregate or if you want greater strength, waterproofing and durability.

Water—For mixing concrete, it's best to use water that is fit to drink. It should not contain any oil, alkali or acid. In hydration, the water and cement in concrete combine chemically to form a paste that binds the aggregates together. It can be thought of as a glue.

The more water added to a given amount of cement, the weaker the concrete will ultimately be. This relationship of water and cement is known to concrete engineers as Abrams' law, or the water/cement ratio (W/C). It is a central factor in the design of the mix. The W/C ratio needs to be adjusted for a large number of variables, including the quantity of water that the aggregates are carrying, and the ambient temperature and humidity at the time of the pour.

The strength of concrete decreases as the W/C ratio increases, as seen in the chart above. The first column expresses the ratio in

weight (pounds of water divided by pounds of cement). The second column gives the same ratio in gallons of water per bag of cement.

W/C (weight)	Gallons per bag	Approx. 28-day strength (psi)
.45	5.0	5,000
.49	5.5	4,500
.53	6.0	4,000
.57	6.5	3,500
.62	7.0	3,000

Strength and quality—The strength of a given sample of concrete is measured by how much compression-loading a test cylinder of concrete 6 in. in diameter by 12 in. high can take before fracturing. The testing is done in a laboratory using a hydraulic piston hooked up to a meter that measures pounds per square inch (psi). Compression-strength figures for concrete refer to tests conducted after 28 days of curing unless otherwise noted. These figures indicate how well a batch of concrete will stand up to vertical and lateral loading. It is also a measure of durability and watertightness. Most engineers require that any load-bearing concrete achieve a minimum 28-day strength of 2,500 psi.

Although most concrete is batched to exact engineering standards at the plant, a lot can happen to affect its quality before, during and after the pour. How much water is added to the concrete after it's initially mixed is a good example of this.

Some drivers will ask if you want the mix stiff or sloppy, and then judge how much water to add, according to the slope (if any) involved in the pour, the angle of the chute from the truck to the forms, the depth of the forms, the amount of rebar, how long the pour is taking, and the weather (hot, dry days call for more water). A thin, watery mix is easier to handle, but much of the water added to the concrete at the job site may not have been figured in the mix design, and the resulting concrete will be less durable and much more subject to cracking.

Slump—To regulate the amount of water in ready-mix, specify slump. If you are working from a set of engineered plans, this may already be listed in the specs. Slump is a measure of the consistency of concrete—the higher the slump, the wetter the mixture. Slump is measured with a 12-in. high truncated metal cone with a base diameter of 8 in. and a top diameter of 4 in. The cone is filled with concrete right off the truck and rodded with a tamping rod. Then the cone is lifted free, inverted and placed beside the sagging pile of concrete for comparison. The distance the

concrete subsides from the top of the cone, measured to the quarter inch, is the slump.

Roadways, industrial floors and any concrete that is consolidated with mechanical vibration requires a 1-in. to 3-in. slump. Most foundations, slabs and walls consolidated by hand methods such as spading can have a slump between 4 in. and 6 in. On most residential pours, taking the time for a slump test isn't feasible. I usually just eyeball the mix. A good 3-in. to 4-in. slump mix will stand in a pile. A 5-in. to 6-in. slump sags into a blob, and a 7-in. to 8-in. slump just flattens out.

Many problems in residential concrete are the result of high slump. Unlike public projects such as bridges and highways, where compressive strength is essential, residential pours seldom suffer failures under a load. Instead, it is surface faults—honeycombing, crazing, dusting, surface cracking and scaling—that cause the problems.

One serious failure of concrete associated with high slump is segregation, which is a re-separation of concrete back into sand and gravel. Bleeding is another kind of separation that can be serious if it occurs on a large scale. Bleeding is the emergence of water on the surface of newly poured concrete. This occurs when the large aggregate settles within the mass, displacing the water in the mix. Heavy bleeding greatly dilutes the cement particles on the surface of the concrete, making it susceptible to abrasion. When this bleed water is troweled into the surface of a slab during finishing, it can result in crazing lines, shallow parallel fissures called plastic shrinkage cracking, a powdery dusting on the surface, and even scaling, which is the flaking of the finished concrete. If you keep slump low, and finish and cure the resulting concrete with care, you can avoid these problems.

The mix—Although all concrete consists of the same basic ingredients, how they are proportioned can make a huge difference in strength and workability. Because of all the variables involved, there are hundreds of possible combinations that will produce concrete with a wide variety of characteristics. As explained on p. 13, it's very important to mention the conditions and requirements of a pour when ordering concrete, so that a mix can be designed or chosen from standard designs that will give you the kind of concrete you need.

The design of a mix is complex because of the interrelationships of the materials. For example, if the size of coarse aggregate is limited in order to pour a thin slab, this will affect the amount of cement needed to reach the necessary compressive strength. The amount of cement used in turn affects the amount of water to be added, as well as the

Mechanical vibrators should be used with caution, and low-slump concrete. Prolonged vibrating can cause the concrete to segregate, bringing the fines—cement and water—to the top, leaving the heavier aggregates below. This can lead to surface failures on a slab.

Photo: Bob Syvanen

size and proportion of fine aggregate. The mix will also have to be adjusted for slump, workability, admixtures and the weather.

Admixtures—There are four kinds of admixtures that can give concrete specific qualities. The first, called air-entraining agents, are known to most builders who work in areas with hard freezes. This admixture is a material that stabilizes bubbles formed by air incorporated in the concrete during the mixing process. The bubbles create tiny voids that act as expansion chambers or shock absorbers, which allow the concrete to withstand freeze-thaw cycles. The amount of air in the mix is a variable of mix design. It is typically 5% to 7% by volume.

Although these air bubbles make the mix slightly weaker, they also have beneficial effects. Air-entrainers increase the workability of the mud (so less water can be used for a given slump), make the mud more resistant to salts, and produce a more durable concrete.

A set-retarder may be added to ready-mix to prolong its setting time by 30% to 60%. If you need extra working time on hot days, tell the dispatcher when you order, and the plant engineer will determine the exact amount according to the weather and the mix.

Concrete companies are also likely to add a water-reducing agent, sometimes called a plasticizer, which may allow as much as a 15% reduction in water content for a given slump. Water-reducing agents can help minimize problems relating to an excess of water, such as segregation, plastic shrinkage cracking, crazing and dusting. It can also increase the concrete's strength and its bond to steel reinforcing rod.

Probably the best-known admixture is calcium chloride. This chemical is an accelerator, used to get an early set in freezing weather. Ideally, concrete should be poured when it's at least 50°F, with the mud maintained in the forms at 70°F. But pours in cold weather are often necessary, and quite common. If concrete freezes while it's setting or during the first few days of curing, it won't gain much strength and problems will develop. Pop-out, scaling and cracking occur when water in the concrete freezes and expands nearly 9% of its liquid volume.

Contrary to myth, accelerators aren't effective as antifreeze. Concrete, even with calcium chloride added, can freeze. Like any accelerator, it will only decrease the setting time. This allows the builder to pour, finish and insulate the concrete before the onset of freezing temperatures. Calcium chloride should be used sparingly and not just for convenience when better scheduling would solve the problem. It attacks aluminum conduit, lowers the resistance of the concrete to sulfates, increases shrinkage, and generally weakens the mix. If you need to use it to beat the weather, limit it to 2% by weight of cement. An effective alternative is to add an extra bag of cement to each cubic yard of concrete.

Working concrete—If you pour a lot of concrete, buy a pair of rubber boots. Leather work boots will rot off your feet after a few dunkings in concrete. The only way I've found to restore the flexibility of the leather is to remove all of the concrete and soak the boots in motor oil for a day—crude but effective.

Gloves are another must. The cement contains lime, an alkali that dries out skin and leaves it cracked and sore. Thick rubber gloves offer good protection, but they are awkward to wear. Lately I've been wearing doctor's disposable examination gloves. They cost less than $10 for fifty pairs, and fit like another layer of skin.

It's a good idea to keep a bottle of vinegar with your concrete finishing tools. Vinegar contains acetic acid, which neutralizes the lime. Wash your hands in it when you quit for the day, and don't rinse it off immediately. You'll smell like a salad bar, but your hands won't be any the worse for wear the next day.

There are several general procedures that should be followed to end up with strong concrete that looks good after the forms are stripped. Concrete should be poured in horizontal layers of 6 in. to 12 in., depending on the stiffness of the mix. Start in the corners of the forms and work toward the middle. Concrete should not be dumped into separate piles and then leveled and worked together. Pouring it near its final place will save your back and keep the mud from separating.

If you are ordering ready-mix concrete,

check the approaches that the transit mixer can make, and calculate how many chute sections you will need. Most trucks carry 10 ft. of chute. You can usually arrange with the dispatcher to have the driver bring another 10 ft. If this doesn't do it, you will have to hire a pumper, build a chute, or truck the wet mud around the site in wheelbarrows. These alternatives are listed in order of preference. Although everybody has had to use a wheelbarrow to complete a pour on occasion, it is slow, risky work.

Most concrete companies will allow between 30 minutes and an hour to empty a full truck (about 8 cu. yd.), before they charge overtime. Money isn't the only issue. Time is also a critical factor. Depending on the air temperature, the mix can agitate in the drum of the truck for up to an hour after batching. After that, more water has to be added, which will weaken the mix. Under average conditions, concrete that is left in a truck for longer than 90 minutes is considered unusable.

Pumpers should be considered for a job where ready-mix is used and some part of the pour is inaccessible by ordinary means. Hillsides, high walls, muddy ground where a fully loaded truck could become stuck, and sites with dense trees or landscaping are all good candidates. In addition to the huge pump trucks with articulating booms or snorkels used for big commercial jobs, there are smaller portable pumping machines and trucks that are suitable for residential jobs. In most areas, they can be hired with an operator for $100 to $200 for an average pour.

Using a pumper can cut down the number of people needed to make the pour, and still give better results. Small pumpers use a 3-in. or 4-in. diameter hose. This requires using pea gravel as large aggregate and adding an extra bag of cement for each yard. This mix is rich and easy to work. If you need to build a chute for the site, use at least ¾-in. plywood and lots of bracing. Pitch it at about a 5-in-12.

Once the mud is in the forms, it needs to be tamped or vibrated to eliminate voids around rebar or against the form faces. Spading the sides of the form just after the mix is placed will minimize honeycombing and sand streaking, and keep aggregates from showing on the surface of a poured wall. You can buy special spading tools—thin, flat pieces of metal mounted on long handles—but 1x4s or long pieces of rebar work fine. Another good technique to eliminate honeycombing is to rap the forms sharply with your hammer, moving up and down the forms. This brings the cement paste out to the surface of the wall to cover the aggregate.

The best tool for settling the concrete into forms is a vibrator, a portable motor with a long, flexible, waterproof shaft. You can rent one for less than $20 a day. Use a vibrator as you are pouring on each level, particularly at the face of walls, corners and around rebar. Plunge the shaft into the mud every few feet along the length of the form, and let it vibrate for 5 to 15 seconds. Prolonged vibrating is not good because it can cause segregation. In fact, a vibrator shouldn't be used on any mix that can be placed and consolidated readily by hand tools.

With a slab, it is common practice to tamp the wet concrete with a Jitterbug®—a tubular-steel frame with a mesh bottom and waist-high handles—to settle the aggregate below the surface (photo, previous page). This aids in floating and finish troweling, and brings excess water to the surface to evaporate.

Curing—Proper curing is essential in achieving high-strength concrete and durable surfaces. As long as it is kept moist and warm—above 80% relative humidity and 70°F is ideal—concrete will harden indefinitely at a diminishing rate, as shown in the chart below. If the humidity drops below 80%, the surface of the concrete begins to lose moisture more rapidly than the interior of the pour. This causes the surface to shrink, and results in a soft, dusty skin that is less resistant to abra-

sion. Surface hairline fissures (crazing) are sure signs that concrete dried out too much during its initial curing. Plastic shrinkage cracks, which seldom have any structural significance but mar the appearance of a finished slab, are also caused by allowing the surface of the concrete to dry out. To get the best cure, the surfaces of the concrete should be kept moist for at least three days. After seven days it will have attained about 60% of its eventual 28-day strength.

Concrete can be kept wet by spraying it lightly with water several times a day; or, in the case of a slab, by maintaining a pond of water on the finished surface. How much moisture it will need depends upon air temperature and humidity. It's impossible to keep concrete too wet during curing. Spreading burlap or straw on the surface and soaking it with water will help hold moisture, as will covering the surface with plastic sheeting.

Membrane curing compounds, which can be sprayed on the surface, are also available. Clear compounds are preferred for surfaces that will be exposed. Black curing compounds have an asphaltic base and are used when staining isn't important. The black compounds will also hold in heat as the concrete cures. White compounds are effective in hot weather to reflect heat from the sun.

In cold weather, concrete slabs should be protected against freezing with straw covered with plastic sheeting or insulating blankets. If temperatures are in the 40°s, maintain this insulation for at least 48 hours. Insulated forms used for columns and walls should remain in place for at least a week. If temperatures are lower, keep the covering on even longer. Since hydration is an exothermic reaction, the primary concern should be holding this heat in. Only if the air temperature drops below 0°F should you supply heat.   □

Trey Loy is a designer and builder who lives in Little River, Calif. He spent five years pouring concrete professionally.

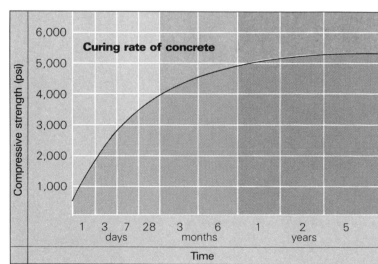

Curing. Concrete acquires most of its strength in the first month after the pour. This graph plots the approximate compressive strength of a six-bag mix using five gallons of water per bag over time, assuming the concrete is maintained at about 70°F.

Photo: Ross Lowell

Figuring and ordering ready-mix concrete

Gulping is the initial reflex for most builders, seasoned or not, when the dispatcher asks the inevitable question, "How many yards you want?" Like it or not, you've got to give an exact figure in cubic yards, no matter how complicated or irregular the pour is. Ordering too much means the driver of the transit mixer will need a place on your job site to dump the excess, which, like the mud inside your forms, will cost you nearly $50 a cubic yard. Ordering too little is even worse. Unless the plant can send another truck with a short load right away (this costs extra), you'll have to cap off your forms, add a keyway, and create a construction joint in a pour that was designed to be monolithic.

The way to prevent these problems is to figure your needs precisely, and then cheat the moment of truth as much as you can. If you have to reserve the concrete several days ahead, give the dispatcher a figure 15% higher than what you think you'll need, and tell him you'll call in the exact number of yards an hour or so before the pour. This way you'll be calculating trench depths and form widths as they exist. If you are figuring a big job, put off estimating the final load until after you have finished pouring from the fully loaded trucks. You won't have a lot of time in which to make that final calculation in order to send it back with the last driver or call it in to the dispatcher, but you will have cut down your margin of error by dealing with only the few yards that remain rather than with the whole job.

On the day of the pour, leave yourself enough time for careful figuring and double-checking. Last-minute concrete panic has a long tradition, but punching the keys on a pocket calculator while you are trying to get the building inspector to sign off the formwork will make for miscalculations every time. Have a helper figure independently how much concrete you'll need, and then compare notes. This should pick up careless errors in math, and the easily made mistake of forgetting to figure in some portion of the pour.

Calculating volume—The only accurate way to calculate most residential concrete jobs is to get down in the trenches and take measurements. Write down the width, depth and length of each trench or form. Having to remove a large root from a footing trench and undersupervising an overeager backhoe operator are just two reasons why the actual measurements may differ from the numbers on your blueprints enough to make a real difference.

Particularly with slabs and trenches, the depth figure can be a compromise based on measurements taken at many points and tempered by the kind of intuition that comes with experience. The same can be true of trench width, depending on the soil.

Once you've noted the measurements for each wall, footing or slab, group them into separate categories, one for each configuration or cross section. If the depth of a footing trench changes, figure this portion of the footing as a separate category, even if the width remains the same. When a configuration has more than four sides, such as a T footing, break it down into separate rectangles, trapezoids or triangles. Keep all this information on a clipboard. I number the categories, draw a small section of each one and fill in the dimensions so I don't get confused. List the different lengths for each cross section below the appropriate drawing, and note its location on a plan drawing of the pour. This really helps later in the pour when you have to calculate the last load in a hurry. You can now add the lengths of all the footings and walls in each category to get the total lineal feet for each configuration.

Concrete is ordered by the cubic yard. Unfortunately, the width and depth of footings, walls and slabs are often in inches, and length is in feet and inches. Calculate each cross section—width x depth—in inches if the increments are small, or in feet and decimal feet if the numbers get too big. Just don't mix the two. To convert your square-inch answers to square feet, divide by 144. List the total area of each cross section on your clipboard. Use a pocket calculator to grind out these numbers.

To get the total concrete needed in each category, multiply the cross-section figure, now in square feet, by the total of all the lengths in the category, which is already in feet. The product will be in cubic feet. Adding all of the totals for the categories together, and dividing by 27—the number of cubic feet in a cubic yard—will yield a total in cubic yards.

You can order concrete in fractional yards, but remember to round off high, not low, to get to the nearest large fraction. I usually order a few extra cubic feet to protect against being short, in addition to the standard 5% allowance for spillage. It's a good idea to have pier holes for decks or retaining-wall footings dug to use any excess concrete.

Placing your order—If there is more than one batch plant near the job site, ask around to see which one other builders like. Since price per yard is usually similar, their impressions will be based on phone contact with the dispatcher, and on pouring with the drivers. These opinions are useful, because the cooperation and expertise of the people in these two positions will determine how easy and successful your pour will be.

Give the dispatcher the day and time you want to see the first concrete truck, and how many yards you'll need. Then describe the mix you want, in enough detail so that the proper concrete will get sent to your job site. There are two established methods—performance and prescription.

When you order with a performance specification, you give the dispatcher a compressive strength in psi, and it's up to the batch plant to supply concrete that will test to that minimum figure in 28 days. Prescription ordering, on the other hand, puts the responsibility on you. It also lets you determine some of the variables in the mix design. You may want to duplicate a mix that worked well for you in the past, or to satisfy a restriction unique to this pour, such as a maximum aggregate size that the pumper can handle.

A minimum prescription tells the dispatcher how many pounds or bags of cement to use with every cubic yard of concrete. When you specify only the cement content, the batch plant will determine all the other variables.

If you are knowledgeable, or if there are engineering specifications on your plans you need to satisfy, carry prescription ordering a step further by specifying slump, maximum size of the coarse aggregate, or the percent of air-entrainment. If not, you should mention any special characteristics of the site or of the pour that will affect batching or delivery. If weather is a problem, ask about admixtures. If you are pouring grade beams on a steep slope, talk about slump. If you are going to use a pump, tell the dispatcher which company, confirm the time of the pour, and have the aggregate and mix adjusted.

Whatever way you order, make sure you get a batch ticket from the driver for each load you receive. This is more than just an invoice that lists the number of yards of concrete. It should also tell you the cement and water content of the mix, the size and amount of aggregates, the amount of air-entrainment, the percentage or weight of admixtures, and the slump.

Maybe most important, give the dispatcher clear directions to your job site and a telephone number where the driver or the batch-plant dispatcher can reach you or someone who can get in touch with you. It's surprisingly easy to lose a truck that weighs 27 tons, but it's more than difficult—and expensive—to deal with its load after it's sat in there a few hours.

—*Paul Spring*

Small-Job Concrete

Site-mixed mud can be batched as accurately as ready-mix, given a strong back and a few guidelines

by Bob Syvanen

Most batch plants charge extra for less than a cubic yard or two of concrete. The service you're likely to get on a small order is pretty minimal, so it often makes sense to mix on site. You can do this by hand in a trough or wheelbarrow, or you can rent a mechanical concrete mixer. You'll be using the same ingredients as the batch plant; if you measure and mix carefully, the quality of the concrete should be at least as good. For folks beyond the range of ready-mix trucks, this is the only way.

Ingredients—Clean water is a must for concrete. Sea water is okay if the concrete won't be reinforced with steel. The pour will attain high early strength, but will not be as strong in the long run. Increase the cement content and reduce the water to recover some of the strength lost to the salt.

Cement should be bought by the 94-lb. bag and kept dry. In most cases, Type I is what you need. For resistance to freeze-thaw cycles in northern climates, buy air-entrained cement, which requires using a portable power mixer, since the air-entraining agent needs vigorous mechanical agitation to be effective.

Large aggregate and sand can be purchased by the cubic yard from quarries, building-supply yards and batch plants. Large aggregate can range from ½ in. to 1½ in. It should be clean, hard, durable gravel or crushed stone such as granite or hard limestone. Most sandstone isn't usable. Crushed stone should be square, triangular or rectangular in shape. Flat, elongated pieces shouldn't be used. Sand, the fine aggregate, should be a mix of coarse and fine grains up to ¼ in.

When ordering cement and aggregates, keep in mind that the amount of concrete you get from mixing the ingredients is not nearly as much as the sum of the volume of those materials. This is because the sand in the mix fills in the voids between the gravel or stone, and the cement nestles in between the particles of sand. A rule of thumb that's sometimes used is to figure the amount of your mix

to be slightly greater than the amount of coarse aggregate you're using. The chart below will get you a little closer, so you don't end up short on the last batch of your pour.

Unless you are pouring just a few cubic feet of concrete, which can be done using dry-packaged pre-mix, have the aggregates and cement delivered. These materials are extremely heavy, and your good old pickup truck can easily get overloaded with a half-yard of wet sand. Most suppliers will deliver with a dump truck that can spot the materials almost anyplace. As near as possible to where you'll be mixing and pouring is best. Lay plas-

tic sheeting, 6 mil or heavier, on the ground for each kind of aggregate. Plywood is even better because it is a harder surface for scooping against with a shovel, but don't plan on using the plywood for anything very important afterward. You can also use old plywood to separate the sand and coarse aggregate piles vertically, so that they can be placed close together without mingling. Stack bags of cement close together, and up off the ground so they don't turn to stone. Cover them with waterproof plastic whether or not it looks like rain. Don't use cement that is so hard it won't crumble in your hand.

It's okay to scavenge aggregates for your concrete as long as you test them to be sure that they are clean and free of fine dust, loam, clay and vegetable matter. The beach is a good place to find clean sand, and old quarries and stream beds often have acceptable gravels. Aggregates taken from tidal areas contain larger quantities of salt, and should be washed with fresh water before being used in concrete. You can test both sand and gravel for dirt or loam by placing them in a glass jar filled with water. Put on the lid, shake the jar, and then wait for the water to clear. If silt covers the gravel or sand, it needs washing.

There are two tests for vegetable matter. For gravel, add a teaspoon of household lye to a cup of water in a glass jar, add the gravel and shake well. If the water turns dark brown, the gravel needs washing. This can be done with a good hosing. Sand is tested by putting it in a clear glass jar with a 3% solution of caustic soda, which can be made by dissolving 1 oz. of sodium hydroxide in a quart of water. If the solution in the jar remains colorless, the sand is in good shape. A straw color is still okay, but anything that resembles brown means finding another source for sand.

Sand shouldn't be rejected because it's holding a lot of water, but you need to know how wet it is in order to adjust the water content of your mix. Although damp sand feels a little wet, it won't leave much moisture on

Cement	+ Sand	+ Gravel	= Concrete
bag	cu. ft.	cu. ft.	cu. ft.
1	+ 1.5	+ 3	= 3.5
1	+ 2	+ 3	= 3.9
1	+ 2	+ 4	= 4.5
1	+ 2.5	+ 5	= 5.4
1	+ 3	+ 5	= 5.8

From *Fine Homebuilding* magazine (February 1983) 13:34-45

your hands, and won't form a ball when squeezed in your fist. It contains about ¼ gal. of water per cu. ft. Wet sand will form a ball, but still won't leave your hands very wet. Most sand falls into this category. It contains about ½ gal. of water in each cu. ft. Very wet sand is obviously dripping wet and holds about ¾ gal. of water per cu. ft.

The mix—The strength and durability of the concrete that comes out of your wheelbarrow or mixer depends on the proportion of the cement to the aggregates, and on the proportion of water to cement. Instruction manuals and construction textbooks often show concrete mixes as a ratio of cement, sand and gravel by volume, such as 1-2-4. The first number always represents the cement content, the second is the small aggregate (sand), and the third is the large aggregate (gravel or rock). The more cement used, the stronger the mix. A rich mix, 1-1½-3, is used for roadbeds and waterproof structures. The 1-2-4 mix is used for industrial floors, roofs and columns. A medium mix (1-2½-5) is used for foundations, walls and piers. A lean mix such as a 1-3-6 is used in less demanding applications.

Volume formulas like the ones above give the proportions of dry ingredients but leave the water content up to you. Start with a trial batch, and use the least water you can to get a workable mix. Add a little at a time, and keep track of how much you used.

I favor mix formulas that specify the water/cement ratio, which is called a paste. A 5-gal. paste contains five gallons of water for every bag of cement. This includes the water contained in the sand. The lower the water figure in relation to the cement, the stronger and more durable the concrete. Sidewalks, driveways and floors require a 5 gal. paste for durability. A 6-gal. paste is good for moderate wear and weathering such as foundations and walls. Where there is no wear, weather exposure or water pressure to deal with, a 7-gal. paste will do. Footings are typically poured with 7-gal. paste concrete.

Listed in the chart above are the formulas I use for 5-gal., 6-gal. and 7-gal. pastes, including volume amounts of aggregates for each mix. Each mix differs from the others not only in the ratio of water to cement, but also in the amount and size of aggregates. These adjustments are compromises between economy, strength, durability, workability and slump (stiffness of the mixture). The engineered mixes batch plants use for ready-mix concrete make the same kind of adjustments. The first formula for each mix lists the ingredients used with a single, 1-cu. ft. bag of cement. The second formula gives the correct amount of each material for mixing one cubic yard of concrete, which is useful for figuring and ordering cement and aggregates.

The amounts of water, cement and aggregates in these formulas are given by volume—gallons and cubic feet. There are other formulas that give proportions by weight, but I don't like them as well because ultimately you are trying to fill up a given space—the forms—

Five-gallon paste

1 bag cement, 4½ gal. water, 1 cu. ft. sand, 1¾ cu. ft. gravel (⅜-in. maximum);

for 1 cu. yd. of mix: 10 bags cement, 10 cu. ft. sand, 17 cu. ft. gravel.

Six-gallon paste

1 bag cement, 5 gal. water, 2¼ cu. ft. sand, 3 cu. ft. gravel (¾-in. maximum);

for 1 cu. yd. of mix: 6¼ bags cement, 14 cu. ft. sand, 19 cu. ft. gravel.

Seven-gallon paste

1 bag cement, 5½ gal. water, 2¾ cu. ft. sand, 4 cu. ft. gravel (1½-in. maximum);

for 1 cu. yd. of mix: 5 bags cement, 14 cu. ft. sand, 20 cu. ft. gravel.

---

with concrete. The formulas above also assume wet sand, with its ½ gal. of water per cu. ft. If you use damp sand, increase the amount of water you add by a quart per cu. ft. of sand. Decrease the water content by a quart for very wet sand. These proportions will yield a fairly stiff mix, depending on the size and shape of your aggregate. But you may need to make adjustments, so mix up a small trial batch first. If the concrete is too soupy, you can correct it by adding aggregates. Don't play with the cement or water content. Instead, add 2½ parts sand with 3 parts gravel in small amounts until the mud stiffens up. For the next batch, be sure to deduct the moisture carried by the extra sand from the total water to be added to the mix. If the test batch is too stiff, use slightly less sand and gravel in the next batch.

Accurate measure—The care that you take in proportioning the mix has everything to do with the quality of your concrete. If you are following a formula for mixing that is given by weight, you will need a bathroom scale for careful weighing. For measuring volume in cubic feet., make a 12-in. by 12-in. by 12-in. frame with no handles or bottom. Place it on a flat surface, fill it level, and lift. Cement is easy to deal with because it comes in 1-cu. ft. bags. You can also make a level mark on the side of your wheelbarrow to indicate the 1 or 2-cu. ft. level. In the case of a ratio mix like 1-2-4, use any convenient measure—a shovel, bucket, or box—but don't let your mind wander when you're counting. For water, mark a large bucket for half-gallons and gallons.

Mixing by hand—A lot of concrete has been mixed by hand, but it is a long, backbreaking job worth avoiding for anything more than a few cubic feet of mud. You can mix in a deep (4 or 5-cu. ft.) wheelbarrow or buy a steel or plastic mortar box (about $70) that holds 6 to 9 cu. ft., or you can make your own mixing tray. A large, shallow plywood box lined with metal so that water won't leak away works pretty well. A flat platform works even better because there are no corners for the shovel or hoe to hang up on. However, mixing must be

done carefully to avoid losing water on this flat surface.

First, load the tray with a measured amount of sand. Spread the correct amount of cement evenly over the sand and mix them together with long push and pull strokes with a hoe or shovel. Work the large aggregate into this mix with the same method. Make a depression in the center of the mix and slowly add the water. Pull the mix toward the water until the dry material is saturated, and then turn the mud over until it reaches a workable smoothness. Use this method even if you are mixing in a wheelbarrow or box. A mortar hoe is useful if your aggregate is no bigger than ¾ in. It looks like a large steel garden hoe with two holes in the blade, and costs about $15 to $20. A square-point shovel turned over so that the back of it faces away from you works too.

Mixing by machine—Machine mixing is easier than hand mixing, but it's still a lot of hard work. Electric or gasoline-powered mixers with a capacity of ½ cu. ft. to 6 cu. ft. can be rented by the day or week. Electric mixers are the least trouble. If your job site doesn't have power, then rent a gasoline-powered model. If you are going to use a mixer for more than two weeks and you do lots of small jobs involving concrete, consider buying one.

Set up the mixer right next to your sand and gravel, and run a water hose there. If you have chosen a shady spot, both you and your concrete will set less quickly. Load the drum of the mixer with all of the large aggregate and about half of the water in the formula. Start up the mixer and add the sand and cement slowly, along with the remainder of the water. Let the mixer run for about three minutes or until the concrete has become uniformly grey. When you are finished mixing for the day, add a couple of shovels of large aggregate and some water and turn the machine on one more time to scour the inside. Emptying the drum and a final rinse with a hard jet of water will leave the mixer clean.

Cold-weather concrete—Most engineers do not want you to pour concrete at air temperatures lower than 40°F. A lot of good loads have been poured when it's colder than this, but it's a bit of extra work. Both the aggregate and the water can be heated to keep the concrete warm while it's being mixed and poured, but don't heat the cement. If you heat just the water, bring it to a boil in a 55-gal. drum or other container and pour it on the aggregates to warm them. If you are heating the aggregates also, keep the temperature of the water below 175°F, or the cement will flash-set when mixed, and you won't be able to get a finish on the concrete. Aggregates can be heated on a tray of heavy sheet metal. Build a makeshift firebox out of large stones or concrete block underneath the tray, and heat the aggregates separately. Take them off the fire when they are hot to the touch. □

Bob Syvanen is consulting editor to Fine Homebuilding. *Photo by the author.*

Gunite Retaining Wall

Sprayed-on concrete does more than just line swimming pools

by Ken Hughes

I am a structural engineer who works in the San Francisco Bay Area, and these days a lot of my clients are building on steep hillside lots. The reason is simple enough—all the good flat lots have been taken. For the designer, hillside lots present the challenge of fashioning a building that takes advantage of the views and a floor plan that works in concert with the terrain. On the other hand, the builder is usually faced with

Ken Hughes is a structural engineer with the firm of Vickerman/Zachary/Miller Engineering and Architecture in Oakland, Calif.

extensive excavation work and an unconventional foundation system. But no matter what type of foundation is eventually constructed, these hillside projects often begin with hefty retaining walls.

The wall discussed in this article holds back the earth above a home built by Servais Construction in the Berkeley hills. This company specializes in building finely crafted houses, both on a contract and a speculation basis. Whenever I get a call from Jim Servais, I know I'd better put on my hiking boots to inspect the lot.

As with most of Servais' projects, I was skepti-

cal when I first saw this site. It was almost too steep to walk. Servais wanted to build a spec house on the property, so it was understood from the outset that we had to approach the project with that in mind. If we couldn't figure out a way to stabilize the earth within budget, we would have to abandon the project.

We began by getting a soils report from Subsurface Consultants of Oakland, Calif. They found the soil to be reasonably stable, with weathered bedrock 4 ft. to 6 ft. below the surface. Given this news, we calculated that some excavation near the center of the site would allow a house to be attractively nestled into the hillside. The vertical cuts into the hill, however, would have to be bolstered by retaining walls.

Retaining-wall design—With any retaining-wall design, the objective is to stabilize a vertical cut in the soil as economically as possible, yet achieve a long-lasting structure that satisfies accepted levels of structural safety. For this project, several retaining walls were required. The largest is 50 ft. long and averaged 7 ft. in height with a steep, upward-sloping backfill. This wall is above both the house and the street, about 100 ft. from the nearest driveway.

By looking at test borings from the site, our soils engineer knew that the retaining walls would have to hold back soil made of sandy clay. The wall footings would be in the transition area, where the sandy clay mingles with the weathered bedrock.

Using this information, I designed two retaining walls for Servais. This way he could run a cost analysis for each design, and pick the more economical solution. Both designs used a conventional continuous-spread footing to resist overturning and sliding. Wall A was a cast-in-place concrete wall. Wall B would consist of concrete blocks, reinforced with steel and completely grouted. We considered two other wall types—a concrete crib wall (precast concrete members stacked together like Lincoln Logs) and one made of pressure-treated timbers. We rejected the crib wall because it would have required another subcontractor to build it, and Servais wanted to keep the cost down by building the wall with his crew. Although there is nothing wrong with pressure-treated wood retaining walls, we vetoed the idea because bank loan officers don't always believe in them.

Modifying the design—When the hillside cuts were made, two things became apparent. First, the 7-ft. vertical cut seemed to hold temporarily without sloughing. This was partly because of a long dry spell prior to excavation. Also, the bedrock turned out to be a little closer to the surface than we expected. We also realized that most of our retaining-wall footing would have to be trenched and placed in bedrock. At this point, we reconsidered the wall's foundation.

We decided to discard the conventional footing because of all the pick-and-shovel excavation it would have required into the stubborn bedrock. Instead, we opted for a reinforced-concrete grade beam atop 18-in. dia. piers, spaced 6 ft. apart (drawing and photo at right). A backhoe could have handled the excavation, but since Servais needed a drilling rig to bore holes for the house foundation anyway, it made sense to avoid the expense of one more heavy-equipment subcontractor.

Footing aside, Servais was not looking forward to building this wall. The labor involved in carrying the concrete blocks up the hill by hand would be time-consuming and expensive, and the alternative of casting the wall in place involved transporting, building, placing and stripping a considerable amount of formwork.

After pondering the blocks versus the pour, Servais called and asked, ''Why can't we build

Drawing: Chuck Lockhart

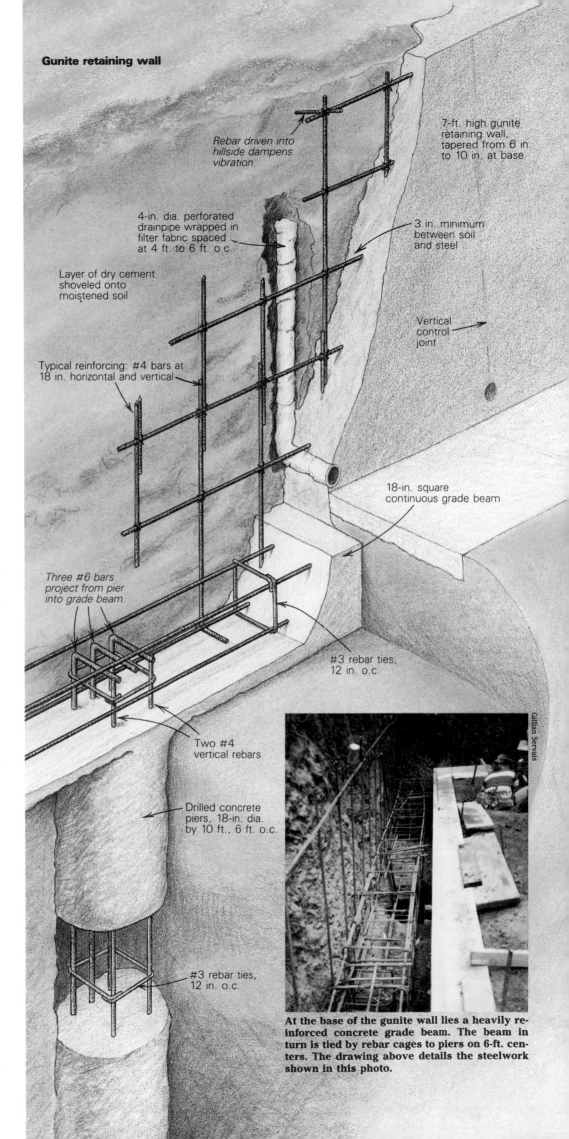

Gunite retaining wall

Rebar driven into hillside dampens vibration.

7-ft. high gunite retaining wall, tapered from 6 in. to 10 in. at base.

4-in. dia. perforated drainpipe wrapped in filter fabric spaced at 4 ft. to 6 ft. o.c.

3 in. minimum between soil and steel

Layer of dry cement shoveled onto moistened soil

Vertical control joint

Typical reinforcing: #4 bars at 18 in. horizontal and vertical

18-in. square continuous grade beam

Three #6 bars project from pier into grade beam.

#3 rebar ties, 12 in. o.c.

Two #4 vertical rebars

Drilled concrete piers, 18-in. dia. by 10 ft., 6 ft. o.c.

#3 rebar ties, 12 in. o.c.

Gillian Servais

At the base of the gunite wall lies a heavily reinforced concrete grade beam. The beam in turn is tied by rebar cages to piers on 6-ft. centers. The drawing above details the steelwork shown in this photo.

Perforated drainpipes let into vertical cuts in the raw earth relieve hydrostatic pressure behind the wall, left. Before the gunite was applied, they were wrapped with filter fabric to keep them from clogging. The white material on the bare earth is portland cement, which helps to prevent sloughing. At the end of the wall, a minimal form turns the corner. Above, gunite is placed in increments as the operator makes a pass from one end of the wall, then back to the other. The contents of the hose are under tremendous pressure, so it takes a firm, steady grip to control the business end of a gunite hose. If it were to whip about out of control, it could easily injure an unlucky by-stander. A few hours after the gunite crew begins its work, the wall is ready for troweling, below.

this wall out of gunite?" The idea made sense. It took advantage of the temporary stability of the vertical cut, which eliminated the need for a back form. It would considerably reduce labor and time, compared to the concrete block or conventional cast-in-place concrete. From an engineering viewpoint, gunite is nearly identical to cast-in-place concrete (see the sidebar at right). The steel reinforcing, concrete strength and wall thickness would not change significantly. By making some changes in the drainage system, I was able to adapt the new foundation to work with a gunite wall.

A drainage system behind a cast-in-place concrete or block wall is typically installed after the wall is in place. Not so with a gunite wall. In this case the drainage network is a series of vertical perforated 4-in. dia. pipes let into cuts in the earth on 4-ft. to 6-ft. centers (photo facing page, left). To keep the gunite from clogging the perforated pipes, the crew wrapped them with Mirafi 140N filter fabric (Mirafi Inc., P.O. Box 240967, Charlotte, N. C. 28224). The drainpipes emerge at the bottom of the finished wall, relieving the lateral hydrostatic pressures on it. Before shooting on the gunite, the crew temporarily plugged the pipe outlets with rags tied to rebar handles.

Wall construction—Once the hillside had been cut, it was important to get the wall built right away. This was a winter project, and a heavy rain could seriously have eroded the exposed hillside. To help maintain the bare earth cut, the crew applied a thin layer of portland cement to the vertical surface. They did this by lightly misting the wall with a hose, and then scattering shovel loads of dry cement across the cut. This trick did two things: it helped to prevent erosion of the dirt, and it created a surface to which the gunite would more readily adhere.

Next, the crew tied the steel reinforcing rods in place. The steel is the same size and spacing as for a poured concrete wall, with one exception. Reinforcing steel in a gunite wall has to be secured to keep it from vibrating as the gunite is blasted into position. To dampen the vibration of the steel, Servais' crew tied wires to the tops of some of the vertical rods, and then wrapped the wires around rebar stakes driven into the hill. Without this bracing, the steel might have bounced around as the gunite was blasted onto the wall, causing already placed gunite to fall out in big chunks. Except for a couple of small forms for the wing walls at the corners, the wall was ready to shoot.

The entire 7-ft. height was placed in one four-hour operation (photo facing page, right). As they sprayed the wall, the gunite crew monitored its thickness by watching thin horizontal "ground wires" tied to the rebar. As they sprayed the wall, the gunite crew monitored its thickness by watching wire depth gauges tied to the rebar. Typically, the gauges are used as guides for a cutting tool that slices the excess gunite off the face of the wall. This results in a true flush surface. With this wall, however, Servais elected to have a more random finish (bottom photo, facing page) in keeping with the Spanish-style stucco house he planned to build.

As soon as the gunite was in place, the drain-

Gunite and shotcrete

Until 1967, the word "gunite" was a proprietary trademark. It is now a generic term used to define the dry-mix shotcrete process. In this procedure, a dry mixture of cement and fine aggregate is pumped through one hose and water through a second. Mixing occurs at a common nozzle where the gunite is ejected at high velocity onto a surface.

"Shotcrete" is a generic term used to describe the pneumatic placement of any concrete through a hose and nozzle at high velocity. While the term properly covers both the wet-mix and dry-mix processes, the word is used most often to describe the pneumatic placement of concrete in a plastic state.

Recent improvements in the pumps that deliver shotcrete to its target have made it the choice over gunite in some circumstances. While a gunite crew can typically move about 30 cu. yd. of material in a day, a comparable shotcrete outfit can pump about 90 cu. yd. Gunite, however, can be trimmed to a smooth surface, while shotcrete leaves a rough finish that is often plastered for cosmetic purposes.

The birth of gunite—Although gunite and shotcrete came into wide use immediately after World War II, gunite dates back to the turn of the century. In 1895, Dr. Carlton Akely, Curator of the Field Museum of Natural Science in Chicago, developed the original cement gun. He was searching for a method to apply mortar over skeletal frames to form the shapes of full-size prehistoric animals. He could not form the necessary convoluted shapes and contours by conventional troweling, so he developed a method to shoot concrete into place with air as the propellant. In a single-chambered pressure vessel, he placed a mixture of sand and cement. Then he pumped compressed air into the chamber, forcing the mixture into a hose. As the sand and cement mixture was ejected from the end of the hose, it passed through a spray of water that hydrated the mixture.

Immediately following World War II, the use of gunite and shotcrete increased tremendously. Builders found numerous applications in all sizes of projects, from swimming pools to tunnel construction.

Although procedures have been refined and equipment improved, the basic process has not changed since it was originally developed. Gunite or shotcrete can be used in lieu of conventional cast-in-place concrete in most instances, the choice being based upon convenience and cost. These processes are particularly cost-effective where formwork is impractical, or thin layers or variable thickness are required. The principles used in the design of cast-in-place concrete structures are also applicable for gunite and shotcrete structures. Compressive strengths of 2,000 psi to 4,000 psi are common, and higher strengths are easy to attain, depending upon the specific mix design.

Although large civil and industrial projects such as dams, tunnels and aqueducts are the most common use for gunite and shotcrete, other modern applications that are becoming more popular include seismic renovations, basement and shear-wall construction in new buildings, and soil nailing. Old masonry buildings can be strengthened to resist seismic forces by applying reinforcing steel and gunite to the face of the brick, thus forming a strong wall attached to the much weaker masonry. This is usually done on the inside face, which allows the exterior rustic brick facade to remain.

Soil nailing is a relatively new procedure that allows construction of very high retaining walls without the need for a footing or vertical piles. This is a common way to stabilize a deep excavation, such as the perimeter basement walls of underground parking structures below high-rise buildings. This process involves reinforcing the earth by drilling and grouting into place an array of tie-back anchors, typically to a 30-ft. depth. The exposed ends of the rebar strands protruding from the anchors are woven into a reinforced gunite or shotcrete wall that forms the vertical surface of the excavation.

Gunite and shotcrete placement are very specialized operations. The quality of the product is highly dependent on the skill of the workers. The American Concrete Institute (Box 19150, Redford Station, Detroit, Mich. 48219) has prepared and made available "Guide to Shotcrete" (ACI-506R-85), which gives detailed guidelines and requirements for successful gunite and shotcrete placements. In addition, the Gunite and Shotcrete Contractors Association (P.O. Box 44077, Sylmar, Calif. 91342) has vast resources of technical data to aid contractors and engineers in the use of gunite and shotcrete.   —*K. H.*

line plugs were pulled, and vertical control joints were struck into the face of the wall. These control joints project upward from each weep hole. Once the masonry starts curing and shrinking, cracks usually start at the weep holes. They are therefore natural areas to direct crack-control joints.

The last step was to apply curing compound to prevent rapid curing and cracking of the surface concrete. Except for achieving its design strength—in this case 2,500 psi—the wall was complete. Excluding excavation and footing construction, most of the construction was completed in one day, and there were almost no forms to strip or backfilling to do.

Construction costs—This gunite wall was built for just over $2,000, or roughly $6 per sq. ft. of surface area, excluding footing construction. By comparison, a similar concrete-block wall would have cost roughly $7 per sq. ft., plus the costs of transporting the blocks uphill and applying a plaster finish. We estimate that a cast-in-place concrete wall for this project would have cost $9 to $10 per sq. ft. Building and setting the forms would have been labor intensive and costly, and likely as not the form lumber would have been hard to reuse. Perhaps even more important, Servais didn't have to agonize over a tall-wall concrete pour into forms that would have been braced on just the downhill side.   □

Building with Ferro-Cement

Freeform concrete walls redefine the interior of a 600-year-old structure on the Italian Riviera

by Jane Speiser

I have been renovating a 14th-century stone house in northern Italy for the past five years. Not only is the house ancient, but half of it had always been used as a barn, and the other half had been uninhabited for 40 years when I bought it. Nearly every floor, wall and roof surface needed work. This sounds daunting, but it allowed me a degree of freedom that I had never had during my renovation commissions for clients in the United States. Basically, I had a shell, the interior of which I could use as a laboratory for structural experimentation. I wanted to create new surfaces compatible with the texture of the original stonework, and also attempt a fusion of sculpture and architecture. It was also important to create the illusion of spaciousness, because the rooms were small—scaled to the size of the 14th-century peasant, whose average height was 5 ft. or so. To do this, I used the ferro-cement construction techniques that I'd learned in San Francisco in the early 1970s, modifying the process by using as much indigenous material as possible. I built an undulating wall to separate the entrance from my darkroom, a staircase, a curved wall surrounding the toilet, a bathtub, a shower, and a scalloped wall between the bathroom and the second bedroom.

In ferro-cement construction, a thin layer of concrete is troweled or sprayed over a form, or armature, which is usually made of steel rebar and wire cloth. Ferro-cement requires much less material than does reinforced concrete. The resulting structure is cheaper and lighter, but just as strong. Elaborate warped planes can be created without resorting to a supporting formwork. The process was originally used by Pier Luigi Nervi and other Italian engineers and architects on vast industrial enclosures and stadiums, whose shells were no more than several inches thick.

Assembling the armature—In order to create such a structure, an armature or skeleton is fabricated from steel rebar ranging from ¼ in. to 1 in. in diameter. I weigh only 108 lb., but I can easily bend ¼-in. or ⅜-in. rebar into elegant forms. Thicker rebar requires a little more strength, but it's rarely used anyway, except in industrial construction.

The rebar is welded or tied in place with lighter-weight galvanized wire. There is a standard rebar tying procedure (as there are standard sailor's knots) that prevents the

The entrance to the author's 14th-century stone house on the Italian Riviera is tucked back in a narrow alleyway.

crossing rods from slipping against each other (drawing, below). As the skeleton takes shape, the rebar is made to intersect, creating as many triangles as possible. Triangulation makes the completed structure—which should be no more than 1½ in. thick—comparable in strength to a standard 6-in. thick reinforced concrete slab.

Tying rebar

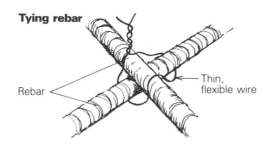

Rebar

Thin, flexible wire

Once the armature is erected it is covered on both sides with ½-in. aviary mesh, which is sewn to the armature with fine galvanized wire. This is the part of the process that demands the most care and accuracy. The mesh must be stretched taut, cut precisely to shape, and attached to the armature every 6 in. If this

part of the work is not done with great care and precision, all of the imperfections in the form, which are hard to detect at this point in the process, will show up larger than life as the cement is applied. Also, if the mesh isn't snugged up securely against the armature, the cement will fall into these voids, mesh and skeleton will separate, and the job will require two or three times the mortar that it should. With 10 years of experience behind me, I always take the time to hand-tailor the mesh to the skeleton.

Applying the mortar—The mortar needs to be a fairly rich mix: plastic enough to apply, but not so runny that it drips. If the mixture is too dry and crumbly, it will not adhere to the mesh. If it is too wet, it will fall through the holes. It is usually applied by two people pressing it into place from both sides at once. It can be applied with a trowel, or—as is often necessary in more intricate structures with curved surfaces—by hand. Wear heavy rubber gloves, because wet concrete is caustic. I usually work up to a height of several feet, then allow the mortar to set up for a couple of hours before continuing. If you try to do too much at once, vibrations in the mesh can dislodge what you've just pressed into place.

A ferro-cement structure usually demands a second application of mortar to smooth out imperfections. I think the best tool at this stage is a large, heavy-duty paint brush, which is more flexible than a trowel and works better on curved surfaces. The ratio of mortar that falls to the ground to that which finds its final resting place on the structure itself has never been calculated scientifically, but even the most skilled masons lose some mortar, particularly when they're working overhead or dealing with warped planes.

Don't worry too much if at first some of the mix does not cling as enthusiastically as you might wish. It will take a bit of practice to improve your batting average. Actually, any waste of time and mortar is offset by the facts that you don't have to erect any formwork, and that the amount of mortar you have to mix is one-fifth of what you would need for poured concrete.

A ferro-cement structure should be allowed to cure for several weeks, and get wetted down each day to keep it from drying out too quickly and cracking. Once it has cured, you will have a strong and permanent enclosure

Illustrations: Roger Barnes

The renovated stone house

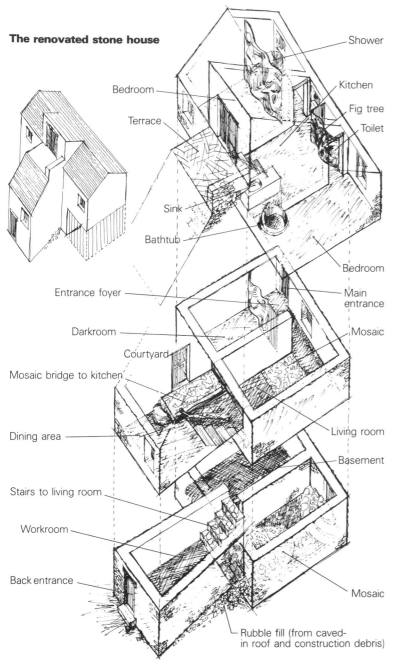

- Shower
- Bedroom
- Terrace
- Kitchen
- Fig tree
- Toilet
- Sink
- Bathtub
- Bedroom
- Entrance foyer
- Main entrance
- Darkroom
- Mosaic
- Courtyard
- Mosaic bridge to kitchen
- Dining area
- Living room
- Stairs to living room
- Basement
- Workroom
- Back entrance
- Mosaic
- Rubble fill (from caved-in roof and construction debris)

This part of the house had been used for centuries as a stable, and the rest had fallen into disrepair. Only the shell and a few heavy beams were still sound. Speiser saw this as an opportunity to experiment with interior and structural design in the renovation. She turned to ferro-cement to maintain a sense of the original textures while at the same time integrating sculpture and architecture.

The house's main stairway has ferro-cement treads, built on an armature of rebar that can be seen wrapped around the old wooden beams.

Speiser set tile in many of her floors. The walkway connects the dining room and the kitchen, and was cast in one day. The tiles were laid in place and leveled as the mortar dried. The tiles in the foreground are a century old.

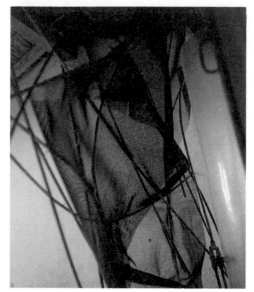

Building up the form. Ferro-cement requires a skeleton or armature which, in non-industrial construction, is usually of ¼-in. or ⅜-in. rebar. These are bent and tied off into as many tri-

angles as possible for maximum strength. Once the armature shaped to the builder's satisfaction, it is put in place, and covered on both sides with a mesh that will accept the several

thin layers of concrete that will be troweled on to form the wall. Reeds, here laid across the bathroom's ceiling joists, are used in northern Italy to support poured concrete slabs.

that resists fire and moisture, and whose surface can be inlaid with tile, painted or simply whitewashed with lime.

This working procedure satisfies requirements for public or industrial structures. However, individual houses—at least in areas where there is no rigid building code—don't have to meet such rigorous requirements, so you can use lighter and less expensive armature materials, which are more enjoyable to work with, and easier to handle than rebar and wire mesh.

Experimental structures—In my house, I used a variety of new and recycled materials, trying to find those that best lend themselves to undulating forms and the labor of a single person. The first thing I built was the wall surrounding the toilet, with its accompanying wood-collage door. For an armature, I used a fig tree that had been cut down in the garden next to my house (photo facing page, top left). It allowed me to create a more organic surface than I could have with rebar.

The basic skeleton was interlaced with reeds from the river that flows through my village. After laying in the reeds, I attached a loose-weave jute covering to the inside and plastered both sides with a lime and sand mortar, which is much stickier than cement, and is therefore easier to apply. It also sets up gradually over several days, so you don't have to rush its application. If you use a pure cement and sand mixture, you can't mix more than you can apply within a few hours, because the mortar will start to set up in the pan. Lime mortar, however, you can mix the night before, so that you can get a good night's sleep between the exhaustion of mixing and the concentration required for the next application.

Once the curved wall was done, I built the bathtub next to the woodstove in the master bedroom (photo facing page, center). I used the standard rebar skeleton, over which I

draped a flat knitted mattress spring. I'm not sure if these exist any more in the United States, but in Italy all beds are made from them. Eventually, they lose their tautness and start to sag too much for the bed, so all the villagers seem to have one they're anxious to get rid of lying out in the backyard. I discovered that the knitted springs' intersecting ringlets of wire form a ready-made double surface that holds cement better than anything else I have used, and can be coaxed into warped planes with the flexibility of a large sweater. These qualities made it perfect for the tub, because it let me use only one layer and required a minimum of cutting, sewing and applying mortar to form a permanent, waterproof, freeform enclosure. I laid tile on the bathtub floor and surfaced its sides with pure cement, which is as watertight as enamel. The end result is rustic but highly serviceable.

After the bathtub, I built the dining-room floor and the walkway between the dining room and kitchen. For the floor I once again used reeds from the river. These reeds range in thickness from 1 in. to 2 in., and they're very strong. When laid and tied in place, they can be walked on (I have jumped on them on occasion) even before the concrete has been poured over them. They've been used for centuries in this part of the world to span floor or roof beams over which, these days, concrete is poured and tiles are laid. They form a strong and—to my mind—elegant support, which is mildewproof and insectproof, and remains in place after the pour. Even the roof of my house—the one part of the house that I hired a mason to work on—was rebuilt using this method.

For the floor, I used the one enormous original floor beam, set two other salvaged beams into the wall, and ran the floor joists—recycled from other parts of the house—over all of them. Many of the oak, olivewood or chestnut joists appeared on the outside to be termite-ridden, but a little scraping revealed a

center that is as strong as steel—too hard to drive a nail into without drilling—and still good for another few centuries. Wood is prohibitively expensive in Italy, so nobody buys new wood when salvaged wood is available.

Over the joists, I tied the reeds into place, poured 2 in. of concrete, and set in place some 100-year-old floor tiles generously given me by a neighbor. The smaller walkway (photo facing page, bottom right) and the multicolored tile surface were cast in one afternoon using rebar and another flat mattress spring for an armature. The tiles were cast in the wet mortar as it was setting up, in one shot, which is the fastest and most permanent way to work, although it takes a certain amount of practice to break the tiles to size and set them in place before the mortar hardens. I find that it's usually possible to do this when the surface to be covered at one time is no larger then two square yards.

For larger areas, like the mosaics in the living room, the entrance and the shower floor, I set the tiles down on a compressed, crumbly, nearly dry mix that has been sprinkled with pure cement that acts essentially as glue. The mosaic is hosed down with a fine spray, so as not to dislodge the tiles. Then I lay a long board over them and pound on it with a rubber mallet as I move it gradually across the whole mosaic. This sinks the tiles evenly into the mortar, which rises around them to form the grout. This technique may sound old-fashioned compared to the instant tile adhesives developed in the United States, but it is inexpensive, and it resists for many decades the settling stresses to which a building in an earthquake zone is subject.

The ferro-cement stairs are attached to a single large beam. Rebar wraps around the beam and is lodged securely between the stones in the wall. A double layer of quite rigid ½-in. square galvanized mesh supports the mortar and tiles. Predictably, my neighbors, when they saw the transparent and apparent-

From *Fine Homebuilding* magazine (February 1983) 13:50-53

Armatures can be made from salvaged or indigenous materials. A fig tree cut by a neighbor becomes this wall's skeleton; reeds from a nearby stream replace galvanized mesh.

ly flimsy armature, all told me that the stairs would never support the weight of a person. It was not until the 200-lb. husband of one of them jumped up and down on the finished product without even causing it to vibrate that they were convinced.

The wall in the entryway (photo top right) is the most visually adventurous of the ferro-cement constructions in my house. Because of its undulating surfaces, the small entrance appears quite large. The effect of curved surfaces on spatial perception has been fairly well documented. Any surface that has textures or planes that negate the customary reference points of rectangular perspective will create the illusion of greater depth.

I built the wall with many different materials, including a reed armature, cloth, and plastic square mesh that is normally used for garden fencing. I discovered recently that garden mesh is cheap, easy to handle (you can cut it with scissors), and gentle on the hands when you bend and attach it. It grasps the cement well and is more than strong enough for any surface that isn't walked upon. Next to the wall is another wood-collage door, a combination of old and new.

In the bathroom wall, which I'm still building, I used the plastic mesh extensively for complicated curves, and the mattress springs for the flatter sections. The end result will be an elaborate scalloping surface that would have been impossible with any other method.

Ferro-cement construction requires a certain amount of skill to master, but it offers an enormous flexibility of form while satisfying the most rigorous building-code standards for permanence and fire resistance. I hope it will attract more advocates in the future, especially among architects who wish to collaborate with sculptors to incorporate freeform spatial divisions into custom construction. □

Jane Speiser is a sculptor and general contractor. Photos by the author.

The elaborately curved wall facing the entrance makes the small space feel bigger by breaking down the expected rectilinear perspective. Speiser used a reed armature, cloth, and garden fence mesh. The door next to the wall is a collage made of wood scraps.

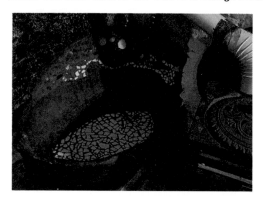

The bathtub, top, next to the bedroom's wood-stove, is ferro-cement, with tiles embedded in the concrete. Flat knitted mattress springs—common cast-offs in Italy—make perfect armatures for such shapes. Above, the cellar stairs.

The completed dining area of the renovated house, as seen from across the walkway in the kitchen. The floor of the dining area is a slab poured over beams and reeds like those in the room's ceiling.

Stemwall Foundations

Use plywood, framing lumber and snap ties to form
the foundation, then build the house out of the forms

by Dan Rockhill

I poured my first concrete foundation on a site tangled with mature, unpruned trees. Furthermore, we were laboring under the watchful eye of an uneasy client. Initially I had no intentions of building the foundation, but our subcontractor threw us a curveball when he suddenly announced that there would be a two to four-week delay before he could pour the footings. As I pondered the delay and our anxious client, I decided to do it myself. My crew and I will never forget that decision.

The site had bedrock fingers that came almost to the surface, making it difficult to drive stakes into the ground. We were pouring a grade beam that would sit directly on the rock, and because it was next to impossible to penetrate it with a stake we stabilized the forms with an elaborate network of crisscrossing braces,

which rendered the interior side of the foundation virtually inaccessible. But this condition didn't concern us, as we were going to let the concrete "seek its own level." All we had to do, according to my friend Richard, was to add plenty of water to the concrete.

It did indeed seek its own level. Soupy concrete oozed from the tiniest holes in the forms, and flooded into the future crawl space through cavernous gaps between the bottom of our forms and the irregular bedrock. To the utter disbelief of the first concrete driver to arrive, we watched the first two or three yards pour into, and then dribble out of, our forms.

Now one thing I did know about contracting, from all the books I had read, was that you were supposed to get the dirt piles as far away from the excavation as possible. This we had done

with diligence during our site prep, not realizing that it would be the only material around to patch the holes between grade and our forms. We started running what seemed like the high hurdles, carrying shovelfuls of dirt from one side of the site to the other. Even the concrete driver pitched in to help, and I know now from experience how unusual that is.

Six hours later, when the last concrete truck left, we collapsed on the grass only to be accosted by our client. She was in a state of delirium because the concrete truck had broken a few twigs on her maple trees as it jockeyed around the site.

Since that dreadful day I have worked on refining a simple system of concrete forms that I can assemble with the help of the most inexperienced crew. The key parts of this system are

From *Fine Homebuilding* magazine (February 1988) 44:32-37

plywood, 2x4s and an ingenious device called a snap tie (drawing, right).

A snap tie is a slender metal rod that connects the opposing walls of a form, holding them apart and locking them together at the same time. Once the concrete is in place, you break off the protruding metal stem and pry out the plastic cone. The resulting crater is easily patched if you're concerned about appearance. Snap ties cost about $.30 apiece, but they ensure accurate work and they are so much easier to use than wood spacers that I wouldn't build a foundation without them. We used about $60 worth of snap ties on the relatively small foundation shown in the photos. We have our own wedges that are used in conjunction with the ties. They cost a nickel apiece to rent, or they can be bought for about $.70 apiece.

I assemble form panels using full sheets of plywood, which are perfect for building 4-ft. high stemwalls atop a separately poured footing. Commonly known as a T-wall system or crawl-space foundation, this kind of foundation is popular in areas with a medium depth of frost penetration (28 in. to 42 in). We use them here in Kansas, where we have to excavate too deep for a slab foundation but not deep enough for a full basement. The resulting 4-ft. high crawl space has the added benefits of being large enough to house mechanical systems and of serving as a handy tornado shelter. Having all the mechanical services in the crawl space frees up valuable square footage above that might otherwise be lost to hot-water tanks or furnaces.

Footings on a vacant site—Most building projects begin with nothing more than a stake driven into the ground to locate a corner of the structure for reference. I always feel a little reverential at this moment, because this act symbolically marries me to the project and quite deliberately initiates construction.

There are as many different kinds of sites as there are buildings, so it is difficult to give any hard-and-fast rules for excavation. Typically, you begin by carving away enough soil for a generous crawl space. Then you should lay out and mark the footings (for details on foundation layout, see Tom Law's article, "Site Layout," on pp. 64-66). Mark the outlines of the foundation trenches with lime so that your backhoe driver can easily see where to cut. Don't use bonemeal to mark the lines. I did once, and the client's dog licked it up as soon as I left the site.

The minimum footing depth is determined by the maximum depth of frost penetration. If you don't know what it is, check with the local building-inspection office. Note that the minimum footing depth can be affected by the quality of the soil. Some soils hold up buildings better than others. Consult with a soils engineer if you have any doubts about the dirt.

A good backhoe operator can save you a lot of headaches by cutting straight trenches to the desired depth. Ask around to find out who the good ones are, and don't base your selection strictly on hourly rates.

Before the backhoe arrives, decide where you want to put the dirt that comes out of the trenches. You need to save some for backfill,

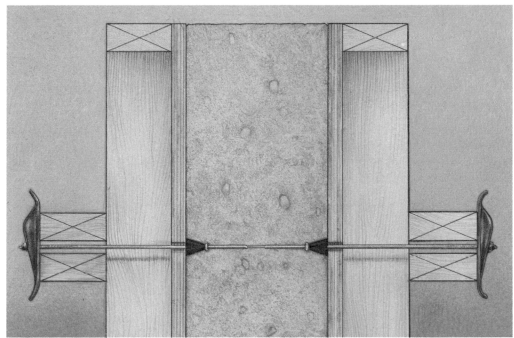

In stable soils, a footing can be poured directly into a trench cut into the earth. But in crumbly soil like this, Rockhill uses framing lumber to make forms for the footings (photo above). Their tops are held together with snap ties. Their sides are bolstered with stakes and tamped earth.

and if you have some good topsoil, set it aside for finish grading. Before digging, be sure that you know where any underground utilities are located. If you don't, I can assure you the backhoe will find them and snap them in a split second. This creates unnecessary delay and danger that can easily be avoided. All utility companies are pleased to come out and mark the positions of their underground lines.

In some parts of the country the soil is stable enough to allow you to "trench pour" a footing. This means that all you do after digging a clean trench is to set the steel, then some grade stakes to control the footing height. Then you can fill the trench with concrete. In soils around here, it's hard to get a perfect trench so I use wood to form the footings (photo above). The side walls of our trenches have to slope back enough to prevent a cave-in, and the trenches have to be wide enough to muscle forms around in.

When the trenches are complete, be certain

that you leave the soil undisturbed beneath the footing. If any soil has been removed and replaced, it must be tamped solid. Around here we can rent gasoline-powered compactors, or "jumping jacks," for about $80 a day.

In most situations, a T-wall footing should be twice as wide as the wall is thick and as deep as the wall is thick. For a standard 8-in. thick stemwall, that means a footing that measures 8 in. by 16 in. I form all the footings out of the dimension lumber that will later be used as joists or headers. When you begin to use the lumber later, be sure to have plenty of sawblades on hand. The concrete dust dulls them quickly. The forms for the footing don't have to be immaculate, as long as they are sturdy and the size of the footing is not compromised. Boards can be lapped or run long at the outside corners in order to minimize cutting.

If I'm using 2x10s for the footing forms, I find it easiest to pour the concrete to the top of the form. On small foundations, the expense of the extra 1¼ in. of concrete is offset by the ease of this operation. If I'm using 2x12s, I snap a chalkline at a height of 8 in. inside the forms and set a few nails along the line as reference points because the concrete will likely obliterate all traces of the chalk.

I have found that the entire job goes a lot faster if I take the time to make my footings dead level. This means carving away obstacles at the bottom of the trench, and making plenty of level checks with the transit. But I think the extra work is worth it because it sets a good precedent for the rest of the job.

To keep the forms at the right level, I drive stakes a couple of feet into the ground and nail through them into the side walls of the form with 16d duplex (double-head) nails. Duplex nails are a lot easier to remove when it comes time to strip the forms. I try to set my stakes about every 4 ft. I also tie the opposing form boards together with a 2x4 collar every 4 ft. or so. The collars impede the pour somewhat, but can be navigated around.

If the site slopes, make sure that the footing stays below the frost line. To do this you will have to step the footing down the slope (for more on how to do this, see "Stepped Foundations" on pp. 67-69). So that you don't weaken the footing, make sure the concrete is at least 12 in. thick where it steps down.

Patch gaps between the stepped forms and the grade with tamped earth before you place the concrete. You can make the "treads" of the step footing as long as you like but not less than 24 in. A 4-ft. tread is good because it maximizes the use of your plywood.

Rebar—Steel gives a backbone to concrete and is a necessary part of your footing. Most 16-in. wide footings use two continuous ½-in. reinforcing bars spaced 8 in. apart down the center of the footing, directly below the inside and outside lines of the future stemwall. The steel will resist tension forces better than the concrete will alone. Rebar comes in 20-ft. lengths, and where sections of it meet they should overlap by at least 12 in. (24 rebar diameters). The overlaps should occur in straight runs of foundation— never at the corners—and should be staggered in parallel runs. Thoroughly secure adjoining pieces with form wire to keep them from working free as the concrete is placed.

To cut rebar you can use either a reciprocating saw fitted with a metal-cutting blade or a hacksaw to cut partway through the steel. Then bend it over your knee to snap the pieces apart. It's easier, though, to use a cutting torch or a cutoff blade in a circular saw. The best cutting method is a rebar cutter/bender. This tool costs about $250, but if you do a lot of rebar work you can easily justify the cost.

Many builders support the rebar in the footing with bricks or broken chunks of block placed about every 10 ft. This holds the steel the required minimum 3 in. above the bottom of the footing, but does not hold it securely in place. I tie rebar to the forms with form wire. To do this, drill two ¼-in. holes, 3 in. and 4 in. up from the bottom of each form board after it is placed,

and thread through some form wire. Weave the wire around the steel (photo facing page, top left) and tie it off outside the form. Then use a nail to twist the wire around the steel until it is snug. I tie off the rebar about every 6 ft. to 8 ft. down the line.

Once the footing forms are complete and the steel is in place, make one last check of the elevations with a transit or a level. Adjust them accordingly, and then remind your crew not to use the forms as a convenient step for climbing out of the crawl space.

Concrete—Around here, I can call the ready-mix dispatcher a couple of hours before I need the concrete and give him a tentative delivery time and quantity. If I have questions about the mix, the dispatcher can usually answer them. Some suppliers even have a field representative who can drop by the job site to answer questions and help verify quantities. The dispatcher will ask you what strength concrete you want, and how much slump (a measure of the consistency of concrete) you want it to have. A typical footing mix should be rated at least at 2,500 psi, with a 5-in. to 7-in. slump.

Depending on what part of the country you're in, the dispatcher will also ask you if you want fly ash or water reducer in the concrete, or if you want it to be air entrained. Air-entrained concrete has tiny air bubbles in it to improve its workability during cold weather. Water reducers allow the concrete to flow more easily without reducing its strength.

Fly ash is a by-product of coal-burning power plants. When used in a concrete mix for footings and walls, it enhances the mix in many ways. It slows down the set, adds more fines to the mix (which makes for smooth walls), helps the concrete to pump easily and strengthens the mix. Best of all, fly ash reduces the overall cost of the concrete.

Fly ash can be used as either a substitute or an additive in a mix design. A 15% fly-ash substitute for portland cement in a five-sack mix (five sacks of cement per cubic yard of concrete) is most common for a 2,500 to 3,000-psi spec. Generally speaking, suppliers pay $70 per ton for portland and only $15 per ton for fly ash. Try it in your walls and footings if fly ash is available in your area. But be careful on flatwork because fly-ash concrete tends to be sticky and hard to finish with a trowel.

While you are waiting for the ready-mix truck to arrive, take a look around the site and make sure the truck can get to as many sides of the forms as possible. If some are inaccessible, make a chute out of lumber to extend the truck's chute. Where the two meet, prop up the ends with a sawhorse.

There is nothing quite like the calm before you hear the diesel working its way toward the job. Take that time to make some decisions about who is going to do what. You will need at least one helper, and a second won't hurt. One worker runs the chute in concert with the driver while others push, pull and poke the wet concrete into the forms. Shovels and lengths of 2x4 are the typical tools for these tasks. Pound the sides of the forms with a rubber mallet every

few feet to discourage any voids from forming in the concrete.

Concrete with a 5-in. slump does not "seek its own level," and it takes a lot of work to move it around the forms. The temptation to add water to the mix is a strong one, but remember that extra water weakens the concrete. According to research conducted by the Portland Cement Association, every gallon of water added to a yard of concrete reduces the strength of the mix by 5%. Some builders order stronger concrete (a six-sack mix instead of a five-sack mix) in anticipation of cutting its strength back with water so that they can move it around the forms more easily. Another way to make the concrete flow better is to order it with a water reducer.

Once the forms are filled with concrete (or filled to the chalkline), use a straight piece of wood to screed it flat. After screeding, use a wood float to level the top of the concrete. The wood float will also roughen the surface, which improves the bond between the footing and the stemwall to come.

While the concrete is still wet, you've got to make a shear connection for the stemwall. This can either be a groove in the top of the footing, a row of bricks set on edge or steel pins made of rebar. I prefer the pins because they are inexpensive and easy to install. I cut rebar into 4-ft. lengths and push them about 5 in. into the leveled footing on 4-ft. centers (photo facing page, top right). Without this precaution, it's conceivable that the backfill pressure could cause the stemwall to slide off the footing.

Form materials—A standard batch of concrete weighs about 140 lb. per cu. ft., and it exerts considerable pressure on the form work— especially at the bottom. To keep walls straight and true, you need a material that is able to resist that kind of pressure. To build 4-ft. stemwalls, I use ¾-in. CDX plywood reinforced with a 2x4 framework. When the foundation is complete, I disassemble the forms and recycle the plywood as subflooring and the 2x4s as wall framing and blocking.

At assembly time, it's best to have all the form materials stacked inside the footing perimeter. I build a jig to crosscut the studs to 45-in. lengths. This dimension allows room for the thickness of the top and bottom plates. Then I make as many 16-ft. panels as I can with continuous 16-ft. plates and two sheets of plywood. The 2x4 frames are built first, and then I nail down the plywood. I space the studs on 16-in. centers, although 24-in. centers would do if the concrete were placed slowly. I tried assembling 2x4 frames with a nail gun, but I found that the nails are very difficult to remove after I'm through with the form work. So I stick to 16d box nails for putting the frames together.

Before I nail the plywood to the frames, I use a jig to locate the ⅝-in. holes for the snap ties. They fall on 16-in. centers between the studs, forming two horizontal rows (photo facing page, bottom left). One row is 12 in. down from the top of the form; the other is 12 in. above the base of the form.

After drilling the holes, I set the plywood atop the frames and nail the plywood off with two or

Footing rebar can be propped up on rocks, bricks or commercial rebar stands called chairs, bolsters or dobies. Rockhill doesn't like to worry about the rebar slipping off its perch during a pour, so he wires them to the forms (above). Rebar pins set into the concrete on 4-ft. centers (right) will help the stemwall resist lateral pressure. When pushing them into place, don't let them go beyond the concrete and into the soil, where they can rust.

Exterior forms are braced with 2x4 tiebacks to keep them aligned as the concrete is placed (above). The short section of form in the foreground is used to complete the run of modular form panels. At outside corners (below), overlapping walers reinforce the forms.

Two rows of snap ties protrude from the exterior half of the stemwall forms. The ties make a handy place to secure the top and bottom courses of rebar. The middle piece is wired to the vertical rebar. With the steel in place, the forms are sprayed with a mixture of motor oil and diesel fuel to keep the concrete from adhering to the plywood.

Preparing for the pour. Wiring rebar in position is a breeze with twister ties and a winder (top left). Metal wedges driven onto the ends of the snap ties press the walers against the stud framework of the forms (top right). As the concrete is distributed around the forms, a worker follows along with the vibrator to eliminate any voids in the concrete (cover). A piece of plywood used as a deflection shield (above) directs the flow of concrete from the chute into the forms. Once the fresh concrete is screeded flat across the tops of the forms, anchor bolts are set so their tops extend about ½ in. beyond the thickness of the plate material (right). Anchor bolts should be at least 1 ft. from the end of a wall and are usually set 4 ft. o. c. The nut is threaded onto the bolt to keep the wet concrete from clogging the threads.

three 8d nails per stud. On most panels, I run the plywood flush with the edges of the perimeter 2x4s. For panels that form inside corners, I let the plywood run 4¼ in. past the outside stud so it can overlap the intersecting panel.

When most of the panels are ready and the footing formwork has been stripped, I snap a chalkline on the footing to mark the outside edge of the stemwall. I erect the exterior forms first, nailing the neighboring panels to one another through the butting studs. Then I brace the panels with stakes and tiebacks placed about 4 ft. o. c. (middle right photo, p. 27). Once I've got the big panels up, I make a custom infill panel to complete the exterior wall of the forms.

The snap ties are inserted when the exterior walls are up. Their stems are sandwiched by two horizontal 2x4 walers (drawing, p. 25). The assembly is then snugged together by driving home the metal wedges. It's okay to let the walers run past outside corners to avoid making long 2x4s into short ones, but be sure to offset butt joints in the walers to avoid weak spots.

Overlap the walers at outside corners and nail them together (bottom right photo, p. 27). As good as snap ties are, they won't hold the corners together. So the intersecting walers are essential for keeping the forms solid at corners.

When all the exterior forms are up, check them for level and square. Because subsequent work will be affected by the accuracy of the stemwalls, their alignment is critical. Check all 90° corners of the foundation using the 3-4-5 method. Level the top plates using a builder's level or a transit rather than a spirit level. Use shim shingles to make the adjustments.

I attach the bottom plates of the leveled and squared forms to the footing with powder-actuated 16d fasteners in every other bay. Then I use a garden sprayer to coat the inside surface of the forms with oil to make it easier to remove them from the concrete. I use 1 gal. of used engine oil with 4 gal. of diesel fuel as a spray mix.

Rebar and interior forms—A typical 4-ft. stemwall will have three pieces of #4 rebar running horizontally on 12-in. centers. One piece runs down the middle; the other two are 1 ft. in from the top and bottom. It's no accident that the snap ties are on the same levels as the top and bottom lengths of rebar because they support these pieces (tie wire holds them in place). The middle rebar is tied to the vertical rebar pins (bottom left photo, p. 27).

Whenever I have a lot of rebar to tie off, I make sure I've got a good supply of twister ties (photo facing page, top left) and a couple of winders. The ties have loops at both ends. The loops fit over the winder's hook, and with a few cranks on the handle the wire is twisted tight. They save a tremendous amount of time.

After the steel is set, check the forms for spots that still need spraying with oil. But be careful not to get any oil on the reinforcing steel, because it will interfere with the bond between the concrete and the steel. Spray the inside surface of the interior forms before you put them in place.

Lifting the interior forms into position differs slightly from raising the exterior forms. At inside corners, one of the forms has a plywood overlap, and you've got to thread the snap ties through the holes in the plywood as the panels are raised. Prop your 16-ft. panels on the footing edge and start to tilt them up until they meet the ends of the snap ties. Thread the lower course of ties through the holes first. After you have done the same with the top course, you have closed the form.

After all the interior walls are up, lock them together with the walers and wedges (photo facing page, top right). Work carefully—the cones will break if you pound the wedges down all the way. Snug is enough. Now use a spirit level to align the tops of the interior and exterior forms. When you're done, secure the bottom plate to the footing with powder-actuated fasteners.

When the forms are up, check the plans and locate blockouts for beam pockets, crawl-space vents, sewage and water-supply lines. You can make these out of dimension lumber, short pieces of plastic pipe or hunks of Styrofoam that have been glued together. For vents, I typically make a box out of pressure-treated 2x stock that has been ripped to a width of 8 in. I tack it to the inside of the forms, where it remains cast in the concrete as a nailer for screening and trim.

None of the walls in the project depicted here is longer than 32 ft., so control joints weren't required. Generally speaking, walls that are longer than 36 ft. require a break in the wall to control cracking. Control-joint theory and practice are beyond the scope of this article. But if you need to know more about them, the Portland Cement Association (5420 Old Orchard Road, Skokie, Ill. 60077) publishes an informative pamphlet on their function and installation.

The pour—A 4-ft. stemwall, 8 in. wide and 10 ft. long, will contain about 1 cu. yd. of concrete. I calculate the amount of concrete I need to fill the forms, and I round up to the nearest half-yard to be on the safe side. Now I am ready to place my order.

Most builders use a five-sack mix for stemwalls, which yields 2,500 to 3,000-psi concrete. I like my walls to be smooth, so I prefer to use a six-sack mix (3,500 to 4,000 psi) and add a little water to it to get a 6-in. to 7-in. slump. I add the extra portland to the mix to offset the loss of strength caused by the additional water.

As the weight of fresh concrete in the forms doubles, the stress on the bottom of the forms quadruples. For this reason, it's a good idea to fill the forms in stages, or lifts. I do 4-ft. walls in two lifts. By the time I get all the way around the perimeter, the concrete has had a chance to take its initial set, taking some of the burden off the forms.

A good driver can regulate the flow off the chute and move the truck ahead slowly to make the pour a pretty slick operation. A plywood sheet about 2 ft. square is handy for diverting the concrete into the form as it comes off the chute (middle photo, facing page). Before the truck arrives, get a concrete vibrator or a rubber hammer, a wooden float and your anchor bolts.

As one worker directs the flow of the concrete into the forms, another should follow, stinging the concrete with a vibrator or rapping the forms with a rubber mallet. This is essential for filling air pockets and eliminating segregation. It also stresses your formwork to the maximum. You can feel it in your feet as the entire assembly resonates from the vibration. Overvibration of the concrete is a common mistake—three to five-second spurts every 3 ft. is plenty.

When you make the second lift be sure to vibrate the joint between the two. Sting the concrete a little extra around blockouts to eliminate the air pockets that typically occur there. As the concrete reaches the top of the form, strike it off level using the edges of the form as a screed. Don't forget to start setting anchor bolts before the concrete sets (bottom photo, facing page).

After the bolts are placed, scrape off any concrete that got on the forms and go home. Let everything sit for at least 24 hours (48 hours is better). I always try to get the concrete in before the weekend so we can start stripping the forms on Monday. It usually takes a day to strip all the forms, pull all the nails and stack the lumber.

I spray the green concrete with a curing compound, which slows the loss of water from the concrete. Both resin and latex-base curing compounds are available. I prefer the latex because it is less likely to clog the sprayer. If you plan to top the concrete with another finish, remove the curing compound with a sandblaster. Curing compound smells terrible but is worth the money because concrete has to cure properly to achieve its rated strength. Ignoring this step can reduce the concrete strength by as much as 50%. Remember, good concrete costs money and bad concrete costs more.

Gravel, insulation and backfill—I like to get gravel into the crawl space as soon as we get the form wood out. I use the smallest size I can afford, keeping in mind the hours someone will spend kneeling on it running mechanicals. Small gravel is comfortable, big is torture.

I also run insulation inside the walls. For a small foundation like this one, I put 4 in. to 6 in. of beadboard on the inside. This insulates the crawl space quite well.

If you backfill too soon, you will surely crack the concrete. After a few days, the concrete will be strong enough to take the compression loads of floor framing, but not the force of an entire truckload of gravel hitting it from the side. I wait to backfill at least until I've got my first-floor wall framing up. Concrete achieves most of its strength in 28 days.

Further information on concrete systems can be obtained from the Portland Cement Association (5420 Old Orchard Road, Skokie, Ill. 60077). Write for their free catalog of brochures. The National Association of Home Builders (15th and M Sts., N.W., Washington, D. C. 20005) has a useful publication entitled "Residential Concrete" ($15 to non-members, $12 to members), and the American Concrete Institute (Box 19150, Redford Station, Detroit, Mich. 48219) also has numerous publications available on the subject. □

Dan Rockhill is a contractor and an associate professor of architecture at the University of Kansas at Lawrence. Photos by the author.

Rubble-Trench Foundations
A simple, effective foundation system
for residential structures

by Elias Velonis

Although it was first used extensively by Frank Lloyd Wright early in the 20th century, the rubble or gravel-trench foundation has largely been ignored by builders since Wright's time—perhaps because it represents a different way of thinking about what it takes to support a house. The conventional poured-concrete or block perimeter wall attempts to solve a building's load-bearing requirements in monolithic fashion by creating a solid, supposedly immovable and leakproof barrier extending from a footing poured below frost line to 8 in. or more above grade. But since freezing water expands 9% by volume with a force of 150 tons per sq. in., monolithic foundations are unlikely to survive in frost country unless they include a footing-level perimeter drain backfilled with washed stone, which carries away water that might collect and freeze under or against the foundation wall.

The two functions of load-bearing and drainage are solved separately with a solid foundation, but the rubble-trench system unites these two functions in a single solution: the house is built on top of a drainage trench of compacted stone that is capped with a poured-concrete grade beam. The grade beam is above the frost line, but the rubble trench extends below it, and the building's weight is carried to the earth by the stones that fill the trench (drawing, facing page, center). The small airspaces around each stone allow groundwater to find its way easily to the perforated drainage pipe at the bottom of the trench. Atop the grade beam, a short stemwall of concrete block, poured concrete or pressure-treated wood is built to support the floor framing. Or you can pour a slab. More about this later.

While this foundation system has been time-tested in many of Wright's houses, acceptance by building officials and the codes they follow is still not assured. In *The Natural House* (Horizon Press, New York, 1954), Wright speaks of what he calls the dry wall footing. "All those footings at Taliesin have been perfectly static. Ever since I discovered the dry wall footing—about 1902—I have been building houses that way.... Occasionally there has been trouble getting the system authorized by building commissions."

The disapproval of a building inspector usually arises from a lack of familiarity with the technique, since the Uniform Building Code states clearly that any system is acceptable as long as it can "support safely the loads imposed." When I first approached our local building inspector with plans for a rubble-trench foundation, he studied them quietly for a moment, ahemmed in good New England fashion, and said, "Yep, that looks as if it oughta work." And so it will, except in what Wright calls "treacherous soils," which I would judge to be any soils with a bearing capacity of less than 1 ton per sq. ft.

Determining the bearing capacity of a soil without engineering analysis is a matter of common sense and experience. If the earth in the trench is dry, seems to be well drained, feels solid when you jump on it, and is a mixture of gravel, rock, sand or clayey sands, it will very likely carry all the weight your house can bear on it. If, on the other hand, your heels sink several inches into soft clay, loose sand or fine silt when you jump into the trench, you'd better consult a soils engineer.

Construction—Assuming you've got stable soil, bulldoze the area of the house level, clearing all topsoil away and saving it for fin-

---

'All those footings at Taliesin have been perfectly static. Ever since I discovered the dry wall footing—about 1902—I have been building houses that way.' —Frank Lloyd Wright

---

ish grading. If you have a sloping site, you will have to cut a level shelf in the hill, graded away from the house on all sides. This will ensure a good path for surface runoff. Lay out your foundation in the conventional manner (see "Site Layout," pp. 64-66), but make sure the batter boards are set up far enough outside the lines of the building that the backhoe will have room to maneuver. Sprinkle a line of lime 4 in. inside the strings that define the building's outer edge. This white line represents the center of the masonry wall that will rise up from the on-grade footing, or grade beam, and it provides the backhoe operator with a centerline to follow with his bucket. Ask the excavator if he has a narrow bucket for the backhoe—16 in. to 20 in. is perfect for

most soils. A wider trench gives you more bearing in softer soils, but it also takes more stone to fill it.

Have the backhoe operator cut the trench with straight sides, as deep as the frost line at the high point and sloping down to one or more outlet trenches along the perimeter (drawing, facing page, bottom). These should run away from the building and out to daylight at a slope of at least 1 in. in 8 ft. If you have a level site, I recommend running trench drains to a drywell, if your water table isn't too high. A drywell is a hole filled with a combination of small (1½-in.) stone and coarser rubble. You can base the depth and diameter of your drywell on the drainage qualities of your soil and the surface runoff you expect. Compute this from average-rainfall data and figures from the site's percolation test.

Clean up all your trenches by hand, making sure that their bottoms are flat and that they slope toward the drain line. Disturbed soil at the bottom of the trench may settle unevenly, so tamp the bottom firm with a pneumatic tamper or the heels of many boots.

Next, pour in a few inches of washed stone, and lay 4-in. dia. perforated PVC drainage pipe on top of it in the foundation and outlet trench. Make sure that the pipe follows the slope without dips that could restrict the flow of water. A ½-in. block taped to the end of a 4-ft. level makes the job of sloping the rigid pipe quite a bit easier. When the bubble reads level, you've got a 1-in. in 8-ft. slope.

I place the perforated pipe with its holes down (that is, at 4 o'clock and 8 o'clock), as I would in laying out a leach field, because as the trench fills with water, I think this orientation gets rid of it quicker. On the other hand, a case could be made for putting the holes up. It would take longer for them to silt up, but this shouldn't be a problem in good soils.

Now begin filling the trench with washed stone, taking care not to disturb the pipe as you cover it. I use 1½-in. stone because it's easy to find and easy to shovel, but larger washed stone is okay, too, as is the occasional clean fieldstone. (This is where the technique gets the name rubble trench.) Tamp the stone every vertical foot or so to make sure it is compact. To this end, I have even driven a loaded dump truck along the filled trench to make sure it was well settled, although this seemed to have little effect.

The outlet trench need not be filled with

From *Fine Homebuilding* magazine (December 1983) 18:66-68

Washed stone is dumped straight from the truck into the foundation and outlet trenches.

stone except for a foot in all directions around the pipe. Cover this stone with hay, burlap or tar paper as a filter, and backfill it with the original soil. If the pipe is running to daylight, be sure to leave its end exposed on a bed of stone. You want it to drain freely, so don't cover it with soil. Cap the end of the pipe with wire mesh to keep out rodents.

The grade beam—After the drains are installed and the trenches filled with stone, you're ready to build the forms for the grade beam. For one-story wood-frame structures, a 16-in. wide by 8-in. deep grade beam with three runs of ½-in. rebar is more than adequate. For a two-story structure, increase the depth of the beam to 10 in. or 12 in., and add two more lengths of rebar in its upper third.

Restring the lines from the batter boards and place the form boards on edge beneath them. I use 2x8s or 2x10s and brace them with stakes every 3 ft. Level the top edges of the form boards all around, nailing in 1x2 spreaders every 4 ft. to 6 ft. across the top to hold the forms in place. To reinforce the corners, use metal strapping or plumber's tape (perforated steel strapping), nailing it around outside corners. Place three runs of ½-in. rebar spaced evenly along the bottom of the beam, wiring them securely at the joints, and stagger these joints around the perimeter. Wire short pieces of rebar across these runs every 6 ft. Bend the lengths of rebar around all corners rather than splicing them there. Lift the rebar about 2 in. off the bottom of the beam with small stones. For a two-story building, prepare two more perimeter runs of rebar, and put them alongside the forms, ready to be dropped in the top third of the beam during the pour.

As the concrete is poured into the form, vi-

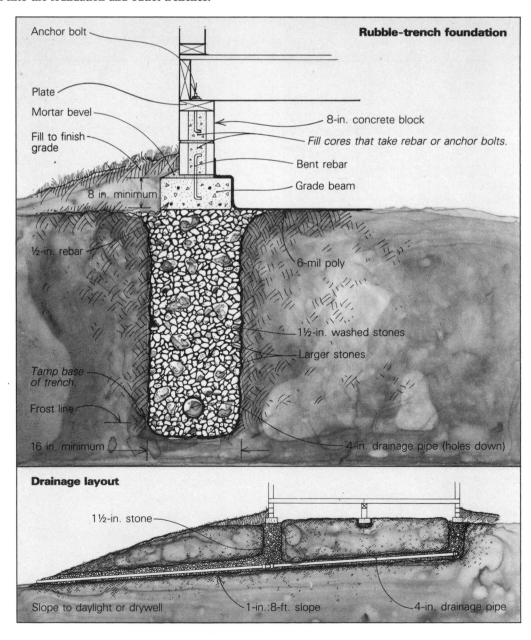

Rubble-trench foundation

Anchor bolt
Plate
Mortar bevel
Fill to finish grade
8 in. minimum
½-in. rebar
Tamp base of trench
Frost line
16 in. minimum

8-in. concrete block
Fill cores that take rebar or anchor bolts.
Bent rebar
Grade beam
6-mil poly
1½-in. washed stones
Larger stones
4-in. drainage pipe (holes down)

Drainage layout

1½-in. stone
Slope to daylight or drywell
1-in.:8-ft. slope
4-in. drainage pipe

Illustrations: Jackie Rogers

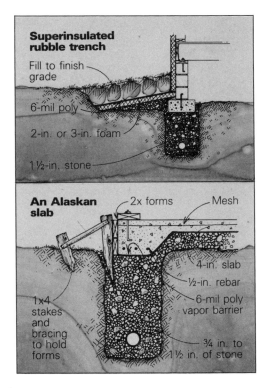

The finished grade beam, with its forms stripped.

Forming and pouring the grade beam. At left, forms for the grade beam are being set up on top of the stone-filled trenches. The 2x10 form boards are held together by steel plumber's strapping and by 1x2 wood stretchers nailed across their tops. Below left, rebar has been placed between the forms, and tied off to spreaders. The transit mixer is discharging its load of concrete, which is being spread and leveled by the crew of Heartwood students.

brate it well with a short piece of 2x4 to get rid of air pockets. Screed along the tops of the forms to get a level surface. If you intend to build a masonry wall, rough up the top of the grade beam with a broom before the concrete cures to ensure a good bond between it and the mortar. In areas where high winds are a problem, set 1-ft. lengths of bent rebar vertically in the top of the grade beam. They will anchor the stemwall. For a pressure-treated stemwall, place anchor bolts for the sill plate every 6 ft., and 1 in. from the end of each plate member.

If your design calls for a slab floor, you can pour the grade beam and slab at the same time, using a turned-down or "Alaskan" slab. For this you need only build outside forms, as shown in the second drawing below. For

Superinsulated rubble trench

Fill to finish grade

6-mil poly

2-in. or 3-in. foam

1½-in. stone

An Alaskan slab

2x forms

Mesh

1x4 stakes and bracing to hold forms

4-in. slab

½-in. rebar

6-mil poly vapor barrier

¾ in. to 1½ in. of stone

houses with wooden floors, however, the height of the stemwall will determine the height of the crawl space, which should be at least 26 in. Since this may make the level of the finished floor higher than you want, raise the grade around the perimeter by 1 ft. or so, which will help slope it away from the house for surface runoff. Leave adequate ventilation ports on the stemwall, and when the building is roofed, cover the earth in the crawl space with 6-mil poly. Place anchor bolts for the plate in the usual manner, and parge the outside of the wall that will be above grade for a clean appearance. You might also add a bevel of mortar between the stemwall and the grade beam, to shed any water that might want to find its way between them.

Insulation alternatives—If you've designed an energy-efficient building and want to extend the insulation down below grade on the exterior, modify the foundation as shown in the first drawing at left. In an area where the frost line is 4 ft. deep, dig the trench only 2 ft. to 2½ ft. below initial grade, as you will be raising the finished grade around the building by 1½ ft. to 2 ft. Fill the trench with stone to 8 in. below initial grade, and then pour the grade beam to the surface. Lay up four courses of block (surface-bonded block walls work very well in this application) and build the floor.

When you're ready to insulate the exterior of the building, glue 2-in. or 3-in. thick panels of rigid foam insulation to the exterior of the stemwall down to the grade beam, then lay more foam panels on a sloping bed (at least 3 in. in 1 ft.) of stone, 2 ft. to 4 ft. around the entire perimeter. This insulation apron preserves a large bubble of relatively warm earth beneath the house, tempering the crawl space with its warmth. Cover this apron with 6-mil poly to protect the foam from water, and backfill up to finish grade. The exposed foam above grade and below the siding should be covered with asbestos board or the equivalent, or parged (see pp. 49-51) with a surface-bonding compound troweled on ⅛ in. thick over a wet coat of Styrofoam adhesive. The adhesive helps eliminate expansion cracks in the surface bonding between panels, but if cracks do appear, they can be caulked.

The only reservation I've heard from other builders about the rubble-trench foundation is that it might settle unevenly. In non-uniform soils this might be a problem, although the reinforcing in the grade beam is ample to span a good deal of uneven settling. We've built four rubble-trench foundations over the last five years, and none has shown the slightest sign of settling, cracking or frost damage. The great advantage of the system, of course, besides speed and relatively low cost, is that instead of building a massive wall underground, you just pour stones into a trench and are free to carry on building above ground. □

Elias Velonis is the founder and co-director of Heartwood Owner-Builder School in Washington, Mass. Photos by the Heartwood staff.

Building a Concrete Bulkhead

In canal country, you can't build a house
unless you can keep the water away from your lot

by Will Rainey

A bulkhead is nothing more than a retaining wall used to prevent erosion of waterfront property. They're pretty common in towns near Houston, where a maze of canals fed by Galveston Bay gives people a chance to live near the water and provides them with a place to keep a boat at the back door.

Bulkheads are basically simple devices. The wall that separates land from canal is built of planks, or "sheet piles," embedded edge to edge in the silty bottom of the canal. These piles are either treated wood timber, precast concrete or metal panels. To keep the tops of the piles from bowing outward from the pressure of retained soil, they're anchored with metal tie rods to timber piles, or concrete "deadmen" buried in the soil.

When I started looking for a waterfront lot several years ago, it became evident that if I purchased one with a bulkhead in good shape, I couldn't afford to build a house on it. So I looked for a lot with a bulkhead in need of repair, or a lot that needed an entirely new one.

After a lengthy search, I found a triangular lot at the end of a cul-de-sac backing up to a canal 90 ft. wide. When the canals were dredged and the original bulkheads installed 20 years ago, the developer had used the most economical material available at the time—wood sheet piles. Now the creosoted piling of full-dimension tongue-and-groove 2x8s was badly deteriorated, and the soil behind the piles was gradually washing into the canal. It could no longer keep the canal at bay. Since the wood had decayed beyond repair, I checked my bank balance and started looking at the problems of installing a new bulkhead.

Choosing a bulkhead material—A bulkhead is not easy to replace, so you want to do the job right the first time. The choice of materials is probably the most important thing to consider since it determines the durability of the structure and the construction techniques used to build it. There's also a significant difference in cost between the various alternatives, and I had to weigh these carefully because my budget was tight. Here's what I found out about the four materials that are commonly used for bulkheads.

Wood. The most popular material for sheet-pile bulkheads is 2x6 or 2x8 tongue-and-groove yellow pine, since it's about the easiest to replace when it gives out and is relatively inexpensive. But it has to be chemically treated to keep marine borers away. These nasty critters are worm-like mollusks, marine termites if you will, that are attracted to wood. The water pollution in Galveston Bay kept them under control for many years, but the water quality is beginning to improve now, and the borers are back.

Creosote is the traditional favorite for treating wood, but I wasn't overly enthusiastic about it— I owned a prime example of its lack of durability. One modern treatment is chromated copper arsenate (CCA), the bluish-green stuff you often see on pressure-treated lumber at your local lumberyard. But don't use lumberyard-grade material for bulkheads because it won't hold up to constant immersion in water. Wood piles should be CCA-treated to marine-grade standards, which call for considerably more CCA to be retained in the wood. Lumberyards don't usually stock the stuff, so you'll have to look for a specialty supplier. A problem with using this kind of material is that the copper in CCA will corrode aluminum in short order. This presented a problem because I originally planned to use aluminum tie rods; steel rods can't be used in timber sheet piles because it's hard to keep the salt water from rusting them away. This was the major reason many of the original bulkheads in the area had failed.

Steel. Steel sheet piles, ¼ in. to 1 in. thick, are the next step up the cost ladder. I figured that with the short lengths and the relatively small quantity I needed, I could buy inexpensive damaged or salvaged steel and cut out enough usable material to do the job. As I looked into it, however, I found that steel and sea water are not compatible.

Ask your local high-school chemistry teacher and you will quickly learn that the chemical reaction between steel and salt water methodically oxidizes and destroys the steel. The process is essentially the same as the one that occurs in your car's battery, where electrical current is produced by the chemical reaction between the acid and the lead plates. In steel exposed to salt water, the product of this chemical reaction is iron oxide—rust. But it can be controlled.

Just as recharging your car battery controls oxide buildup on the lead plates, a trickle of current from a DC power source, applied directly to the steel piles, can minimize the formation of rust. Don't try it yourself, though. Contact an engineer who specializes in the subject. Another way to control the chemical reaction between steel and salt water is to weld small aluminum or zinc bars—anodes—to the steel below the water line. These anodes will be oxidized instead of the steel, and so are often referred to as "sacrificial" anodes. They usually have to be replaced every five or six years, and that can be a nuisance. The third way to protect the steel is to keep it from contacting the salt water in the first place by coating it with a fusion-bonded epoxy (more on this later) or industrial-grade epoxy paints. But any coating will fail unless it covers 100% of the steel—even a pinhole gap in the coverage will quickly lead to problems.

Concrete. Concrete poured in place is rarely used for bulkheads because it takes special expertise to pour concrete directly in water, particularly in salt water. Precast piles 4 ft. or 5 ft. wide are used instead. Concrete bulkheads are more expensive than wood or steel, partly because you have to rent a crane to position the heavy panels. But the fabrication work is labor intensive, and I figured the costs would be manageable if I did a lot of the work myself.

Even though concrete is a durable material, constant immersion in salt water can be tough on it. Conventionally reinforced concrete used in salt water tends to crack and spall as the salts penetrate the concrete and attack the rebar. The rust swells as it forms, and the expansion causes the concrete to crack. Once a crack has formed, it opens a direct path for the salt water to reach the steel rebar, accelerating the deterioration. This problem can be solved in two ways. To prevent salt from reaching the rebar, you have to take special care to seal all surfaces of the concrete. There are products on the market that can do this (one is SIL-ACT Silane, made by Advanced Chemical Technologies Co., 2601 Northwest Expy., Oklahoma City, Okla. 73112), though they can be expensive.

The alternative is to protect rebar from corrosion by factory-coating it with fusion-bonded epoxy before it's put in the concrete. Epoxy-coated rebar was originally developed for use in cold-climate bridges that had to withstand frequent doses of de-icing salts. In fusion bonding, the metal to be coated is cleaned, heated and sprayed with a powdered epoxy. The hot steel melts the powder, causing it to bond to the metal for a very effective protective coating. One company that does this work is ABC Coating Co. Inc., PO Box 9693, Tulsa, Okla. 74157.

Aluminum. Aluminum is considerably more expensive than any of the other materials, but there are some labor tradeoffs because it's easier to handle and install. It's available in about the same thicknesses as steel piles, but to buy it at a reasonable cost I would have had to buy ten times the quantity I needed. There wasn't much chance for shopping around, either—I could find only one company in the U. S. that manufactures it (Ravens Metal Products, Inc., Marine Division, PO Box 1835, Sanford, Fla. 32771).

Concrete: a good compromise—Though none of the four materials available for bulkheads was perfect, I decided that concrete was a reasonable compromise in terms of expense, durability and ease of installation. But it did present some problems with logistics.

I might have been able to purchase precast concrete piles with a larger order, but I decided to pour my own to save money. Because of the small size of my lot, I had to figure out a way to pour them and still have room to bring a small crane in to set them in place. The solution was to pour them flat on the ground and make each successive pour on top of previous ones, gradually building stacks of concrete slabs. Unfortunately, this plan meant that I couldn't waterproof the concrete, since I couldn't get to the bottom of each slab once it was poured. So I decided to use the fusion-bonded epoxy rebar.

As an additional precaution, I figured on making the slabs relatively thick in order to put more concrete between the rebar and the water. To withstand the loads at my lot, I needed piles that were only 5 in. thick. But to protect the rebar, I designed 8-in. thick piles, with rebar midway between the two faces, and called for a denser concrete mix than is usually used.

Building the forms—All told, I had to pour 27 14-ft. long piles to protect the 110 ft. of canal frontage: 25 of them 4 ft. wide, and two of them 5 ft. wide. But because I poured each pile on top of the one below, I didn't have to worry about enclosing the forms completely—all I needed to provide were the edges of the forms and a good solid surface to start stacking everything on. Once I got started, I simply separated the pours with plastic sheeting. The first pour was made on a bed of sand that had been compacted with a gas-driven compactor—I call them ground pounders—and screeded level. A layer of 4-mil polyethylene over the sand served as the bottom of the first pour.

I built the forms from 2x10s ripped down to a full 8 in., the thickness of the finished piles. To help them withstand the pressure of the wet concrete, I nailed a pair of 2x4s on edge to the outside face of each formboard, creating a C-shaped member in section, and extended them to form a sort of finger joint at the corners of the forms (top photo, facing page). A ⅜-in. dia. bolt run vertically through the joint pinned adjoining sides of each form together, and made it easy to disassemble later on. I ripped another 2x10 with two angled cuts to form both the tongue and the groove of the finished piles (drawing, facing page), and nailed those to the inside edges of the forms. To the bottom edge of the forms I nailed a small chamfer strip, which supported the forms as they were placed on top of the previously poured pile. At the bottom of each form, in one corner, I used a 2x block to form a "snipe" (top photo, facing page). The effect was to "cut off" a corner of the finished pile. As each finished pile was jetted into place, this helped to push it tight against the previously installed pile. If you've ever driven single-bevel wood stakes into the ground, you know how this works. Because the first pile installed had nothing to push against, it had a snipe on both lower corners.

The #4 rebar was tied with steel wire and hung from the sides of the forms with copper wire—steel wire is more prone to rust, and stirrups would have provided a path for corrosion to enter the concrete, defeating my other attempts to protect the rebar. To provide a tie between the top of the piles and the concrete cap that I would eventually pour over them, I drilled holes in the top end of each form and stuck the ends of the rebar right through. I put two weep holes toward the top of each pile, forming them from ¾-in. OD PVC pipe. Weep holes are important because they allow excess moisture on the land side of the bulkhead to drain to the canal. They also allow the water table to stay at approximately the same level as the canal.

Before the piles were ready to pour, though, there was one more element to be added to the forms, one that became the key to installing them. Since the soil in the canal was light, mostly silt and sand, I could "jet" the piles into place. Jetting is a technique that uses pressurized water to blast the silt and soil out from under piles as they're being lowered into place. The alternative is to pound them into place with a pile driver, an expensive proposition.

I borrowed an idea from heavy construction for the jetting. Each pile I poured had a built-in jetting device: a 1½-in. OD PVC pipe manifold cast into the bottom (photo facing page, left). This was connected to a pair of PVC pipes lead-

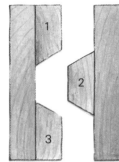

A 2x10 ripped twice at an angle created strips for tongue and groove of pile forms.

Forming the piles. The concrete forms for each bulkhead pile were strengthened with 2x4s on edge (above), and pinned together at the corners with a single bolt. Wood spreader bars help to keep the long sides of the forms from bowing out, while plastic sheeting separates individual pours from the ones beneath. A 2x10 ripped twice at an angle (drawing, left) formed the molds for the tongue and groove. The key to installing the bulkhead was a manifold of PVC pipe that was cast into the bottom of each concrete pile (photo left). Using a technique called jetting, a high-pressure water pump forced canal water through the manifold to loosen sand and silt beneath the pile, which gradually sank 7 ft. into the canal bottom from its own weight. Reinforcing each pile is a grid of fusion-bonded epoxy-coated rebar.

ing to the top of the pile, where a steel fitting connected them to a flexible fire hose and a high-pressure pump. In effect, the system would allow each pile to dig its own hole.

Designing and testing the concrete—The success of the bulkhead depended in part on the characteristics of the concrete I used. A strong, dense concrete provides additional corrosion protection for the rebar. In standard vertical pours, the weight of the concrete itself increases density. But since I was pouring the slabs flat, I had to get the density some other way. I started with a 4,000-psi mix at 28 days. To keep the amount of water in the mix to a very low level without making the concrete difficult to work with, I specified an additive called a "high-range water reducer." It's simply a chemical that makes the concrete behave like it has a lot of water in it, allowing the mix to flow more readily. After 30 min. or so, the effects of the additive dissipates, and the concrete returns to its normal state. Adding water to do the same thing would weaken the final strength of the concrete. About 3 gal. of the additive in a 7-yd. load of concrete turned what would have been pretty stiff 2-in. slump concrete into flowing 8-in. slump concrete, and made finishing the concrete fairly easy. To give me a little more working time during the pour, I also called for the addition of a retarding additive to slow the setup time. The use of additive combinations is a good way to customize a concrete mix, but anything you put in the concrete should be from

the same company since one company's product isn't always compatible with another's.

When the strength of concrete is particularly important, it's a good idea to take samples from each load and test them as the job progresses. On large commercial or industrial projects this is fairly common, and there's a standard procedure for the testing. As the concrete is poured from the truck, a small portion is placed into a cylindrical form about 6 in. in dia. and 12 in. tall. The concrete is allowed to cure at the job site, and is then taken to an independent laboratory where the cylinder is compressed until it fails (you'll find a lab in just about any large city). The ultimate strength of the concrete can be determined from the results of the test.

I took test cylinders from each load, but in my project the results of the test wouldn't be available until long after the pour was complete, too late to make any corrections. I made the tests mostly to keep the ready-mix folks honest. And I made a point of buying the test cylinder forms from them so they'd know I was keeping close track of the mix. As it turned out, the concrete was plenty strong. Some of the cylinders broke at 6,000 psi, and none at less than 4,000 psi.

Making the pour—Designing the concrete mix was one thing, but getting the correct mix was quite another. The first load showed up without the water-reducing admixture, so I had to make the pour with stiff 2-in. slump concrete. Fortunately the forms were on the ground and I didn't have any trouble getting the concrete out

of the truck. The next two loads were missing the admixture, too, so I made a trip to the ready-mix company and picked up the admixture myself. I added it to the remaining loads when the trucks showed up, mixing it in the truck for about five minutes before making the pour. As the concrete was being placed, I used a vibrator to help consolidate it.

As each slab was poured, I finished it by hand with a wood float and covered it with polyethylene to slow curing and increase strength.

Placing the bulkhead piles—When all the piles were complete and in stacks five deep, it was time to get them into the canal. Rather than go to the trouble of removing the old wood piles, I decided to leave them in place and run the new piles just in front of them.

The process of installing the bulkhead piles took about three-and-a-half days (with several hours lost to pump failures), and depended on the muscle of a rented 14-ton crane (photo, p. 33). As each pile was lifted from the stack, I connected the manifold fittings on top to a 10-hp water pump, using a long length of 1½-in. dia. fire hose. A second length of hose stretched from the pump to the canal for water intake. The pile was then hoisted into the canal, and I fired up the pump. As it sucked canal water and forced it through the manifold inside the pile, the force of 140 gal./min. at 100 psi blasted away silt and sand from beneath. This large volume of water fluidized the sand and allowed the pile to sink under its own weight. When each

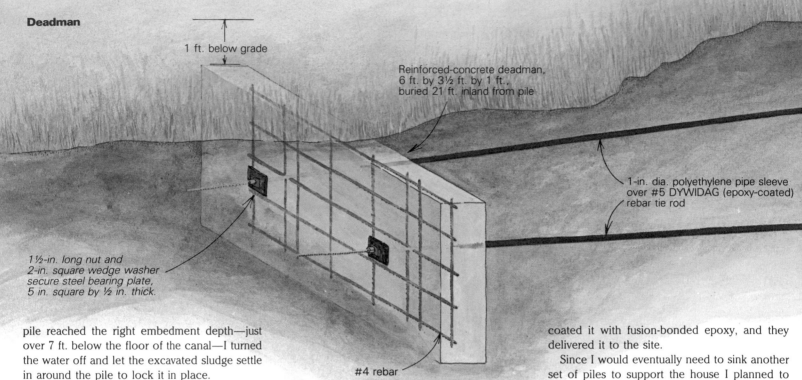

Deadman

1 ft. below grade

Reinforced-concrete deadman, 6 ft. by 3½ ft. by 1 ft., buried 21 ft. inland from pile

1-in. dia. polyethylene pipe sleeve over #5 DYWIDAG (epoxy-coated) rebar tie rod

1½-in. long nut and 2-in. square wedge washer secure steel bearing plate, 5 in. square by ½ in. thick.

#4 rebar

pile reached the right embedment depth—just over 7 ft. below the floor of the canal—I turned the water off and let the excavated sludge settle in around the pile to lock it in place.

Though this system worked pretty well, it was sometimes difficult to keep the pile plumb as it was being lowered into place. This was solved by connecting a third 1½-in. OD hose to the pump and fitting it with a 15-ft. length of ¾-in. metal pipe that I used as a hand jet. With this I could blow sand away from either side of the pile to keep it in line and plumb.

When all 27 piles were in place and jetted to the proper depth, I placed 12-in. wide strips of filter fabric against the land side of each joint to keep sand and soil from washing through the cracks. The material is a non-woven, needle-punched polypropylene, and the force of the sand backfill kept it in place. Sand is the best backfill because it drains quickly and doesn't expand or contract as its moisture content changes.

There comes a time in nearly any project when a little bit of forethought will lead to big savings in time and effort later on, and that point on this job came when the backfill reached a level just short of the weep holes. I knew that later on I'd have to attach forms to the piles for the continuous concrete cap I planned to pour along their top edges. So before any more fill was added, I threaded the looped end of a length of ¼-in. dia. polypropylene rope through each weep hole. The loop was anchored on the land side of the piling with a 20d nail, sort of like a cotter pin through the

end of a bolt. Later on this rope would tie the forms tightly against the piles.

There was one more task to complete before the backfilling could resume. I made a long tube of filter fabric and draped it along the length of the bulkhead just behind the weep holes. Periodic gaps that I had left in the tube's seam allowed me to shovel it full of gravel, and the combination serves as a drain behind the weep holes. The filter fabric will keep the gravel channel open and prevent the sand backfill from washing through the weep holes as the tide goes in and out.

Installing the tie rods—Various materials can be used for these crucial elements, but because I chose concrete for the piles, steel made the most sense for the tie rods. I chose DYWIDAG rebar (DYWIDAG Systems International, USA, Inc., 301 Marmon Drive, Lemont, Ill. 60439). This grade-60 product has helical deformations along its length instead of the ribbed deformations typical of standard rebar. These not only offer a solid connection to the concrete, but also allow DYWIDAG to be easily threaded into couplings that enable it to be extended. I had the rebar shipped directly to a company that

coated it with fusion-bonded epoxy, and they delivered it to the site.

Since I would eventually need to sink another set of piles to support the house I planned to build on the lot, I carefully located the bulkhead tie rods so as not to interfere with the future house piles. With a rented trenching machine I gouged a ditch for each rod, angling it into the ground on a slope away from the bulkhead. The trencher also started holes for the deadmen that would anchor the tie rods. A bit of effort applied to a good shovel helped me to finish the holes, 4½ ft. deep by 6 ft. long, and 1 ft. wide.

Before putting the rods in place, I slipped polyethylene pipe over each one from the bend all the way to the end as extra protection for the epoxy-coated steel. The ends of the pipes were sealed with duct tape to keep out any dirt or water. The bent end of each was placed in the cap forms, and the rod was simply laid in the trench with its other end extending through the hole for the deadman. The tie rods were then covered over with soil, leaving the rod extending through the hole. I lined the holes with sheet plastic (to prevent the soil from absorbing moisture from the concrete before it could cure), tied a skeleton of rebar together and hung it in each hole by tying it to the exposed section of the tie rods. Then I simply filled the holes with concrete to anchor the ends of the rods.

Forming the cap—Next came the poured concrete cap for the piles. Forming the cap on the land side was fairly easy—the backfill

The finished bulkhead has a 2x12 wood bumper, bolted into the concrete cap as a replaceable guard against damage by small boats tied up alongside.

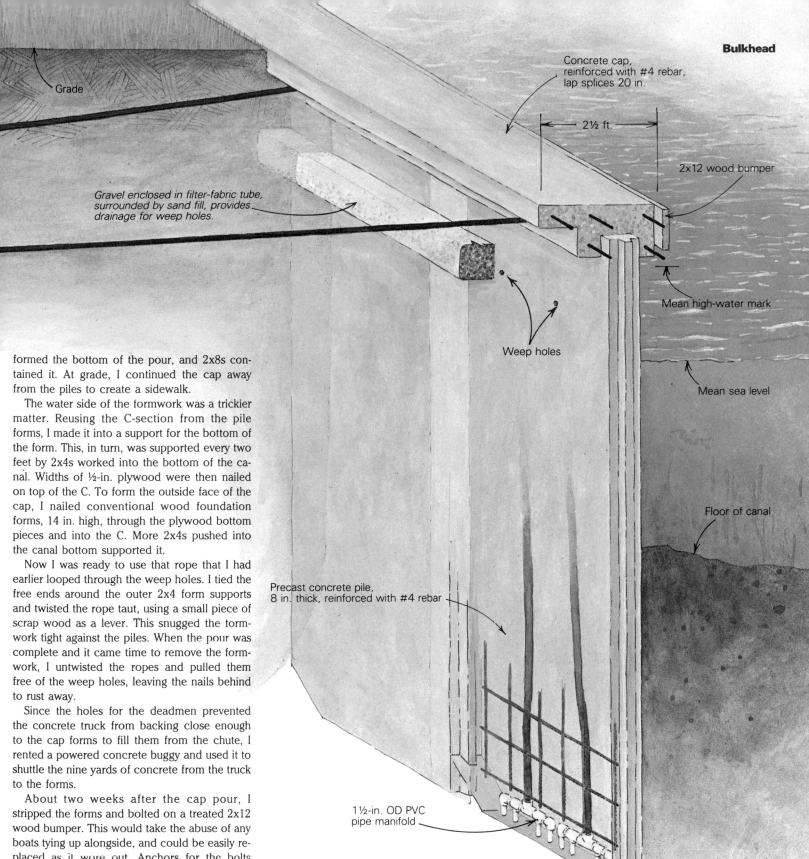

Grade

Gravel enclosed in filter-fabric tube, surrounded by sand fill, provides drainage for weep holes.

Concrete cap, reinforced with #4 rebar, lap splices 20 in.

← 2½ ft. →

2x12 wood bumper

Weep holes

Mean high-water mark

Mean sea level

Floor of canal

Precast concrete pile, 8 in. thick, reinforced with #4 rebar

1½-in. OD PVC pipe manifold

formed the bottom of the pour, and 2x8s contained it. At grade, I continued the cap away from the piles to create a sidewalk.

The water side of the formwork was a trickier matter. Reusing the C-section from the pile forms, I made it into a support for the bottom of the form. This, in turn, was supported every two feet by 2x4s worked into the bottom of the canal. Widths of ½-in. plywood were then nailed on top of the C. To form the outside face of the cap, I nailed conventional wood foundation forms, 14 in. high, through the plywood bottom pieces and into the C. More 2x4s pushed into the canal bottom supported it.

Now I was ready to use that rope that I had earlier looped through the weep holes. I tied the free ends around the outer 2x4 form supports and twisted the rope taut, using a small piece of scrap wood as a lever. This snugged the formwork tight against the piles. When the pour was complete and it came time to remove the formwork, I untwisted the ropes and pulled them free of the weep holes, leaving the nails behind to rust away.

Since the holes for the deadmen prevented the concrete truck from backing close enough to the cap forms to fill them from the chute, I rented a powered concrete buggy and used it to shuttle the nine yards of concrete from the truck to the forms.

About two weeks after the cap pour, I stripped the forms and bolted on a treated 2x12 wood bumper. This would take the abuse of any boats tying up alongside, and could be easily replaced as it wore out. Anchors for the bolts were made from lengths of solid drilled and tapped plastic rod that had been bolted to the cap forms just before pouring.

Finishing up—After pouring nearly 55 yards of concrete for the project, the only task remaining was to tighten the tie-rod connection between piles and deadmen. I uncovered the ends of the tie rods behind the deadmen. With a hacksaw blade, I cut the plastic pipe surrounding the rebar and removed the excess pipe to expose the end of the tie rod. I doped the outside face of the deadmen with roofing cement and slipped a square steel bearing plate over the

end of the rebar, seating it into the tar-like goo. A wedge washer accommodated any misalignment between the tie rods and the deadmen, and a nut threaded on the rebar locked the assembly in place. I needed some sort of pressure plate for hydraulic jacks to bear against, so I drilled a hole through the walls of a piece of 2-in. square, ¼-in. wall steel tubing and slipped it over the rebar, holding it in place with another nut threaded on the rebar. Then, with a pair of two-ton hydraulic jacks placed horizontally between the deadman and the steel tubing, I pulled the tie rods tight and turned the first nuts down snug against the bearing plate and wedge wash-

er. After covering the exposed tie-rod ends with concrete, I covered the deadmen with sand, and the job was done. Now time and nature will judge the effectiveness of my efforts at creating a "permanent" structure. □

Will Rainey is a licensed professional engineer employed by a consulting firm in Houston, and a weekend builder. Photos by the author.

Concrete and Masonry Fasteners

A survey of anchors reveals some surprising alternatives to lead and plastic

by Kevin Ireton

"**I**f a horse can't eat it, then I don't like it." That's what Richie Allen, then a first baseman for the Philadelphia Phillies, said when Astro-Turf was installed at the Houston Astrodome. As a carpenter, I've always had similar feelings about concrete. If my saw can't cut it and if I can't hammer a nail into it, then I don't like it. But just as Allen had to play on AstroTurf, I've had to work with concrete.

I've never been comfortable fastening material to masonry—attaching a ledger board to a foundation wall to support a deck, installing a threshold on a concrete slab or hanging kitchen cabinets on a brick wall. Like most residential carpenters, I usually chose lead anchors, also called lag shields. These were the only anchors I knew about, and the only ones available in most hardware stores and lumberyards, except for plastic anchors, which I always thought were pretty wimpy. Powder-actuated fasteners were not appropriate either, because I didn't want to risk splitting the wood, or blowing out the concrete and therefore having to use additional fasteners in one place.

I recently discovered that there's a whole industry out there devoted to concrete and masonry fasteners, with trade associations, testing and regulatory agencies, manufacturers, importers, distributors, and lots of nifty fasteners (some of the manufacturers, and the types of fasteners they make, are listed in the chart on p. 43). But most of the applications for these fasteners fall into the domain of commercial/industrial construction, and that is where the marketing efforts are aimed. Many residential

Concrete screws work like woodscrews. After drilling a pilot hole (top), the drill bit is retracted into the shaft of the installation tool, a Phillips-head bit is inserted and the screw is driven into the hole (middle), cutting its own threads in the concrete (bottom).

builders don't know about all the different masonry anchors available or where to buy them, which is a shame.

Drills and drilling—The hardest part of any anchor installation is drilling into concrete. I've drilled plenty of holes with my standard electric drill. But if you have to drill a lot of holes at one time, especially in the larger sizes (over ¼ in.), buying a hammer drill is a good investment. These heavy-duty tools combine a high-speed reciprocating action with the drilling operation. The impact of the drill bit works to shatter any aggregate that gets in the way, and the reciprocation helps remove dust and debris.

Most tool manufacturers now make ⅜-in. hammer drills, which can drill up to ½-in. dia. holes in concrete, more than big enough for most residential jobs. Look for a hammer drill that's variable speed and reversible, and be sure to get a side handle and depth gauge.

Selecting the right drill bit is even more important than what drill you use. Part of my distaste for drilling into concrete stems from buying cheap bits, ruining them in the first few holes, then continuing to use them. In order to bore into concrete or masonry, a drill bit needs to have a carbide tip, and carbide-tipped tools are expensive. For years I balked at paying the price for mid-range and top-of-the-line carbide drill bits. Now I know better.

If you're going to use the bit in a hammer drill, be sure to buy one that's designed for this use. Carbide bits for standard rotary drills won't stand up to the impact of a hammer drill.

Most masonry drill bits are coated for corrosion resistance. In general the bright finishes, like zinc, are there to appeal to consumers and

Carbide-tipped masonry bits

Fast-spiral flutes

Regular twist flutes

Lead anchors

Woodscrew anchor

Lag-screw expansion shield

All fasteners are shown actual size.

begin to flake off after a few uses. Bits for commercial use usually have a black finish, so buy these if you can. They're probably better bits.

Stay away from drill bits with fast-spiral flutes (facing page, bottom), which look more like a bolt thread than a twist bit. They're the cheapest bits to manufacture and to buy, so they abound in hardware stores. Often you'll want to drill, in one operation, through wood into concrete, and the fast-spiral bits don't work well in wood. The flutes are too small to auger large chips of wood out of the hole.

Once you've got a decent drill bit, the next priority is not to ruin it. Drill slowly and let the drill and bit do the work. Don't lean all your weight on the drill to speed things up—this will overheat the bit and destroy its temper.

If progress stops partway into a hole, it means that you've hit a piece of aggregate or rebar. Pull the bit out and look at the tip. If there are metal shavings on it, you hit rebar. Don't try to drill through it, you'll just hurt the bit. Cut your losses and drill somewhere else.

If there aren't any metal shavings on the bit, you probably hit a piece of aggregate. Put an awl or drift punch in the hole and hit it with a hammer. This will fracture the stone, and then you can continue drilling. Although it's more important with some anchors than it is with others, it's always a good idea to clean out the hole before you set the anchor.

Lead shields and anchors—For years lead was the cornerstone of the masonry-fastener industry. Its malleability lets these anchors conform to the rough surface of concrete or masonry. Unfortunately, that same malleability limits their holding power and also makes them susceptible to working loose through vibration. A lot of lead anchors are still available today, including single and double-wedge expansion shields, lag-screw expansion shields and wood-screw anchors (facing page, bottom). The latter two are what you'll get when you go into most hardware stores and ask for concrete anchors.

The big drawback to lead anchors is that they require hole spotting. This means you have to drill a screw or bolt-size hole through whatever you're mounting, hold the piece in place on the concrete or masonry surface, and mark the hole. Then you put the piece down and drill a larger hole in the concrete for the lead anchor. And since drill bits tend to skate around on concrete before boring in, it's easy to end up with a misaligned anchor.

One of the manufacturers I spoke with referred to lead anchors as dinosaurs, but, he said, they're still made because they're so popular in consumer markets. Tradition makes them sought after, though, not design. There are better anchors out there.

Flush anchors—Also called drop-ins, these machine-thread anchors are made of lead, steel and a variety of alloys. The steel anchors (below left) are some of the strongest available. Most operate by means of internal wedges that expand the outer sleeve. Some require a setting tool to install, others use the machine bolt to expand them in the hole. Hole depth is critical with machine-thread anchors, but this also means the anchors install flush with the surface, which is an advantage in certain circumstances.

Wedge and sleeve anchors—In addition to being much stronger than lead, the major advantage of these anchors is that you can drill one hole, the same size, through both materials you're fastening. If you were hanging a cabinet on a concrete wall, for instance, you would simply drill through the mounting rail into the concrete, insert the anchor and tighten it.

Both wedge and sleeve anchors are made of steel—a resilient material, like concrete itself—and achieve their strength by actually compressing the concrete. They work by means of a threaded rod with a cone-shaped tip (below right). The tip end is inserted into the hole, and a nut or screwhead is tightened on the other end. With wedge anchors, the cone-shaped tip pulls against a short sleeve with barbs in it that wedge in place near the bottom of the hole. They are extremely strong and impossible to remove once installed. They're often used in highway bridge renovation to anchor guardrails.

Wedge anchors can be used to anchor sill plates to foundation walls. I know at least one builder who prefers them to poured-in-place anchor bolts because they allow him to put an anchor wherever needed to straighten crooked plates. You can get a box of 25 ½-in. by 5½-in. wedge anchors for about $25 to $30.

With sleeve anchors, the cone-shaped tip pulls against a long metal sleeve with slots cut in it. The slots cause the sleeve to flare outward and exert considerable pressure against the sides of the hole. These anchors aren't as strong as wedge anchors, but have the advantage of being available with different head styles. You can get them with a stud bolt and nut, a hex head, a round head or with a flat slotted head.

Sleeve anchors provide a larger bearing surface than wedge anchors and so will work in a wider variety of applications. In their technical notes on brick construction, the Brick Institute of America (11490 Commerce Park Dr., Reston, Va. 22091) rates sleeve anchors as the most versatile for brick masonry.

Occasionally wedge and sleeve anchors will

Setting tool

Wedge and sleeve anchors

Flush anchors (machine-thread)

Lead expansion shield

Lead double-wedge expansion shield

Steel drop-in anchor

Sleeve anchor

Sleeve anchor

Wedge anchor

spin in the hole without grabbing, usually because the hole is too big. If this happens, use a hammer or a flat bar to pry outward on the anchor while continuing to turn the nut or screwhead. You have to be careful not to tighten wedge and sleeve anchors too much or you'll strip them. Most manufacturers recommend turning the nut or screwhead three to five turns after the anchor first engages.

Chemical fastening systems—The newest and fastest-growing species of fasteners is the adhesive chemical anchor, which comes in two forms, caulking-type tubes and glass capsules. Both involve setting a threaded rod, or sometimes just a piece of rebar, into a hole half filled with a mixture of resin, quartz aggregate and a hardening catalyst. Curing time varies from ½ hour to 24 hours, depending on the temperature of the concrete. The resulting bond is incredibly strong because the epoxy-like mixture flows all around the threaded steel and into the pores of the surrounding concrete or masonry.

One of the chief advantages of chemical anchors is that they don't stress the base material into which they're set. They can be used near

Chemical anchoring. In hollow walls, a tubular screen contains the adhesive.

In solid masonry, glass capsules are shattered by a bolt, which releases their chemicals.

the edge of a concrete wall or slab, where an expansion bolt would be liable to break the concrete. Chemical anchors are also highly resistant to corrosion and vibration.

To install chemical anchors, you have to drill a hole to the specified depth, then clean out all the debris, using a vacuum cleaner, compressed air, or a blow-out bulb. Cleaning out the debris is a very important step.

Tube systems work in a couple of ways. The simplest has all the ingredients in a single tube that fits a standard caulking gun (below left). Just before using it, you mix the ingredients by moving an internal agitator, either by hand or with an electric drill.

A much more sophisticated system, called Polly-All (Ackerman Johnson Fastening Systems) involves a pneumatic caulking gun that holds two tubes, one with resin, the other with catalyst. As you squeeze the trigger the chemicals are automatically dispensed in the proper amount and mix outside the tubes.

The equipment for this system is expensive (over $450 for the pneumatic caulking gun alone), but since the epoxy is mixed outside the tubes you can use a portion of a tube and save the rest. It is worthwhile if you are doing a lot of chemical anchoring. The manufacturer is working on a less expensive, manual version of the gun.

Both systems allow you to squirt epoxy wherever you need it, in whatever amount you need. With the use of a tubular screen, they will also work in hollow walls (top photo at left), like concrete block. The tubular screen contains the epoxy just enough to keep it from running out of the hole, and also lets it mushroom inside the hollow space for a more secure anchor.

The other type of chemical anchor comes in long capsules made of glass (below right). They're filled with a pre-measured amount of

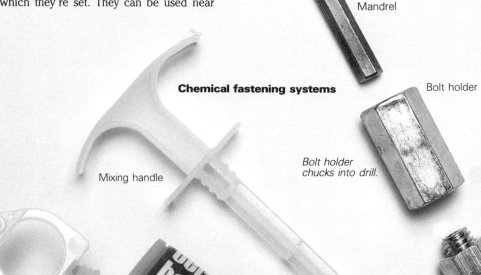

Chemical fastening systems

Mandrel

Bolt holder

Bolt holder chucks into drill.

Mixing handle

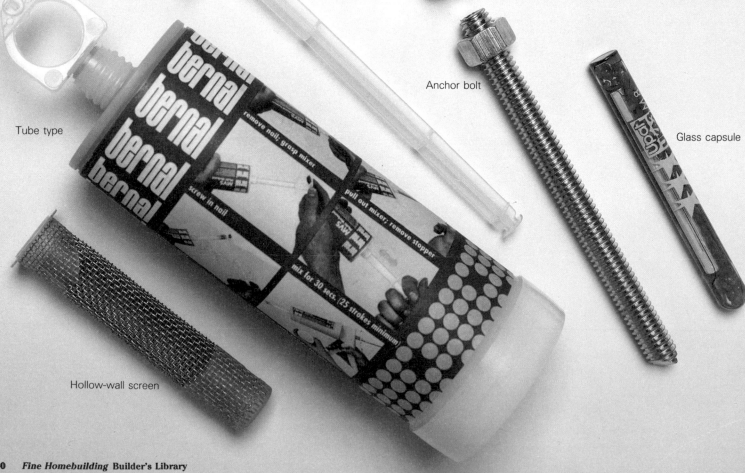

Tube type

Anchor bolt

Glass capsule

Hollow-wall screen

elastic monomer resin, quartz aggregate and a hardening catalyst separated in its own little vial. After drilling the hole to the specified length and cleaning it carefully, you insert the capsule. Then you chuck a sharpened piece of threaded rod or rebar into your drill (using a special attachment sold by the manufacturers) and slowly drive it into the capsule, which shatters. The glass shards become part of the aggregate. After the rod is fully inserted, let it spin for an extra second or two. This ensures that the glass is ground up and that all the ingredients are well mixed. It's important with all the chemical anchors not to move the insert while the epoxy is curing.

Glass capsules are sold in boxes of 10. A 4-in. capsule and the appropriate threaded rod (sold separately) cost about $3.

Concrete screws—About 10 years ago, Buildex (Illinois Tool Works, Inc.) developed the TAPCON, a blue-coated screw made of steel so hard that it can be screwed into concrete. You still have to drill a pilot hole for these screws, a precise one in fact, but you don't need any kind of anchor. The screw cuts its own threads in the concrete. TAPCONs became so popular that Buildex couldn't keep up with the demand, so they licensed other companies to manufacture and market them under the TAPCON trademark.

Since then, many companies have developed their own concrete screws (below left), and for good reason. Concrete screws are stronger than nearly all the lead and plasic anchors used in residential construction. They're faster and easier to install. They're reusable and fireproof. The only reason more builders don't use them is that most builders don't know about them.

Concrete screws are expensive. The 2¼-in. #14s that I bought at my local hardware store cost $0.60 each. But compared to the cost of other anchors, and considering how quickly they can be installed, they're worth it. Builders who use them buy them by the box, which undoubtedly works out to be cheaper than what I paid. Some manufacturers even throw in a drill bit with each box of 100 screws.

Concrete screws are available in lengths ranging from 1¼ in. to 4 in. and in either hex head or Phillips head. The hex heads, and the drivers they're installed with, hold up a little better under the high torque these screws are subjected to. High torque also means that if you've got a choice, you should use a screw gun to install them (because of the clutch), not a regular drill.

You can get special installation tools for concrete screws that have a drill bit, an integral hex-head driver and a Phillips-head bit attachment all in one (below left). You're probably better off using a hammer drill and a screw gun, if you've got both. If you don't, the installation tool will save a lot of time.

Concrete screws will work in block and brick (into the unit itself, not the mortar joint), as will most of the fasteners discussed in this article. But since these materials aren't as strong as concrete, fasteners anchored into them won't produce shear and withdrawal values as high as those into concrete. Also since bricks range from very hard and brittle to soft and crumbly, none of the fastener manufacturers list strength values for anchoring into them.

Toggle bolts—When you're anchoring into concrete block or hollow brick, you should drill into one of the webs if you can. But if your drill bit penetrates one of the voids, consider using a toggle bolt (below right). These are long thin bolts with large wing-shaped mechanisms called toggles that screw onto them. The toggle is spring loaded and folds flat against the bolt so that it can be slipped through a hole in the hollow block. Once inside, the toggle springs open and can be tightened against the block.

One problem with toggle bolts is that you have to drill a hole much bigger than the bolt in

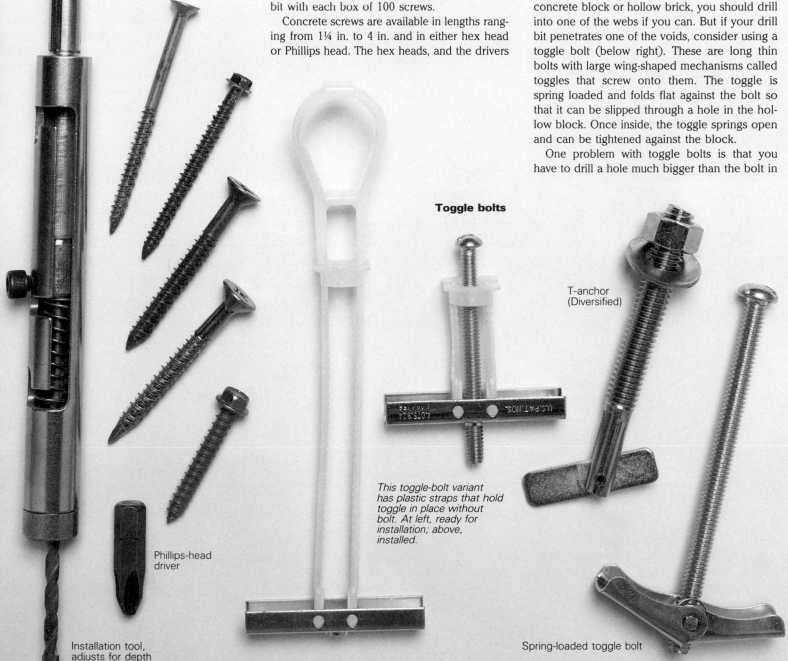

Concrete screws

Phillips-head driver

Installation tool, adjusts for depth

Toggle bolts

This toggle-bolt variant has plastic straps that hold toggle in place without bolt. At left, ready for installation; above, installed.

T-anchor (Diversified)

Spring-loaded toggle bolt

order to get the toggle in. I discovered some T-anchors, made by Diversified Fastening Systems, Inc., that don't have that problem. These have a flat piece of steel pinned through the end of the bolt. The bolts have a slot cut in the end into which the flat piece of steel can be pivoted parallel to the shank of the bolt. In this position, the T-anchor will fit through a hole the same size as the bolt itself. Because the flat piece of steel is pinned a little off center, gravity opens the T-anchor once it clears the hole.

Another nifty variation on the toggle bolt was first manufactured by Unifast under the trade name Kaptoggle, though a similar version is now also made by Toggler. With this toggle bolt you still have to drill a hole that's bigger than the bolt. But the advantage with this fastener is that the toggle is installed separately and will stay in place without the bolt, which means you can remove or change things without the toggle falling down inside the block.

The toggle comes attached to a pair of ribbed plastic straps. You insert the toggle into the hole, while holding onto the straps. Once it's inside it opens up against the block, just as the other toggles do. Still holding onto the straps, you slide a plastic cap down the straps until it engages the hole, and its flange is against the block. Then you simply break off the straps, and the cap and toggle are locked in place.

Hammer-driven fasteners—The two most significant features of these fasteners are that installation is usually simpler and removal more difficult (an advantage where vandalism is concerned). With one type of hammer-driven fastener, you still have to drill a hole in the concrete or masonry. But once the hole is drilled, installation is quick. All you do is insert the fastener

and hammer an integral pin down the center of it, which expands the sleeve inside the hole.

These fasteners come in a wide variety of styles, including steel stud anchors, zinc alloy and aluminum pin anchors, and nylon nail anchors (below). All the ones I tested drove in easily and seemed to hold well. I especially liked the zinc alloy and aluminum anchors. Their strength and cost fall nicely between the plastic and steel. And there was something satisfying about the way the pin and sleeve head meshed together like a well-peened rivet.

One recent innovation in hammer-driven fasteners, the Rawl SPIKE (The Rawlplug Co.), looks like a blunt nail with a crook near the end. It's made of grade 8 hardened spring steel and results in great holding power, much stronger than most plastic and lead anchors, especially in withdrawal. You have to drill a hole first, but then you just drive it in. Like the concrete screw, the Rawl SPIKE is simple and effective.

Some companies make an installation tool for hammer-driven fasteners. It looks like a motorcycle throttle grip and has a steel piston running down the middle of it. You insert a pin or threaded stud in the bottom, place the unit against the wall, and hit the piston with a hammer. One of the problems with driving nails into concrete is that they're very unforgiving of a glancing blow. If you don't hit the nail squarely on the head, it will bend. Because of its piston, the installation tool precludes this problem. But after pounding in a few of these with only limited success, I decided I'd rather take the time to drill. You're more certain of a secure anchor if you use one that requires a predrilled hole, and drilling is a less intense operation than swinging a hammer with all your might.

Cut nails and spiral masonry nails round out the hammer-driven category of fasteners. These are tough to drive (a 22-oz. or 28-oz. hammer helps), and they don't always hold. You have to be careful not to overdrive them; one blow too

many will fracture the concrete, loosening the nail. But nails are fast and cheap. In basements all over the country, paneling is installed over furring strips nailed to concrete walls. If you use nails on brick masonry, be sure to nail into a mortar joint and not into the bricks themselves.

One trick that helps with all fastenings to concrete and masonry, but that is particularly useful with concrete nails, is also to use construction adhesive. Apply it to one surface, the furring strip for instance, press the two surfaces together, then pull them apart for a minute. This helps spread the adhesive, increases its initial tack, and also lets you see any low spots on the wall.

Plastic anchors—The truth is that nearly all the anchors I've listed will work for most residential applications. Because they were developed to support the greater loads found in commercial and industrial construction, most are far stronger than you'll need around a residential project. Among the biggest loads I can imagine supporting with masonry fasteners are kitchen cabinets. So I asked several manufacturers what they would recommend for this application. All of them said plastic anchors would work fine.

As with all the other fasteners, plastic anchors come in a variety of shapes, sizes and colors. There are cylindrical vinyl plugs, toggles, conical screw anchors and ribbed anchors, also called prong anchors (facing page). They all require hole spotting and drilling a bigger hole for the anchor than for the screw.

The plastic toggle, pioneered by Toggler, will work in both hollow-wall and solid-wall applications. Its wings fold parallel to the shank so that it can be inserted into a hole. The wings open up inside a hollow wall. In a solid wall, they remain folded and act as an expansion anchor.

What does it all mean?—While researching concrete and masonry fasteners, I wrote to the Specialty Tools and Fasteners Distributors Association (Box 44, Elm Grove, Wis. 53122) and then to 39 different manufacturers, asking for

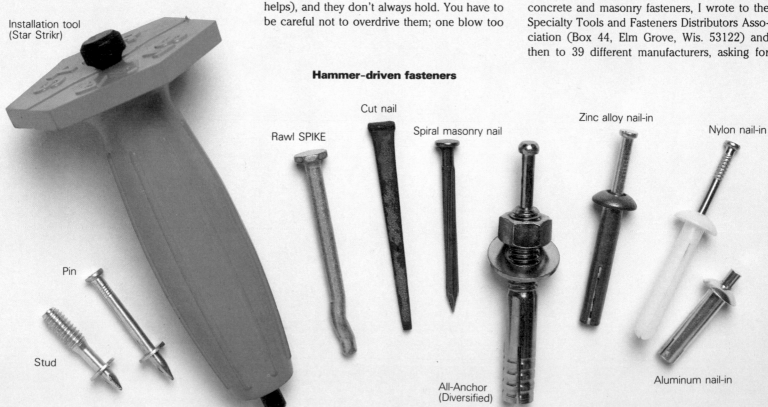

Installation tool
(Star Strikr)

Pin

Stud

Hammer-driven fasteners

Rawl SPIKE

Cut nail

Spiral masonry nail

All-Anchor
(Diversified)

Zinc alloy nail-in

Nylon nail-in

Aluminum nail-in

product literature and samples. I visited a half-dozen hardware stores and lumberyards, and made a lot of phone calls. At the last minute, I was still discovering anchors that I'd never seen before. And I won't be surprised to learn, after this article is published, that I've left out a few.

Nonetheless, some of the fasteners I've discussed ought to be a regular part of every builder's hardware arsenal, and the others are good to know about for those oddball situations that come up now and then.

I found no reason to recommend one brand of fastener over another. Most of the manufacturers have their products tested by independent laboratories and submit the results for approval by groups like the International Conference of Building Officials.

Your best chance of finding a large selection of concrete and masonry anchors in stock is at an industrial-supply house. If you don't have one near you, your local hardware store or building-supply yard can probably order what you need through a distributor. I turned up one industrial-supply company that sells mail order: Phillips Brothers Supply (2525 Kensington Ave., Buffalo, N. Y. 14226). All the companies mentioned in this article offer a variety of fasteners (see the chart at right), and if you contact them directly, they'll supply technical assistance and help you track down a supplier in your area.

If all else fails, and you can't find any of these fasteners, you can do what one custom builder in Idaho does: epoxy redwood plugs into holes drilled in concrete and use woodscrews. □

Plastic anchors

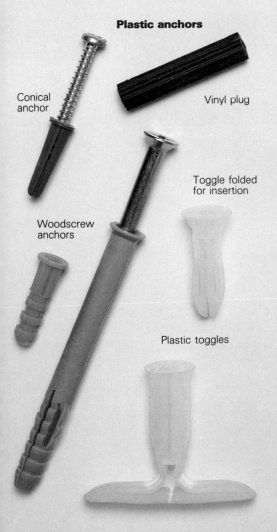

Conical anchor

Vinyl plug

Woodscrew anchors

Toggle folded for insertion

Plastic toggles

Sources of supply — Fasteners

Manufacturers	Lead anchors	Flush anchors	Wedge and sleeve	Chemical	Concrete screws	Toggle bolts	Hammer driven	Plastic anchors
Ackerman Johnson Fastening Systems, Inc. 136 Official Rd. Addison, Ill. 60101 (312) 543-2797	•	•	•	•	•	•	•	•
Albert Berner GmbH & Co. KG Moore Construction Fastening Products, Inc. 250 Barber Ave. Worcester, Mass. 01606 (617) 853-3991	•	•	•	•	•			
Barrett Manufacturing Co. 4124 W. Parker Ave. Chicago, Ill. 60639-2173 (800) 621-7522	•	•	•		•	•	•	
Buildex 1349 W. Brynmawr Ave. Itasca, Ill. 60143 (312) 595-3500					•	•		
Celtite, Inc. 150 Carley Court Georgetown, Ky. 40324 (800)626-2948				•				
Diversified Fastening Systems, Inc. P.O. Box 339 Charles City, Iowa 50616 (515) 228-1162		•			•	•		
Elco Industries, Inc. 1111 Samuelson Rd. P.O. Box 7009 Rockford, Ill. 61125 (815) 397-5151				•				
Hilti Fastening Systems P.O. Box 21148 5400 South 122nd East Ave. Tulsa, Okla. 74121 (918) 252-6000			•	•	•	•	•	•
Mechanical Plastics Corp. P.O. Box 328 Pleasantville, N. Y. 10570 (914) 769-8450						•		•
Ramset Fastening Systems 2100 Golf Rd. Suite 460, West Bldg. Rolling Meadows, Ill. 60008 (312) 640-0770		•		•	•	•	•	
The Rawlplug Co., Inc. 200 Petersville Rd. New Rochelle, N. Y. 10802 (914) 235-6300	•	•	•	•	•	•	•	•
Semco Plastic Co., Inc. 1366 Kingsland Ave. St. Louis, Mo. 63133 (800) 325-0622						•		•
Star Expansion Co. Mountainville, N. Y. 10953 (914) 534-2511	•	•	•		•			
Unifast Industries 45 Gilpin Ave. Hauppauge, N. Y. 11788 (516) 348-0260	•	•		•	•			
Upat Essve Inc. 1849 Peeler Rd., Suite D Atlanta, Ga. 30338 (404) 392-1699		•	•	•		•		•
U. S. Anchor 1531 N. W. 12th Ave. Pompano Beach, Fla. 33069 (305) 782-2221	•	•	•	•	•			•
Wej-It Anchoring Systems P.O. Box 521120 Tulsa, Okla. 74152 (918) 743-1030	•	•	•		•		•	

Editor's note: This is not a definitive list of manufacturers, but includes the major companies and any others that responded to our request for information and samples of their products.

Building a Block Foundation

How to pour the footing and lay the concrete block, and what to do about waterproofing, drainage and insulation

by Dick Kreh

Foundation walls are often the most neglect-ed part of a structure. But they are actually the most structurally important element of a house. They support the weight of the build-ing by distributing its entire load over a large area. Apart from structural requirements, foundations have to be waterproofed, insulat-ed and properly drained.

Although the depth of a foundation wall may vary according to the specific needs of the site or building, the footings must always be below the frost line. If they're not, the foundation will heave in cold weather as the frozen earth swells, and then settle in warm weather when the ground softens. This shift-

ing can crack foundations, rack framing, and make for wavy floors and sagging roofs.

Concrete blocks are composed of portland cement, a fine aggregate and water. They have been a popular choice for foundations be-cause they're not too expensive, they go up in a straightforward way, and they're available everywhere. Block foundations provide ade-quate compressive strength and resistance to fire and moisture. They don't require form-work, and they're not expensive to maintain.

All standard blocks are 8 in. high and 16 in. long—including the usual ⅜-in. thick mortar head and bed joints. But they come in differ-ent widths. The size given for a block always

refers to its width. The size you need depends on the vertical loads and lateral stresses that the wall will have to withstand, but as a rule, most concrete-block foundations are built of 10-in. or 12-in. block.

Footings—After the foundation area has been laid out and excavated, the concrete footings are poured. Footings should be about twice as wide as the block wall they will sup-port. A 12-in. concrete-block foundation wall, for example, should have a 24-in. wide foot-ing. The average depth for the footing, unless there is a special problem, is 8 in.

Concrete footings for homes or small struc-

From *Fine Homebuilding* magazine (June 1983) 15:44-48

Chalklines snapped on the cured footing guide the masons in laying up the first course of concrete block for a foundation wall. The blocks are stacked around the site to minimize legwork, yet allow the masons enough room to work comfortably.

tures need a compressive strength of 2,500 pounds per square inch (psi). You can order a footing mix either by specifying a five-bag mix, which means that there are five bags of portland cement to each cubic yard, or by asking for a prescription mix—one that is ordered by giving a psi rating. Some architects and local building codes require you to state the prescription mix when you order. Either way, footing concrete is a little less expensive than regular finishing concrete, which usually contains at least six bags of portland cement to the cubic yard. The six-bag mix is richer and easier to trowel, but isn't needed for most footings. For more on mixing and ordering concrete, see pp. 8-13 and 14-15.

There are two types of footings—trench footings and formed footings. If the area where the walls are to be built is relatively free of rock, the simplest solution is to dig a trench, and use it as a form. Keep the top of the concrete footings level by driving short lengths of rebar to the proper elevation. Don't use wooden stakes because later they'll rot and leave voids in your footing. You'll need a transit level or water level to get the rods at the right height. After you install the level rods even with the top of the proposed footing, pour concrete in the trench, and trowel it flush with the tops of the rods. Some building codes require that these stakes be removed before the concrete sets up.

If the ground is rocky, you may have to set up wooden forms and brace them for the pour. I've saved some money in this situation by ordering the floor joists for the first floor and using them to build the forms. This won't damage the joists and will save you a lot of money. When the concrete has set, I remove the boards and clean them off with a wire brush and water. The sooner you remove the forms, the easier it will be to clean them.

After the footings have cured for at least 24 hours, drive nails at the corners of the foundation. To find the corner points, use a transit level or drop a plumb line from the layout lines that are strung to your batter boards at the top of the foundation (for more on laying out foundations, see "Site Layout," pp. 64-66). Next, snap a chalkline between the corner nails on the footings to mark the wall lines. Stack the blocks around the inside of the foundation. Leave at least 2 ft. of working space between the footing and your stacks of block. Also, allow room for a traffic lane so the workers can get back and forth with mortar and scaffolding.

Mortar mix for block—For the average block foundation, use masonry cement, which is sold in 70-lb. bags. You have to supply sand and water. Masonry cement is made by many companies. Brand name doesn't matter much,

but you will need to choose between mixes of different strength. The average strength, for general masonry work, is universally classified as Type N. Unless you ask for a special type, you'll always get Type N. I get Type N masonry cement unless there is a severe moisture condition or stress, in which case I would use Type M, which is much stronger. The correct proportions of sand and water are important to get full-strength mortar. Like concrete, mortar reaches testing strength in 28 days, under normal weather conditions.

To mix the mortar, use one part masonry cement to three parts sand, with enough water to blend the ingredients into a workable mixture. Mortar for concrete block should be a little stiffer than for brickwork, because of the greater weight of the blocks. You will have to experiment a little to get it right. The mortar must be able to support the weight of the block without sinking.

The mixing water should be reasonably clean and free from mud, silt or organic matter. Drinking water makes good mortar. Order washed building sand from your supplier. It's sold by the ton.

The following will help you estimate the amount of mortar you'll need: One bag of masonry cement when mixed with sand and water will lay about 28 concrete blocks. Eight bags of masonry cement, on the average, will require one ton of building sand. Remember that if you have the sand dumped on the ground, some will be lost since you can't pick it all up with the shovel. For each three tons, allow about a half-ton for waste.

Laying out the first course—Assuming the footing is level, begin by troweling down a bed of mortar and laying one block on the corner. Tap it down until it is the correct height (8 in.), level and plumb.

A block wall built of either 10-in. or 12-in. block requires a special L-shaped corner block, which will bond half over the one beneath. The point is to avoid a continuous vertical mortar joint at the corner. Now lay the adjoining block. It will fit against the L-shaped corner block, forming the correct half-bond, as shown in the drawing and photo at right.

When the second L-shaped corner block is laid over the one beneath in the opposite direction, the bond of the wall is established. On each succeeding course the L corner block will be reversed.

Once the first corner is laid out, measure the first course out to the opposite corner. It's best for the entire course to be laid in whole blocks. You can do this simply by using a steel tape, marking off increments of 48 in., which is three blocks including their mortar head joints. Or you can slide a 4-ft. level along

the footing and mark off 48-in. lengths. In some cases, of course, dimensions will require your using a partial block in each course, but it's best to avoid this wherever possible. If a piece of block must be used, lay it in the center of the wall or where a window or partition will be, so it is not as noticeable. After the bond is marked on the footing, a block is laid on the opposite corner and also lined up. Then you attach a mason's line to the outside corner and run it to the opposite corner point. This is called "ranging" the wall.

Sometimes there are steps in the footings because of a changing grade line. The lowest areas should always be built up first to a point

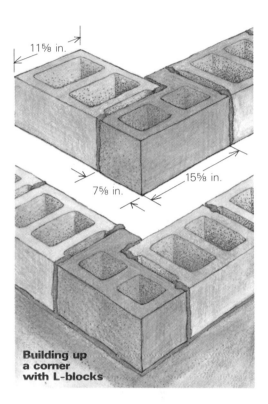

11⅝ in.

15⅝ in.

7⅝ in.

Building up a corner with L-blocks

Laying the first course begins at the corners. Once the four corners are laid and aligned, the entire bottom course is laid directly atop the footing. Most foundation walls are built from block 10 in. or 12 in. wide, and special L-shaped corner blocks, like the one shown here, have to be ordered. Only 8-in. wide blocks can be laid up without L-shaped corner blocks.

Illustrations: Christopher Clapp

where a level, continuous course of block runs through from one corner to the other. Steps in footings should be in increments of 8 in. so that courses of block work out evenly.

After one course of block is laid completely around the foundation to establish the bond and wall lines, it's time to build up the four corners. But before you begin laying block, you should make a story pole, sometimes called a course rod. Do this by selecting a fairly straight wooden pole and marking it off every 8 in. from the bottom to the height of the top of the foundation wall.

Any special elevations or features, such as window heads, door heads, sills and beam pockets, should be marked on the pole to coincide with the 8-in. increments wherever possible. After checking all your pencil marks, make them permanent by kerfing the pole lightly with a saw. Then cut the pole off even with the top of the foundation wall and number the courses of block from the bottom to the top so you don't find yourself using it upside down.

Now you can start laying up the corners so that you end up with only one block at the level of the top course. Successive courses are racked back half a block shorter than the previous ones (photo top left), so trowel on only enough mortar to bed the blocks in a given course. If the local code or your specifications call for using wire reinforcement in the joints, leave at least 6 in. of wire extending over the block. At the corners, cut one strand of the wire, and bend the other at 90°, rather than butting two sections together and having a break in the reinforcement.

Check the height of the blockwork periodically with the story pole. The courses of block should line up even with the kerfs. Once the corners are laid up, you can begin to fill in the wall between. Keep the courses level by laying them to a line stretched between the corners. Keep the corners plumb by checking every course with a spirit level.

Using manufactured corner poles—So far, I've described laying a foundation using the traditional method of leveling and plumbing. But in recent years, manufactured metal corner-pole guides have become popular with builders. They guide the laying up of each course and require less skill than the old way. They work like this. The corner poles are set on the wall once the first course of block is laid out. They are plumbed, then braced in position. Each pole has course heights engraved on it. Line blocks are attached to the poles on opposite corners at the desired course height, and the wall is laid to the line. There is no doubt that the use of manufac-

After the first course is laid, the corners are built up to the topmost course. Above left, a mason checks course heights against a story pole, which is graduated in 8-in. increments.

Reinforced concrete lintels, left, are used to tie the main foundation to walls that are laid at a higher level, such as porch foundations or garage walls.

tured corner-pole guides has increased the mason's productivity without adversely affecting the quality of the work.

If you have to tie a porch or garage wall into a main foundation at a higher elevation, lay a concrete-block lintel in mortar from the corner of the wall being built to the footing at the higher elevation (photo bottom left). Then lay blocks on the lintel to form the wall. This saves time and materials in an area that doesn't require a full-basement foundation.

Stepping the wall at grade line—As you build up to the natural grade line of the earth, you can set the front of the wall back about 4 in. to form a shelf for a brick veneer, if the plans call for it. This is done by switching to narrower block—from 12-in. block to 8-in. block, or from 10-in. block to 6-in. block. The inside of the wall stays in the same plane.

Making the last course solid—On some jobs, specifications require that the last course of block be solid to help distribute the weight of the structure above and to close off the holes. You need only grout the voids in the top course of block. Broken bits of block wedged into the voids in the course below will keep the concrete from falling through. The sill plates will rest on this top course, and the floor joists on top of the plate.

The sill plate has to be bolted down to the top of the foundation wall. So you have to grout anchor bolts into the top of the wall every 4 ft. or 5 ft. These bolts should have an L-bend on the bottom and be mortared in fully so they don't pull out when the nut is tightened against the sill plate. They should extend about 2 in. out of the top of the wall. In some parts of the country, building codes require that the walls include a steel-reinforced, poured-in-place concrete bond beam in every fourth course.

Waterproofing the foundation—The traditional method of waterproofing a concrete block foundation is to parge (stucco) on two coats of mortar and then to apply a tar compound on top of that. This double protection works well, unless there is a severe drainage problem, and the soil is liable to hold a lot of water for a long time.

There are various mortar mixes you can use to parge the foundation. I recommend using a mix of one part portland cement to one-half part hydrated lime to three parts washed sand. This is a little richer than standard masonry cement and is known as type S mortar. The mix should be plastic or workable enough to trowel on the wall freely. Many mortars on the market that have waterproofers in them are all right to use. However, no two builders I know seem to agree on a mix, and most have worked out their own formulas.

Prepare the foundation wall for parging by scraping off mortar drips left on the block. Next, dampen the wall with a fine spray of water from a garden hose or a tank-type garden sprayer. Don't soak the wall, just moisten it. This prevents the parging mortar from drying

Photos: Dick Kreh

Troweling technique

Laying up concrete blocks with speed and precision takes a lot of practice. But it's chiefly a matter of learning several tricks, developing trowel skills and performing repetitive motions for several days. A journeyman mason can lay an average of 200 10-in. or 12-in. blocks in eight hours. A non-professional, working carefully and after practicing the techniques shown here, ought to be able to lay half that many. If you've never laid block before, what follows will show you the basic steps involved in laying up a block wall.

1. First, mix the mortar to the correct stiffness to support the weight of the block. Then apply mortar for the bed joints by picking up a trowelful from the mortarboard and setting it on the trowel with a downward jar of the wrist. Then swipe the mortar onto the outside edges of the top of the block with a quick downward motion, as shown.

2. Apply the mortar head joint pretty much the same way. Set the block on its end, pick up some mortar on the trowel, set it on the trowel with a downward jerk and then swipe it on the top edges of the block (both sides).

3. After buttering both edges of the block with mortar, press the inside edge of the mortar in the head joints down at an angle. This prevents the mortar from falling off when the block is picked up and laid in the wall.

4. Lay the block on the mortar bed close to the line, tapping down with the blade of the trowel until the block is level with the top edge of the line. Tap the block in the center so you won't chip and smear the face with mortar. Use a hammer if the block does not settle easily into place.

5. The mortar in the head joint should squeeze out to form a full joint at the edge if you've buttered it right. The face of the block should be laid about 1/16 in. back from the line to keep the wall from bowing out. You can judge this by eyeballing a little light between the line and the block.

6. Remove the excess mortar that's oozing out of the joint with the trowel held slightly at an angle so you don't smear the face of the block with mud. Return the excess mortar to the mortarboard.

Check the height of the blockwork by holding the story pole on the base and reading the figure to the top of the block. Courses should be increments of 8 in.

Finishing the joints—Different types of joint finishes can be achieved with different tools. The most popular by far is the concave or half-round joint, which you make by running the jointing tool through the head joints first, and then through the bed joints to form a straight, continuous horizontal joint. If you buy this jointing tool, be sure that you get a convex jointer. These are available in sled-runner type or in a smaller pocket size. I like the sled runner because it makes a straighter joint.

After the mortar has dried enough so it won't smear (about a half-hour), brush the joints lightly to remove any remaining particles of mortar.   —*D.K.*

1

2

3

4

5

6

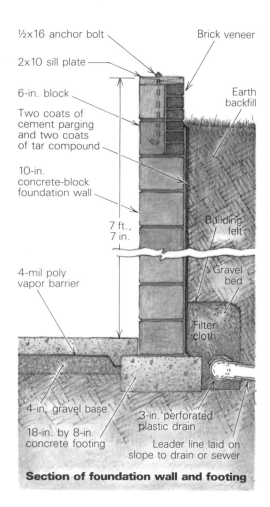

½x16 anchor bolt

Brick veneer

2x10 sill plate

6-in. block

Two coats of cement parging and two coats of tar compound

10-in. concrete-block foundation wall

Earth backfill

Building felt

7 ft., 7 in.

Gravel bed

4-mil poly vapor barrier

Filter cloth

4-in. gravel base

18-in. by 8-in. concrete footing

3-in. perforated plastic drain

Leader line laid on slope to drain or sewer

Section of foundation wall and footing

The completed foundation has been sealed with two ¼-in. thick parging coats and topped with an application of tar compound, which finishes the waterproofing. Backfilling should happen only after the first floor is framed and the walls framed up, so the added weight of the structure will stiffen the walls and make them less liable to bulge from the pressure of the earth.

out too quickly and allows it to cure slowly and create a better bond with the wall.

Start parging on the first coat from the bottom of the wall to the top, about ¼ in. thick. A plastering or cement-finishing trowel is excellent for this. After troweling on the parging, scratch the surface with an old broom or a tool made for this purpose. Let the mortar dry for about 24 hours or until the next day, and repeat the process for the second coat. Dampen the wall between coats for a good bond. Trowel the final coat smooth, and let it dry for another 24 hours.

To complete the waterproofing job, spread on two coats of tar compound (photo, top). You can do this with a brush or roller if the weather is warm. Many builders in my area use a product called Hydrocide 700B (Sonneborn Building Products, 57-46 Flushing Ave., Maspeth, N.Y. 11378). It comes in 5-gal. containers and is available from most building-supply dealers. I like it because it stays a little tacky and seals the wall very well. It's gooey, though, so wear old clothes and gloves when you're applying it. Kerosene will get it off your hands and tools when the job is done.

Drain tile—Most codes require some type of drain tile or pipe around the foundation to divert water build-up and to help keep the basement dry. The design of the drain-tile system is important. Generally, drain tile or pipe is installed around the exterior wall of the foundation, below the wall but above the bottom of the footing, as described below.

Begin by spreading a bed of crushed stone

or gravel around the foundation next to the wall. Lay the drain tile or perforated plastic pipe on top of this bed. The bottom of the drain pipe should never be lower than the bottom of the footing, or it won't work properly. Lay filter cloth over the drain pipe to keep mud and dirt from blocking the holes. Then place another 4-in. to 6-in. layer of crushed stone or gravel over the pipe, as shown in the drawing above.

The water collected by the drain tile has to flow away from the foundation. One way to make this happen is to drain the water to a natural drain away from the foundation area by installing a leader line on a slope that is lower than the drain pipe. The other method is to drain the water under the wall of the foundation and into a sump pit inside the basement. The water that collects is then pumped into a pipe up to grade or street level and allowed to drain away there naturally.

A third method, which has worked well for contractors in my area, is to put the drain tile inside the foundation on a bed of crushed stone, just beneath the finished concrete floor, which will be poured after the drain tile is in place. One-inch plastic pipe is installed about 6 ft. o.c. through the wall at the bottom of the head joints in the first course of block. When the foundation wall is done, crushed stone is spread around the exterior edges as before, but no drain pipes are needed.

The idea is that any water that builds up outside the foundation wall will drain through and into the drain tiles. In addition, the water inside the foundation area will also flow into

the drain tile and into the sump pit in the basement floor, where it can be pumped out to grade level to drain away, as mentioned above.

Insulating the foundation—In recent years, the use of rigid insulation applied to the exterior of the foundation wall has helped to reduce dampness and heat loss. This is especially important in the construction of earth-sheltered homes. There are a number of products that will do a good job. Generally, rigid insulation is applied to the waterproofed wall with a mastic adhesive that's spread on the back of the foam board. Use a mastic or caulking that does not have an asphalt base. Most panel adhesives will work. The building-supply dealer who sells the insulation will know the proper adhesive to use for a specific type of insulation board. Also, there are granular and other types of insulation that can be poured into the block cells.

After all of the foundation work has been completed, the backfilling of earth should be done with great care so that the walls don't get pushed out of plumb. It is always better to wait until the first floor is framed up before backfilling. This weight resting on the foundation helps prevent cracking of the walls, and the framing material will brace the block wall and make it more rigid. If the walls are cracked or pushed in from backfilling, the only cure is very expensive—excavate again and replace the walls. □

Dick Kreh is an author, mason and industrial-arts teacher in Frederick, Md.

Insulating and Parging Foundations

Covering concrete walls with rigid foam insulation and troweling on stucco requires experience with the materials

by Bob Syvanen

If you've got the idea that a builder's skill is an unchanging body of knowledge passed down through the generations, think for a minute about insulating a foundation from the exterior. Even in cold climates, what you used to see between the bottom of the siding and the grade was the bare concrete foundation wall. But these days, with estimates of heat lost in a house through the foundation running as high as 30%, what looks like concrete is more likely parging, or stucco, applied over rigid foam insulation.

Insulating the outside of foundations has been a problem for a lot of builders, including me, because many of the materials and methods are new. Although rigid foam-board insulation doesn't look like much of a problem, it isn't as simple as it first appears. Polystyrene is the insulating material most often used. It comes in 2-ft. wide panels and handles like plywood, but it's a lot lighter. You can cut it with anything from a knife to a table saw. But polystyrene foam is produced in two forms: expanded and extruded. Expanded polystyrene (EPS), also known as beadboard, is more susceptible to soaking up moisture than its extruded cousin, say the researchers on one side of this controversy. This could lead to a considerable loss in R-value. Although I used expanded polystyrene on the job shown here, I think the extruded version is probably the better bet despite its higher cost.

There are more than 100 makers of EPS; but extruded polystyrene is made in the U. S. by only three companies: Dow Chemical (Midland, Mich. 48640), whose blue-tinted Styrofoam is often called blueboard; Minnesota Diversified Products (1901 13th St. N. E., New Brighton, Minn. 55112), whose yellow product is trademarked Certifoam; and U. S. Gypsum (101 S. Wacker Dr., Chicago, Ill. 60606), the makers of pink Foamular.

Whichever brand you use, the process of applying it is the same, and so are the problems. For instance, asphalt-based products dissolve most foams, so my usual method of waterproofing a foundation is suddenly out

the window. And to complicate things even more, polystyrene needs to be protected above grade from impact as well as from deterioration by ultraviolet light from the sun.

I wanted a protective coating that was easy to install, good looking and long lasting. There are many commercial systems—fiberglass panels, super stucco mixes, and even a rigid insulation with a factory-applied coating that can be attached to concrete forms before pouring—but I wanted to use materials that were more traditional.

I first used asbestos board cemented on the foam, but it is fragile, hard to repair, impossible to glue, and required a lot of fitting time at corners, doors and windows. I also tried a latex-cement product applied directly on the foam. Unfortunately, it didn't age well. In fact, I have repaired not only the job I did with it, but several others in my area.

I finally settled on covering the foam with cement-stucco, called parging where I live. When stucco is used for exterior wall finish on a house, it is usually done in three coats like plaster. I was determined to come up with a single application process. Although parging and surface-bonding mixes can be applied directly to the insulation, I don't trust the bond, and want a thicker parging for durability. This means using some kind of lath.

On the first parging job, I used small-mesh chicken wire. I stretched it over ⅜-in. wood lath at 12 in. o. c. both horizontally and vertically to hold it off the surface of the insulation. Chicken wire wasn't the answer. Although my mason got the chicken wire to support the cement out of sheer stubbornness, the diamond-shaped pattern showed through a little, and there were some shrinkage cracks.

I refined the system by using metal lath,

and by reducing the thickness of the wood lath to ¼ in. This worked much better for the mason, but the lath strips were still tedious to install. Next I eliminated the wood lath and applied the metal lath directly to the foam. What I ended up with is a protective coating that is long lasting, attractive and relatively easy to install. Although it is a little expensive, after seeing some of the jobs using cheaper materials other local builders and I have done, I think it's worth the cost.

Since the insulation and lath-work usually fall to the carpenter or contractor who is on the site every day, the only sub I use is my mason, who is much faster and neater than I am with a trowel. I am used to paying anywhere from $2 to $4 a square foot for parging, although conditions vary enough that both the mason and I get the best deal when I use him on a time-and-materials basis. Since a bag of masonry cement covers 20 to 30 sq. ft. of wall, most of the expense is in the labor.

Installing rigid foam insulation—A partially earth-sheltered, passive-solar house I just completed gave me a good chance to try my new system. The plans called for its concrete walls to be insulated with two layers of 2-in. foam. One wall is 7 ft. 10 in. high, and the other three are 2 ft. high. The parging was to cover the first 2 ft. below the mudsill on all of them. Some folks also use insulation laid horizontally below grade, but I simply ran my panels down to the footings.

The first thing I needed was a good adhesive, since there shouldn't be any give in the plane of the insulation panels if the parging is going to last. But the high wall is also below grade and part of the living space, so it had to be well waterproofed. Since asphalt-based products can't be used with foam, I looked

Installing foam insulation. First, a waterproofing agent that also serves as a mastic is spread directly on the concrete. Temporary braces hold the foam panels in place while the mastic dries. Two-by-four nailers are used between the two 2-in. layers. The horizontal nailer is 27 in. down from the sill—the width of the metal lath that will be applied next.

From *Fine Homebuilding* magazine (December 1983) 18:33-35

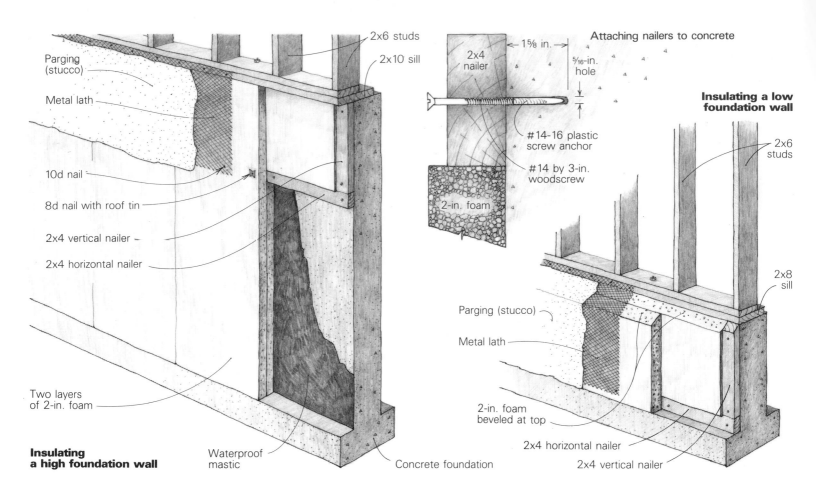

Insulating a high foundation wall — Parging (stucco); Metal lath; 10d nail; 8d nail with roof tin; 2x4 vertical nailer; 2x4 horizontal nailer; Two layers of 2-in. foam; Waterproof mastic; Concrete foundation; 2x6 studs; 2x10 sill

Attaching nailers to concrete — 2x4 nailer; 1⅝ in.; ⁵⁄₁₆-in. hole; #14-16 plastic screw anchor; #14 by 3-in. woodscrew; 2-in. foam

Insulating a low foundation wall — 2x6 studs; 2x8 sill; Parging (stucco); Metal lath; 2-in. foam beveled at top; 2x4 horizontal nailer; 2x4 vertical nailer

around for something else. What I found was a mastic, Karnak 920 (Karnak Chemical Corp. 330 Central Ave., Clark, N.J. 07066), which is marketed as both an adhesive and a waterproofing agent. Theoretically, you trowel the mastic waterproofing on the concrete and then press the foam panels in place. But the walls had enough irregularity that the panels contacted the mastic in only a few places, and they fell off about as fast as I put them on. I then found out that the foam has to be applied before the mastic skins over. This is enough time to apply just one or two panels and brace them with sticks, 2x4s, stones or buckets (photo previous page). When the foam was applied in this way, the adhesive held.

In this case, I installed 2x4 pressure-treated wood nailers with the first layer of foam panels in order to get nailing for the second layer. On the 7-ft. 10-in. wall, I began by placing a horizontal nailer 27 in. down from the sill. It was used to attach both the second layer of insulation and the lath, which comes 27 in. wide. Next, I attached vertical nailers above the horizontal at 24 in. o. c. because the panels are 2 ft. wide. I also filled in with nailers at windows and corners to catch the edges of the panels. A quick, easy way to fasten these 2x4s to concrete is to hold the nailer in place and drill through the wood into the concrete. Using a hammer-drill makes this almost fun.

For 2x4s, I use a #14-16 screw anchor, 1½ in. long, with a #14 by 3-in. flat-head woodscrew. Use a piece of tape on a ⁵⁄₁₆-in. masonry bit, at 3 in. from the tip, to limit the hole depth. Most hammer-drills have an attached depth guide. If the hole is too shallow, the tip of the woodscrew won't hit the con-

crete before snugging up the 2x4. If the hole is too deep, the screw won't grab the anchor.

After drilling, insert a plastic screw anchor into the hole in the nailer and turn a woodscrew a few turns into it. Then hammer the screw-and-anchor combination through the nailer into the concrete. Last, screw the woodscrew home (drawing, above center).

On the 2-ft. wall, I cut the first layer of foam to fit between the footing and the horizontal nailer and installed it before the nailers. This way, I could wedge the nailers between the foam and the footing while I fastened them.

On the high wall (drawing, above left), I cantilevered the mudsill over the concrete by the depth of the foam so that I could nail the metal lath directly to its top edge. This meant that the second vertical layer of foam tucked up underneath it, flush with its outside edge. This layer is held in place with 8d nails wherever there are nailers. The 8d nail reaches through to the nailer, and when given an extra tap, the foam compresses and snugs up the panel nicely. I use a roofing tin on each of these nails to increase its bearing surface. This is a stamped 2-in. by 2-in. flat metal plate with a hole in the center, and is typically used to hold down roofing felt on windy days. You can buy them from a roofing-supply yard or make your own by cutting out sheet-metal squares. I have also seen pins and plastic shields manufactured for this purpose.

The 2-ft. wall was insulated in a similar manner, but here the sill is flush with the outside face of the foundation wall (drawing above right). Since the foam projects past the sill, I beveled the top edge of the foam at a 45° angle. The vertical 2x4 nailers were also bev-

eled at the top before I fastened them to the concrete. The corner nailers are beveled from each direction (photo facing page, top left). To make a neat bevel cut in the foam, I snapped a chalkline the length of the foundation on the face of the panels and sawed along it with a bread knife. This bevel design worked well here because the finished grade was to come at the bottom of the bevel.

Installing metal lath—The galvanized metal lath I use measures 27 in. by 96 in. It is sold in single sheets or in bundles of ten. There are two things to keep in mind when you're working with metal lath. First, the diamond-mesh pattern is formed on an angle between the front and back of a sheet. This means that the dividing wire that is roughly horizontal forms a small lip or cup at the bottom of each hole. Make sure these cups are facing up to catch and hold the parging. It will work both ways, but things go better if the cups are up. The other thing to remember is that metal lath is sharp. I don't think I have ever worked with the stuff without cutting myself. The cuts are not bad, just annoying. Wearing gloves helps, but I find that more annoying than the cuts.

On the high wall, I nailed the mesh to the top edge of the sill. The bottom of the mesh nails through the foam into the horizontal nailer 27 in. below. I used leftover 3d shingle nails at the top, and 10d commons on the bottom and along the edges wherever I had a nailer. The 2-ft. walls were fastened similarly, but because of the beveled top, I had to bend the lath before I nailed it in place.

For corners, expanded corner bead—the plasterer's version of a metal sheetrock cor-

Photos: Bob Syvanen; Illustration: Frances Ashforth

ner—is the best way to go because it forms a neat, stiff straight line. But I didn't have any on the job, so I pre-bent the lath at 90° before installing it (photo top right). Bending sheet metal, particularly metal lath, on the job site isn't hard if you think of how it's done in the shop and duplicate the procedure. The shop uses a brake, which is a cast-iron table and a bar that folds the sheet metal over the edge of the table. On site, I sandwich the sheet between two 2x boards and "break," or fold, the piece that sticks out over the bottom 2x using a scrap block about 2 ft. long. Nailing the sheet metal to the top of the bottom 2x keeps it from creeping out as the bend progresses. This system is particularly good for metal lath because you don't have to handle the material constantly as you bend it.

Parging—Parging is not impossible for a novice, but a good finish takes experience. The first job I did turned out okay, but there was lots of room for improvement, so I went to school by watching mason John Hilley.

The parging he uses is a one-coat stucco with a steel-trowel finish. Other finishes might work better, but I am satisfied with this one. The mix he uses is 16 shovels of sand per bag of masonry cement. He doesn't have any trouble using up a batch that size before it begins to set. Masonry cement is a mix of portland cement, hydrated lime and additives that combine with water and sand to form mortar or stucco. For a parging mix, use Type M for higher compressive strength and greater resistance to water.

Large expanses of stucco are usually worked with darbies and floats. For foundations, though, a standard mason's trowel is easier. The mud is picked up on the bottom surface of the trowel and immediately applied to the lath. The free hand assists by pushing against the top face of the trowel, forcing the mud into the mesh. It is a quick process—pick up, apply, press. With each pressing motion, the excess cement gets pulled along with the sliding trowel (photo center right).

At the same time that the parging is applied, it should be roughly surfaced, to establish an even thickness. As with brick jointing and slab work, compressing the material is what finishing is all about. This requires a bit of pressure, but it should be with good control. Use two hands on the trowel, one on the handle and the other on the flat of the blade, and keep your arms straight.

The finishing is done when the shine leaves the surface of the parging. The trowel is dipped into water, shaken once to get the excess off, then pressed against the surface of the stucco using both hands (photo bottom right). Try to get a smooth finish in just a few strokes so you don't overwork the cement.

A cloudy, cool day is best for parging because the mix can be worked longer before it sets. If the parged wall is in direct sun, mist the surface with a pump-up garden sprayer filled with water to keep the surface of the stucco from drying out too fast, which will cause shrinking and surface cracks. □

Preparation. This foundation corner (top left) is ready for lath. Cutting the double bevel on the top of the corner nailer to match the bevel on the two layers of foam requires much less work later when the stucco is finish-troweled. An 8d nail and its roof tin, which acts like a large nailhead, are just visible at the bottom of the photo. This same corner is ready for parging once it is wrapped in metal lath (top right), which is pre-bent on a brake. The lath is nailed to the sill at the top, and through the second layer of polystyrene into the horizontal 2x4 nailer at the bottom.

Parging. Mason John Hilley forces the stucco mud into the lath (above), using two hands and the weight of his body. Just one coat of parging is used, but it is troweled twice. The first time is a rough troweling. When the shine disappears, the surface is smoothed with the same trowel, dipped frequently in water. This finish process is also done with one hand on the face of the trowel for direct pressure (right), and with arms held straight for good control.

Insulated Masonry Walls

Concrete block comes of age
with new technology for energy efficiency

by Bion D. Howard

Masonry construction is a large portion of commercial building, but masonry's share in residential construction has dwindled to less than 10% since World War II, according to the National Association of Home Builders. Only recently have concrete-block walls started to make a comeback in residential construction. This is largely a result of the development of new insulation systems for masonry construction, and of new data that link thermal mass with energy-efficient performance of buildings. Part of my work at the National Concrete Masonry Association (NCMA, Box 781, Herndon, Va. 22070) is to study and evaluate new ways of constructing better insulated masonry walls.

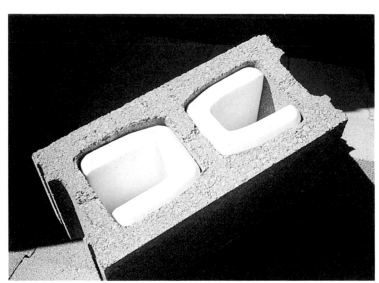

Insulation inserts. **Foam insulation inserts like these can be friction-fit into conventional block cores before the block is laid up. They will increase the insulative value of an 8-in. block to at least R-5.**

Mass and insulation—Before examining options for insulating masonry walls, it's important to put the current emphasis on high R-values into perspective. The recent interest in superinsulation has focused a lot of attention on resistance insulation, chiefly in the form of fiberglass batts. The notion that larger R-values are the only means of achieving better thermal protection emerged as a fabulous marketing tool for the insulation industry. Unfortunately, this isn't the entire story. The missing parts of the equation are the heat-storage capacity and radiative properties of the walls, which can cause the building envelope to perform differently than steady-state calculations predict. New ways of insulating masonry walls can provide the best of both worlds: reducing conductive heat loss using insulation, and improving heat-retention capacity with mass.

The main types of masonry-wall insulation systems are as follows: lightweight concrete made with insulating aggregates; loose fill insulation that is poured into the concrete-block cores as the wall is built, or "pour-foam" installed in the same way; formed plastic insulation inserts that can be fit into the cores of conventional concrete block, or concrete blocks with specially designed cores; cavity walls where two separate *wythes* (a wythe is a single vertical masonry wall) of masonry are built, creating an air gap that is partially filled with insulation; exterior application of insulation covered with weatherproof coatings after wall construc-

tion; and composite construction units using high-strength adhesive or mechanical bonding of two masonry wythes on either side of a layer of insulation.

These insulation systems for masonry walls can be used alone or in various combinations. Thermal-mass research shows that the greatest energy savings are possible when the insulation is located outside the masonry wall. But this isn't always the best option because of cost or design constraints. Integral insulation systems, which have grown in number since the energy crisis, are an alternative choice and can often be more economical than exterior insulation on masonry walls.

Lightweight, insulated concretes—One method of "pre-insulating" concrete masonry is to change the properties of the concrete itself. Lightweight concrete was invented for shipbuilding during World War I. It is made by using a lightweight aggregate such as expanded polystyrene or glasses made from polymer or mineral bases. The lightweight aggregate decreases the density of the concrete, thereby boosting its insulative value.

As might be expected, some strength is sacrificed with the use of lightweight aggregates. The Sparfil Corporation (5 Veronica St., Box 235, Coburg, Ont., Canada K9A 4K5) spent several years on costly experimental development before it was able to claim that it had a block that was both highly insulated and lightweight. The

Sparfil block relies on expanded polystyrene for insulative value. An 8-in. block is rated at R-6 and weighs just 27 lb. (a standard 8-in. concrete block weighs about 35 lb.). For still greater R-value, these blocks have staggered cores that can be filled with foam inserts. If these are used, a finished wall of 12-in. block can approach R-25. Sparfil walls (top drawing, facing page) are constructed by the surface-bonding technique, using high-strength, fiber-reinforced cement coatings on both sides of the dry-stacked block masonry. Vertical steel reinforcement can also be added to these walls through the cores. The main advantages of mortarless assembly are reduced labor costs, equal or better strength than unreinforced mortared assemblies and very low air-infiltration rates. The surface-bonding compound also creates a sound moisture barrier.

Pour-in core insulation—The principal loose fill materials for block and brick walls are Perlite (Perlite Institute, 6268 W. Jericho Tpk., Commack, N. Y. 11725) and Zonolite (W.R. Grace Corp., Construction Products Division, 62 Whittemore St., Cambridge, Mass. 02140). These are loose beads of expanded polystyrene. Another type of pour-in insulation, called Poly-C (Upjohn Corp., Chemicals and Plastics Research, 555 Alaska Ave., Torrance, Calif. 90503) is poured into the cores as a liquid. A chemical reaction causes it to expand and fill the cores. Frothane (Therma-froth Systems, Inc., 99 Collier St., Suite 300, Binghamton, N. Y. 13901), Kasko K1-10 (Kasko Industries, Inc., 301 West Hills Rd., New Canaan, Conn. 06840), Air-Krete (Air Krete, Inc., Box 380, Weedsport, N. Y. 13166) and Thermal-Krete (Omni-Tech Energy Products, Inc., 1515 Michigan Ave. N.E., Grand Rapids, Mich. 49530-2085) are other brands of insulation that can be "foamed" into the voids in the block cores.

Both Perlite and Zonolite insulations are silicone treated for water repellency, and are fireproof. The plastic-based insulating fills are not fireproof, but are isolated in the masonry cores. You should check local fire codes before installing plastic insulations inside block since there may be special regulations in your area. Today's

From *Fine Homebuilding* magazine (February 1986) 31:38-41

foam-insulation products claim to be free of urea-formaldehyde emissions, but it's wise to get this in writing from the contractor or supplier, so that a document may be passed on to new owners. Some states require urea-formaldehyde foam-insulation certificates before homes can change hands or be occupied, even though the ban on urea-formaldehyde foam has been lifted.

Formed plastic insulation inserts—Masons, mason contractors and builders seem to like using new concrete blocks with plastic insulation inserts that have been installed at the block plant. Most of these blocks are made from conventional concrete and arrive on site with insulation already in place and ready to lay up. No added labor or insulation is required to reach R-7 to R-10 values.

The Korfil block (photo facing page) comes with a C-shaped insert that is friction-fit into the block cores (Korfil Inc., Box 123, Chicopee, Mass. 01014). Bend Industries (2929 Paradise Dr., West Bend, Wis. 53095) and the Miller Material Co. (Box 1067, Kansas City, Mo. 64141) have developed several similar insert systems (middle drawing, right). Cast and "split" architectural facings are available on some of these insulated blocks.

The Insul Block Corp. (55 Circuit Ave., West Springfield, Mass. 01089) and Formbloc Inc. (Box 546, Concord, N. H. 03301) manufacture cross-web, flat panels of foam that fit into modified concrete blocks. In these systems, the foam insulation extends the full length and nearly the full depth of the block, providing a more complete thermal barrier. In order to accommodate the maximum amount of foam insulation, the blocks are designed with cut-down block cross webs, as shown in the bottom drawing at right.

Korfil has taken the cut-down web method a step further in a new insulated block that should approach R-10. Called NCMA Korfil High-R, the block uses two polystyrene insulating panels per unit. These panels are slightly different sizes and overlap the mortar joint areas to reduce air infiltration and moisture migration. Horizontal reinforcing rods can be used with channels already formed during the block-making process. Full-scale structural testing of High-R walls is now underway at NCMA's research laboratory, and the system should be available on the market in the near future.

Sparfil makes inserts to add to its insulated lightweight blocks. The Sparfil cores are long and narrow, and they're staggered in three rows to reduce thermal bridging. This approach elongates the normally linear flow of heat through the unit, delaying heat loss.

Essenco's E/Block (Essenco Inc., 834 Eagle Dr., Bensenville, Ill. 60106) uses a phenol-based high-R foam that's inserted in special mini-cores in the block (top drawing, p. 55). Only small areas of thermal bridging occur at the block ends. The block is manufactured with a recess at each end that can be filled with site-applied insulation. Essenco claims E/Block also can attain 20% to 30% higher compressive strength than regular block. High ratings on fire, sound and moisture resistance are also claimed. Early in 1985, these blocks were exhibited at the Na-

Drawings: Frances Ashforth

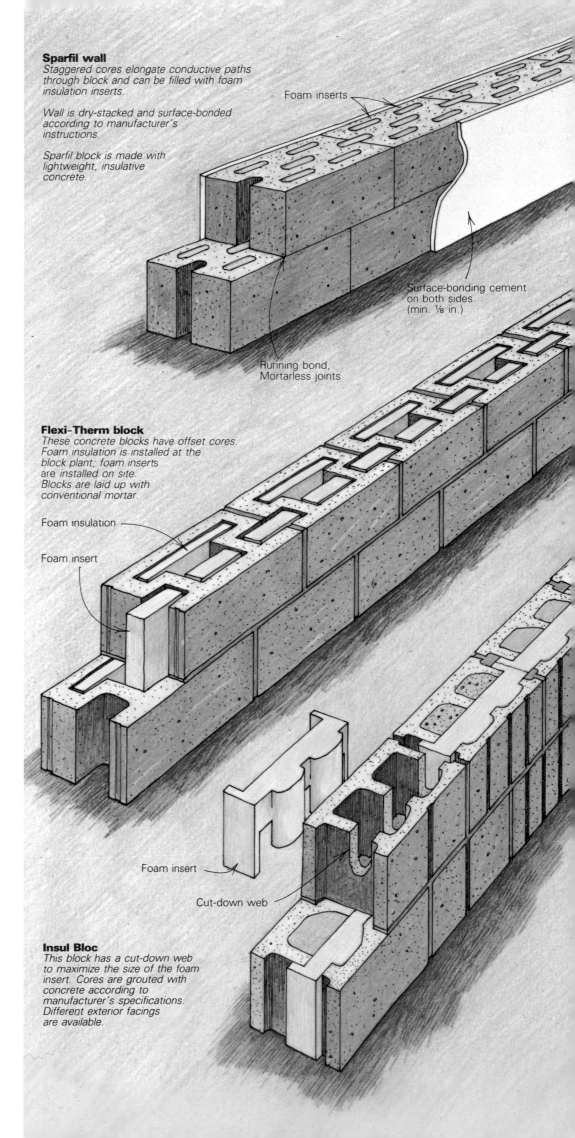

Sparfil wall
Staggered cores elongate conductive paths through block and can be filled with foam insulation inserts.

Wall is dry-stacked and surface-bonded according to manufacturer's instructions.

Sparfil block is made with lightweight, insulative concrete.

Foam inserts

Surface-bonding cement on both sides (min. ⅛ in.)

Running bond, Mortarless joints

Flexi-Therm block
These concrete blocks have offset cores. Foam insulation is installed at the block plant; foam inserts are installed on site. Blocks are laid up with conventional mortar.

Foam insulation

Foam insert

Foam insert

Cut-down web

Insul Bloc
This block has a cut-down web to maximize the size of the foam insert. Cores are grouted with concrete according to manufacturer's specifications. Different exterior facings are available.

Cavity walls. Two separate masonry wythes, one of brick and the other of concrete block, create an airspace that is filled with rigid insulation. The width of the cavity can vary from ¾ in. to 6 in. or more, depending on how much insulation is desired.

tional Concrete Masonry Association Concrete Industries Exhibition in Atlanta and met with good reviews by industry experts.

Cavity walls—Cavity walls saw their first use nearly 70 years ago, and we know about them primarily because of more recent demolition work. The brick and block cavity wall (photo above) has several definite advantages over single-wythe masonry. In cavity-wall masonry construction, a continuous air space is left inside the wall between two masonry wythes.

A cavity wall takes more time and material to lay up than a single-wythe wall, but it's the best system in many ways, providing insulation space, thermal mass, sound reduction, fire resistance, and control over moisture and air infiltration.

The cavity wall is built with weep holes in the bottom of the exterior wythe to give moisture an escape route away from the insulation layer and interior masonry wall. Larger screened holes, built into the exterior wythe just under the eaves, can help to vent the wall and can eliminate the need for a vapor retarder.

Cavity-wall construction is good for hot, humid and stormy climates where moisture is a problem and outdoor heat must be repelled. Under these conditions, reflective insulations like foil-faced boards perform well when used in the cavity. If a foil-faced insulation board is used in a cavity wall, the air space must be kept even,

and mortar droppings should be removed from the back of the outer wythe so that they don't touch the insulation.

The air space, or cavity, can be made as wide as 6 in. for low-rise construction, according to recent NCMA structural analyses. This provides enough room for up to 4 in. of foam-board insulation, available in a variety of R-values. A 2-in. air gap (on the exterior side of the cavity) is recommended with a 6-in. cavity. Pour-in insulation can also be used, so the cavity wall is a fairly flexible system. Wall ties 6 in. and longer are now available through construction jobbers, but many current building codes allow a maximum 4-in. cavity, and recommend ¾ in. as the minimum air gap. The cavity-wall builder may need to have a structural engineer perform calculations to support the use of a wider cavity, especially in multi-story construction.

In cooler climates, the vapor barrier in a cavity wall should be located closer to the interior conditioned space. If furring and drywall are going to be used against the exposed inner masonry, polyethylene film can be installed either against the masonry or between the furring and the drywall. Alternatively, the new airtight drywall approach (ADA) could be used, where the interior drywall provides the air barrier. With this system, vapor transmission is reduced by eliminating air leakage through the use of gaskets and careful workmanship.

Composites and composite walls—Composite masonry is typified by rigid insulation sandwiched between brick and block masonry units, with no airspace between the materials (middle drawing, facing page). This arrangement is achieved with high-tech adhesives and mechanical fasteners. It is also possible to use factory-applied expanding polyurethane foams, which are highly adhesive. This system is now in use in England. The advantage of composite wall construction over cavity wall construction is that the wall lays up in one step just as you'd lay up pre-insulated block. A wide variety of exterior appearances can be obtained with less labor than it takes to build a cavity wall.

In most cases, however, composite masonry units are laid up with mortar. In mortared wall systems, the mortar joint will affect the overall R-value of the wall. For example, a wall built with 8-in. by 8-in. by 16-in. composite masonry units rated at R-12 would have an overall R-value of R-8.2 because the R-1.6 mortar joints account for 7% of the wall surface area.

Several composite masonry-wall systems are currently under development. Experiments show it is possible to produce composite units on modified conventional concrete-block machines. Composite masonry units could use low-density insulating concrete, and integral insulation would provide full thermal breaks if the wall were surface-bonded rather than mortared. It is also probably feasible to mix a phase-change material in with the inner side's concrete, enhancing the thermal characteristics of the wall.

Exterior insulation of block walls—This approach can be used for retrofits as well as for new construction (see article on pp. 49-51). The development of exterior insulation systems for concrete block makes it possible to transform aging concrete and "cinder-block" buildings into good-looking, energy-efficient structures. Exterior insulation retrofits can be done with little or no disturbance to those living in the house. The drawing at the bottom of the page shows a typical exterior insulation system.

The typical block wall built before the energy crisis has an R-value of between R-3 and R-7. Adding 1 in. of R-5 exterior insulation and covering this with a protective stucco finish more than doubles the insulative value of the block wall. In addition, the wall mass can better interact with the HVAC system and the indoor environment by storing heat.

Exterior insulation systems (insulation and protective covering) can cost between $3 to $7 per sq. ft. installed. The cost varies with the thickness of the insulation, the size of the building and the particular detailing requirements. You also have to take into account any replacement of windows and doors that should be done as part of the energy upgrade. NCMA's pamphlet TEK 134 describes the exterior insulation of block walls in detail and provides cost data, fuel conversion factors and a map of suitable block exterior insulation levels for the various parts of the U. S. More information is also available from the Exterior Insulation Manufacturers Association (1133 15th St. N.W., Washington, D.C. 20005). EIMA is now setting industry standards, compiling case histories on exterior insulated projects and coordinating industry information.

Cost versus performance—A good way to evaluate insulated masonry wall systems is to compare them with conventional concrete-block construction. Let's assume the unit cost of a conventional, uninsulated 8-in. block wall to be 1.0, and recognize that this wall will be rated at R-0.11 to R-0.29 per in., depending on concrete density. Advanced systems like Sparfil will cost nearly 2.5 units and provide 0.69 to 2.88 R-per-inch, depending on the concrete mix and the insert installation.

Core insert insulations will cost at least 1.4 units and provide an R-per-inch of 0.40 to 0.56. Loose-fill insulations installed on site (Perlite, Zonolite) provide a broader range of R-per-inch, 0.33 to 0.70, at only slightly higher cost.

The cut-down, web-type insulated block appears to be most cost-effective option. Korfil and Insul-Bloc are now producing and/or developing such products. The composite systems, most of which are still experimental, should provide R-per-inch of 0.73 to 3.75 at a cost of 1.7 to 2.0 units. □

Bion Howard is a technical advisor and energy engineer for the National Concrete Masonry Association in Herndon, Va.

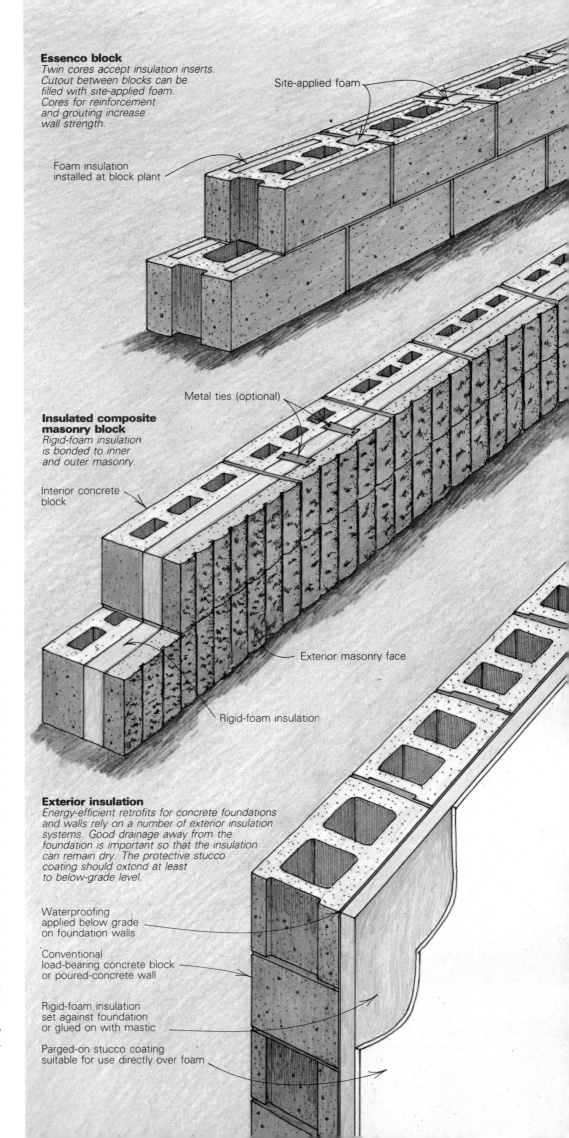

Essenco block
Twin cores accept insulation inserts. Cutout between blocks can be filled with site-applied foam. Cores for reinforcement and grouting increase wall strength.

Site-applied foam

Foam insulation installed at block plant

Insulated composite masonry block
Rigid-foam insulation is bonded to inner and outer masonry.

Metal ties (optional)

Interior concrete block

Exterior masonry face

Rigid-foam insulation

Exterior insulation
Energy-efficient retrofits for concrete foundations and walls rely on a number of exterior insulation systems. Good drainage away from the foundation is important so that the insulation can remain dry. The protective stucco coating should extend at least to below-grade level.

Waterproofing applied below grade on foundation walls

Conventional load-bearing concrete block or poured-concrete wall

Rigid-foam insulation set against foundation or glued on with mastic

Parged-on stucco coating suitable for use directly over foam

Dry-Stack Block

Precision-ground concrete blocks make it easy to build a wall

by Rob Thallon

Designers have been trying for years to develop a mortarless concrete-block system that could be used by unskilled builders. The concrete blocks in use today look quite uniform, but their dimensions actually vary so much that mortar is necessary not just to hold them together, but also to make up for their irregular sizes. Mixing and applying the mortar to the joints in a block wall require skill and time (see article on pp. 44-47), and the process accounts for 20% to 30% of the material and labor in a masonry project. Manufacturers have recently developed mortarless, interlocking block for industrial and commercial buildings. I use it in house construction. It's called dry-stack block, and it can be laid up as easily as the plastic toy blocks in a Lego set.

Dry-stack blocks look very much like ordinary concrete blocks, but they are consistently a full 16 in. long and 8 in. high (regular blocks are an inexact ⅜ in. less in each direction to allow for the mortar joint). During the manufacturing process, the dry-stack blocks I use are sent through a machine that grinds the top and bottom surfaces to a tolerance of 0.005 in. These parallel, exact and smoothly ground surfaces are what allow the block to be laid up so regularly without mortar.

Most dry-stack blocks have interlocking tongues and grooves at their ends to help align and secure them during placement. Besides standard blocks, there are also bond blocks for bond beams (these have knockouts to accept horizontal rebar), and half blocks. Special corner blocks are manufactured without tongues for finished outside corners (drawing, facing page, center). Where the block remains exposed, its edges are usually chamfered to create a hand-tooled corner that's less likely to chip. It is also possible to have the face of the block ground and sealed to create a smooth, marble-like appearance.

There are three essential differences between the ordinary mortar-laid block and the dry-stack. First, the dry-stack method uses mortar only at the joint between the footing and the first course of block. This mortar joint at the base lets you set the first course absolutely level. Second, ordinary block is usually grouted (filled with concrete when the wall is complete) in only the cells containing reinforcing steel (rebar), while dry-stack blocks are usually grouted in every cell. This locks the blocks in place, and also fills the bond beams completely without having to pour them individually (drawing, facing page, top).

Third, you have to be careful with ordinary block walls to be sure that fallen mortar (as distinct from grout) doesn't hang up in the rebar or clog the bond-beam channels. This usually means that you have to build the wall in 4-ft. vertical increments so that the grout completely fills the appropriate cells. A dry-stack wall, however, can be grouted all at one time because there is no mortar to clog the steel or to plug up the cavities. Grouting tall dry-stack walls all at once can save a lot of time, especially if you use a concrete pumper.

When the dry-stack system was first introduced in the Eugene, Ore., area, about half the projects were questioned by the building department. The building official wanted to see calculations proving that the dry-stack system is as strong as a regular block-and-mortar wall. This is reasonably easy to demonstrate by showing that the compressive strength of the block is greater than that of mortar.

Residential applications—I had seen dry-stack block used successfully on several houses before I had the opportunity to try the system myself. I had designed a house for a steep site, with a complex foundation and several retaining walls. It looked as though using dry-stack blocks would allow a significant saving on labor. In addition, my client wanted a warm-colored block, and not having to use

From *Fine Homebuilding* magazine (August 1983) 16:54-57

Reaching as high as 12 ft., the finished block-work is ready for the carpenters. A quarry-tile feature strip is visible just below the top course at locations where the walls will act as foundation for the house. The brownish-red blocks used above grade are special order.

mortar meant we wouldn't have to mix colored mortar to match the block.

Before ordering the block, I asked the supplier about various coloring agents, but everything they showed me gave the blocks a bland uniform color—they looked phony. As an alternative, the manufacturer (Willamette-Greystone Inc., P.O. Box 7816, Eugene, Ore. 97403) suggested using scoria, a brownish-red volcanic aggregate found in Oregon's Cascade Mountains. This seemed to be just what I wanted, so I ordered a special run of blocks.

When the blocks finally arrived at the building site, I was surprised and disappointed. Instead of the rich, red-brown color I had expected, the blocks were pink. Evidently a slurry of scoria dust and cement had come to the surface as the blocks were extruded and vibrated during the manufacturing process. We eventually remedied the problem by sandblasting the finished wall.

With the footings poured and blocks on hand, we began building the walls. Our crew consisted of an experienced block mason and two laborers. I worked part time. The mason and I were anxious to see just how easily the dry-stack block could be laid up—he from a professional's point of view, and I from the perspective of a novice. On this job, we used almost 3,000 blocks and finished the foundation walls and three large retaining walls in about two weeks. The mason estimated that it would have taken four weeks using regular block and mortar.

First course—Getting the first course level is the most important part of the whole process. If you don't get it right, you'll be fighting your mistakes for the rest of the job. So the first rule is to have good footings, flat and within ¼ in. of level.

Mark the corners of your building on the footing, just as you would for an ordinary block-and-mortar wall, and check for square. It's a good idea to lay out at least one wall on the footing without mortar to test the blocks for length. We found that our blocks varied enough in length to accumulate a ½-in. error in a 20-ft. run if we didn't pay attention. By laying out the blocks dry, we could see how big a gap we had to leave between blocks to make things come out even.

After setting the corner blocks in mortar, we stretched out the mason's line and got down to laying the first course. We found that the work proceeded more easily than we expected, because the vertical joints don't require any special attention. This is a boon for the inexperienced mason. All you need to do is to lay two tracks of mortar along the footing, set the block on the mortar and level in both directions (drawing, right). The smooth surface of the blocks makes leveling easy. As

Dry-stack and conventional block walls compared

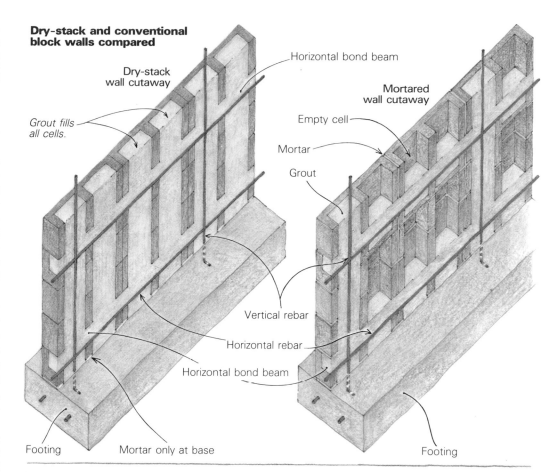

Dry-stack wall cutaway

Horizontal bond beam

Mortared wall cutaway

Empty cell

Mortar

Grout

Grout fills all cells.

Vertical rebar

Horizontal rebar

Horizontal bond beam

Footing

Mortar only at base

Footing

Four types of dry-stack block

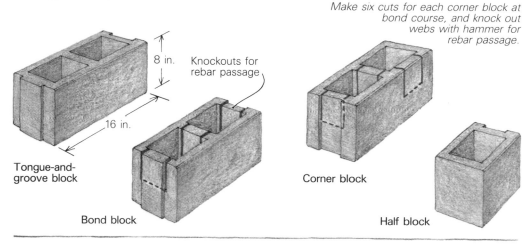

Make six cuts for each corner block at bond course, and knock out webs with hammer for rebar passage.

8 in.

16 in.

Knockouts for rebar passage

Tongue-and-groove block

Bond block

Corner block

Half block

Laying the first course

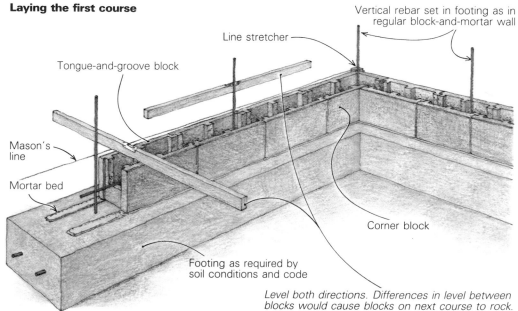

Vertical rebar set in footing as in regular block-and-mortar wall

Line stretcher

Tongue-and-groove block

Mason's line

Mortar bed

Corner block

Footing as required by soil conditions and code

Level both directions. Differences in level between blocks would cause blocks on next course to rock.

Does it pay? As the chart shows, the dry-stack method costs more for the block itself and needs 30% more grout, but it requires virtually no mortar and saves on labor for grout and laying block. In the project from which the figures were taken, about 10% more was spent for materials, but 24% was saved overall by using dry-stack blocks.

	Dry-stack block walls (actual cost)	Standard block-and-mortar walls (estimated cost)
Materials		
Mortar	$ 30	$ 300
Block	2,847*	2,477
Grout	(32 yd.) 1,481	(20 yd.) 920
Steel (2500 ln. ft.)	461	461
Subtotal	$4,819	$ 4,158
Labor		
Laying block	$2,700	$ 5,400**
Grout-pump truck	165	(2 lifts) 330
Grout labor	200	300
Subtotal	$3,065	$ 6,030
Total	$7,884	$10,188

*2,414 8-in. regular, 345 8-in. half, 175 12-in. regular. **Based on the mason's estimate of cost-per-square foot at about $2, a conservative figure. The 1979 Western Edition Building Cost File quotes a figure of $3 per square foot.

A quarry-tile inlay makes a thin red stripe around the house. The blocks have been sandblasted and scaled, revealing their volcanic aggregate.

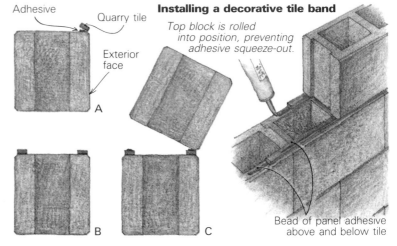

Installing a decorative tile band

Adhesive

Quarry tile

Exterior face

A

B

C

Top block is rolled into position, preventing adhesive squeeze-out.

Bead of panel adhesive above and below tile

we worked, we checked for length every 4 ft. or so and either tightened or loosened the joints slightly to come out even at the corner.

As block walls grow—Once the first course was laid, we built up the corners, carefully plumbed, as a guide for subsequent courses. The weight of the top blocks kept the lower blocks from moving and allowed us to stretch a mason's line tight from corner to corner as a guide. We kept the line about a string's width from the wall, so that accumulating error from successive blocks wouldn't force the string slightly out of line as they touched it. We found that we could set the blocks into place so rapidly that moving the string line became a significant part of the work.

In fact, the work sometimes went so fast that, in our enthusiasm, we made mistakes. The beauty of a mortarless system is the ease with which such mistakes can be corrected. At one point we dismantled a large portion of a 6-ft. tall fireplace footing and ash dump that had gone awry, and put it back together again in less than an hour.

We had some high walls on this project. One, which we built with 12-in. block, was more than 12 ft. tall, and we had several 8-in.

block walls over 8 ft. tall. Walls this high can get pretty wobbly before they are grouted, so we spread a double bead of panel adhesive between every fourth course for stability.

Bond beams—The only major cutting we had to do was notching the corner blocks at bond courses to let in the rebar. We used a mason's saw for this, and for the other minor cutting chores required for vents, bolts and the like. On small jobs, a circular saw with a carborundum blade would work fine.

The rebar required for a dry-stack wall is the same as that required for a block wall with mortar. The minimum requirements are listed in local building codes. For retaining walls up to 4 ft. high or 12 ft. long, we used one #4 bar at 32 in. o.c. vertically and one #4 bar at 24 in. o.c. horizontally (every third course with 8-in. blocks). Beyond these limits we had the wall engineered. Masonry suppliers have brochures listing rebar requirements compiled by the National Concrete Masonry Association.

To form an opening for a window or door, we supported a 2x8 formboard at the appropriate level with temporary posts, then laid up a steel-reinforced and grouted bond beam to serve as a lintel.

A tile feature strip—Just before the last course of block was laid, we installed a narrow tile feature strip. I wanted a thin band of color built into the wall itself to complement the horizontal water-table band at the top of the wall (photo above) so I had a local tile shop split 4x8 red quarry tiles into four lengthwise sections. I sandwiched these between two courses of block. The two pieces with finished edges were used on the exposed side of the wall, and the other two pieces on the hidden side.

We dusted off the top surface of the block and then laid a bead of panel adhesive about 10 ft. long near the inside edge of the tile. I aligned the tiles carefully along the chamfered edge of the block (drawing, above) and rolled them back onto the adhesive, forcing the excess toward the center of the block. This process was then repeated less carefully on the inside of the block and finally on top of the tiles with the final block.

The panel adhesive turned out to be an indispensable part of building the walls. It held all the tiles in position, bonded the top two courses together, and made the wall very rigid. This added rigidity was especially important during the grout pour.

Dry-stack block suppliers

If you're ready to build with dry-stack blocks, you need a local supplier because the blocks are just too heavy to ship economically more than about 150 miles. But finding a local supplier can be frustrating. We called all the block companies listed by the National Concrete Masonry Association as sources for interlocking block and ground-surface block, and asked each company if it could supply the mortarless units. It seems that relatively few builders ask for dry-stacks, so there aren't many made. As a result, the blocks can be hard to find.

The blocks we did find vary in price, design and dimensional tolerances. As a general rule, the closer the tolerances, the more expensive the block. In addition to the basic 8-in. high by 16-in. long by 7⅝-in. wide wall block, each supplier had a full line of sash, corner and bond blocks.

The McIBS Co. is the most active force in the mortarless-block industry. This company has perfected a special liner that can be used in conventional block-molding machines. The blocks made with these liners are double tongue-and-groove on their ends, tops and bottoms, and they maintain dimensions within 0.030 in. This close tolerance, coupled with the tongue-and-groove arrangement on all the hidden surfaces of the block, allows a wall made with them to be sealed with products designed for conventional block walls.

Of the mortarless blocks we found, McIBS are the most widely available, and the most expensive—from 30% to 80% more than conventional blocks. They are currently available in California, Colorado, Illinois, Indiana, Missouri, Nevada, Texas and Wisconsin, with several more states soon to join the list. Write to McIBS Inc., 130 S. Bemiston, St. Louis, Mo. 63105 for more information. For $1, they'll send you a brochure detailing their products and how they are used.

In Texas, builders can find interlocking blocks with tolerances held to ⅛ in. at two places: Valley Builder's Supply, Inc., P.O. Drawer Z, Pharr, Tex. 78577; and the Barrett Co., Rt. 3, Box 211 BI, San Antonio, Tex. 78218. Both suppliers make blocks with tongue-and-groove joints cast into their ends. They cost only about 3% more than conventional blocks. Both companies recommend surface-bonding the finished wall (see article on pp. 60-63). Similar blocks are available in Minnesota at Charles Friedheim Co., 3601 Park Center Blvd., Minneapolis, Minn. 55416, and in Iowa at the Marquart Block Co. 110 Dunham Place, Waterloo, Iowa 50704.

Oklahoma builders can find dry-stack blocks at the Harter Concrete Products Co., 1628 W. Main St., Oklahoma City, Okla. 73106. Harter grinds its blocks to 1/32-in. tolerance, and then cuts a ¾-in. deep slot in the top and bottom of each block to accept a plastic spline. The splines help to align the blocks vertically. The cost is currently about $1.30 per unit for the basic block, and surface-bonding the finished wall is also recommended.

Yet another type of dry-stack block is made by the Buehner Block Co., 2800 S.W. Temple, Salt Lake City, Utah 84115. The block is 8 in. by 16 in., but only 5⅝ in. wide. The blocks hold to ⅛-in. tolerance, and use an interlocking system of plastic rings for alignment during placement.

Dry-stack blocks end up in a wide variety of projects—from houses to roadside sound barriers, from racketball courts to Holiday Inns. They might even become the universal building component that Frank Lloyd Wright and his son Lloyd envisioned in the 1920s. The blocks are simple to use, and they offer the thermal storage capacity that is an essential element of passive-solar design. *—Charles Miller*

Grouting—Before we scheduled delivery, we calculated the amount of grout we needed with the following formulas:

for 6-in. block:
number of full blocks/110 = cu. yd. grout;

for 8-in. block:
number of full blocks /90 = cu. yd. grout;

for 12-in. block:
number of full blocks/50 = cu. yd. grout.

On our job, for example, the calculation was: (2,414 + 345/2 [half blocks])/90 + 175/50 = 32.2 cu. yd. grout.

Because we needed so much grout (about 60 tons), we decided to hire a concrete pumper to get the grout from the trucks to the walls. We completed in about two hours a job that would have taken at least two days if we had done it by hand.

The only problem we encountered was on one of the tall walls. When we filled the cells, the weight of the mud blew out the side of one of the lower blocks, which was probably already cracked. Grout spurted out all over the place. We were able to repair the block and then refill the wall, a little at a time. If you have a wall 8 ft. high or taller, I recommend grouting it up to about 5 ft., filling the shorter walls, and then returning to top off the tall walls after the first pour has had time to set up a bit.

Cleaning—You will inevitably slop some grout over the sides of the block. It's easy to rough-clean the surface by scraping it within 24 hours after the grout is poured. If you have the chance, clean the walls immediately with a light water spray and a soft brush. This can save a lot of hard work later on. To remove the grout stains completely, use a masonry cleaner like muriatic acid.

On this particular job, we wanted to remove both the stains and the pink slurry that formed the surface of our dry-stack blocks. We decided to sandblast only after trying several chemical cleaners without success. The blasting produced the desired results and cost only $320 for the whole job.

Waterproofing—You waterproof dry-stack block the same way as regular block. Below grade, we used Thoroseal, a water-base sealer, which we brushed on in two coats. Some prefer to apply one coat with a trowel, but this requires more skill. I'm sure that any of the asphalt-base sealers recommended by masonry suppliers would also work.

We sprayed the exposed walls with a two-part application of clear acrylic sealer. First, we applied a coat of relatively inexpensive Stone Glamour, then we sprayed on a finish coat of Mex-Seal for the reflective surface we wanted. These sealers bring out the blocks' color much as an oil enhances wood, and protect the block from the deteriorating effects of water penetration. For longest life, an exposed block wall should be resealed every five years or so, depending on the severity of the thaw-and-freeze cycles in your climate.

We didn't do anything to seal the exposed cracks between the blocks, even though we worried about the problems they might cause. I was afraid that capillary action would pull water through the cracks and cause moisture problems inside the house, and that the moisture in the cracks might freeze and fracture the blocks.

We resolved the first problem by sealing the inside of the walls with Thoroseal wherever they enclosed living space. We decided to ignore the second potential problem because the climate here isn't very severe. In the three years since we finished the walls, there has been no cracking.

In a climate where the combination of moisture and freezing is liable to cause problems, I would seal the exposed joints with clear silicone caulk. The caulk could be spread between blocks as the wall is laid up, or applied to the grooves between the blocks' chamfered edges after the wall is assembled. Either one of these procedures would increase construction time and expense, but the job would still be quicker, cheaper and easier than laying up a wall with mortar.

Cost comparison—When we finished the project, all of us who had worked on it were impressed with the dry-stack block. The mason was sure we had cut our labor time significantly by using the dry-stacks (chart, facing page), and he thought that they would result in a 50% labor saving on an average project. The laborers liked it because they got to lay some block themselves, which broke the drudgery of their usual lot—lugging heavy objects around the site all day.

What's wrong with this system?—Availability, that's what. Dry-stack blocks are so heavy that long-distance shipping is prohibitively expensive. Consequently, the blocks have to be manufactured close to their point of use. Although makers of standard concrete block are liberally scattered around the country, relatively few have the grinding or molding equipment necessary for making dry-stacks (see the sidebar above). And unfortunately, there isn't a comprehensive list of manufacturers that make the blocks. So if you're interested in the dry-stack system, get out the Yellow Pages and do some dialing. □

Rob Thallon is a partner in the architectural firm of Thallon and Edrington in Eugene, Ore.

Surface-Bonded Block

A strong, fast and inexpensive alternative to
poured-concrete or block-in-mortar walls

by Paul Hanke

Pouring concrete walls is a difficult and risky business, and I don't recommend it for the inexperienced. Even professionals sometimes have forms let go, creating various degrees of disaster and pandemonium on the site. Laying up block with mortar has drawbacks, too. It is time-consuming, it takes practice, and the result isn't especially strong.

Surface-bonded block, on the other hand, suits owner-builders to a tee, and can be a less expensive alternative for professionals. It is a method of laying concrete blocks without mortar, then troweling both wall surfaces with a

Paul Hanke is a designer and draftsman at Northern Owner Builder, in Plainfield, Vt.

portland-cement coating laced with chopped fiberglass for strength. Built on standard footings, surface-bonded block walls can be used below and above grade, for foundation walls and for finished living spaces. The method is fast and reliable. It requires no particular skill, and the finished wall is stronger than a block-in-mortar wall.

Surface bonding was originally developed as a low-cost construction technique for self-help housing. A USDA booklet on the subject (Information Bulletin No. 374, now out of print) shows a 12-year-old boy doing a successful job after 15 minutes of practice. Even professional masons are reported to be 70% more productive using this method than laying up block in

the conventional way. The USDA estimates that stacking and bonding 100 blocks would take a person an average of 7.4 hours. Several years ago, two friends of mine, Chapin and Donna Kaynor, built an earth-sheltered house using this technique. It took their crew of four inexperienced people, some of whom worked only part time, less than five days to stack and bond about 1,200 blocks.

Strength and cost—Stacked blocks coated with bonding mix have an average tensile strength (ability to withstand longitudinal stress) of from 300 psi to 500 psi, according to lab tests conducted by the USDA and the University of Georgia. This is about equal to the

From *Fine Homebuilding* magazine (December 1982) 12:34-37

strength of unreinforced concrete, and is six times stronger than block laid up with ordinary mortar joints. Mortar has very little adhesive power, and virtually no tensile strength. Its main purpose is to level blocks between courses. Because the weakest part of conventional block walls is the bond between block and mortar, these joints tend to crack, making water seepage a problem. A surface-bonded wall, with its seamless outer coating, is much more watertight (though the coating alone should not be relied upon below grade).

Having a block-in-mortar wall built costs about twice the price of the materials, plus footing and reinforcing. In our area, concrete foundation walls currently cost around $95 to $105 per cubic yard poured in place, including formwork and labor. A typical full basement accounts for about 5% of the cost of a house, or over $3,200 for the average $65,000 home. The builder using surface-bonded block can save as much as 35% to 40% of this figure.

Estimating materials—To build surface-bonded walls, you will need standard hollow-core concrete block, surface-bonding mix, galvanized corrugated brick-ties for shims (the Kaynors used about 250 for 1,200 blocks), threaded steel rod and connectors, a few sacks of mortar, and the rebar and concrete for footings. Order 8x8x16 block (about 65¢ each) for walls above grade or foundation walls that will extend less than 5 ft. below grade. Order 8x12x16 block (about $1.05 each) for a foundation wall deeper than 5 ft. Use the USDA table at right to estimate the number of blocks you need. Be sure to add extra block for reinforcing pilasters, the column-like buttresses used to strengthen the walls (discussed below), and half-length block, which you may need for the door and window openings. Order 5% to 10% extra to make up for waste and breakage. Have your block delivered to the center of your work area if at all possible, or deposited in strategic piles around the perimeter. To save time and effort, don't carry those heavy blocks any farther than you have to.

As the table shows, nominal 8x8x16 block is actually 7⅝ in. by 7⅝ in. by 15⅝ in. to allow for ⅜-in. mortar joints; so you can't figure in exact 16-in. modules when laying block dry. Having to calculate with fractional numbers would be a real headache, but estimating tables supplied by the USDA or surface-bonding mix manufacturers greatly simplify the task.

Bonding mix, which comes dry and includes the chopped fiberglass strands, is sold in bags of various sizes. A 50-lb. sack will cover about 50 sq. ft. of wall. Check the exact coverage when you order, and allow about 10% for waste from broken sacks, mixing and troweling. A 50-lb. sack of grey-colored mix currently costs about $14 in Vermont. White is about $17 per sack. You can also mix your own, as explained at right. In addition to the bonding mix, get enough sacks of mortar to lay the first course of block, plus a few extra sacks to use in spots where you need to shim more than ⅛ in.

I recommend using ⅜-in. threaded steel rod to connect the sill or top plate at the top of the

wall to the footing below. It is available in 2-ft. and 3-ft. lengths at hardware stores. Threaded rod is expensive, but it makes a secure connection. You'll also need connectors to join the lengths of rod, and nuts and washers to secure the wood sill to the top of the wall. The rod isn't for concrete reinforcement, but to tie footing, foundation and framing together to resist uplift forces. The block cores that contain the rod don't require filling with concrete.

The alternative to running threaded rod all the way up through the wall is to fill the cores at the top of the wall with concrete two or three courses deep every 4 ft., and embed standard ½-in. J-bolts. Use screening to keep the grout from falling all the way to the footing, or stuff fiberglass insulation down the block core. This method works, but it will not provide a continuous connection from footing to sill, and will not resist uplift.

Footings—As a general rule, footings should be twice as wide as the wall above, and as deep as the wall will be thick. A standard 8-in. thick wall calls for a 16-in. wide footing. Pour 24-in. wide footings for either a 12-in. thick wall or a two-story house.

The bottom of the footings should be at least 12 in. below the frost line, and almost anyone can safely pour them. You can pour into shallow forms or directly into trenches of the proper size, provided that their sides and bottoms are of firm, undisturbed soil. Place two No. 4 (½-in. dia.) lengths of rebar near the bottom of a 16-in. wide footing. A 24-in. wide footing will require three lengths of rebar. Check codes for the rebar requirements in your area. Remember to widen the footings for pilasters.

Although you can mix your own concrete for footings, ready mix concrete delivered to your site is best. Insert the lengths of threaded rod vertically into the concrete at the corners and pilaster locations, on both sides of all the door and window openings, and every 4 ft. to 6 ft. along the wall, as shown in the drawing on the bottom of the next page.

Stacking block—After your footings are poured and have been allowed to cure, you can begin on the walls. Using your batter boards and strings (see "Site Layout," pp. 64-66), drop plumb lines to establish the outside corners. Use the table to determine exact wall lengths, and allow an extra ¼ in. per 10 ft. for irregularities in the blocks. Measure the diagonals to be sure that your corners are square, and adjust if you need to. Snap chalklines from corner to corner as guides, and then lay and level the first course of block in a bed of thick mortar. Check the top of the first course with a 4-ft. level as you go. If a block is too high, tap it down with the butt end of your trowel; if it's too low, remove the block, add more mortar and reset. Don't put mortar in the vertical joints between blocks; just butt them tightly against each other. Some skill is required here. Take your time and do a good job.

The rest of the wall is simply stacked dry in a standard running bond—each block overlapping half the block beneath. Begin by

Mixing your own

Here is the USDA formula for preparing 25 lb. of your own bonding mix. A friend who investigated this option concluded that it costs about 65% as much as a comparable commercial mix.

19½ lb. portland cement (78% by weight), white or type I grey, which is more common. This is the glue that holds things together. It comes in 94-lb. sacks.

3¾ lb. hydrated lime (15%) for increased workability. It comes in 50-lb. sacks.

1 lb. glass-fiber filament (4%), chopped into ½-in. lengths. Use type E fiber or, better yet alkali-resistant type K fiber, available from plastic and chemical-supply dealers, building-material dealers or boatyards.

½ lb. calcium chloride flakes or crystals (2%), to speed setup time and harden the mix. It's available from agricultural-chemical supply houses. Calcium chloride is also used for salting roads.

¼ lb. calcium stearate (1%), wettable technical grade, makes the mix more waterproof. You can obtain it from chemical distributors.

Since the bonding mix sets rapidly after water and calcium chloride have been added, do not make more than a 25-lb. batch at one time (dry weight).

Begin by mixing the powdered ingredients, except for the calcium chloride. Add the glass fiber, and remix only enough to distribute the fibers well. Overmixing breaks the fibers into individual filaments, which makes application difficult. Be sure to wear a proper respirator. The chemicals are very corrosive, and you don't want to breathe fiberglass, either.

Mix the calcium chloride with 1 gal. of water, and slowly add this solution to the dry ingredients. Mix thoroughly. Add about ½ gal. more water, until the mix is the right consistency—creamy, yet thick enough for troweling. A mix that's too thick is hard to apply and may not bond properly. —P.H.

Wall and opening dimensions for surface bonding

Number of blocks	Length of wall or width of openings		Number of courses	Height of wall or openings	
1	1 ft.	3⅝ in.	1	0 ft.	7⅝ in.
2	2	7¼	2	1	3¼
3	3	10⅞	3	1	10⅞
4	5	2½	4	2	6½
5	6	6⅛	5	3	2⅛
6	7	9¾	6	3	9¾
7	9	1⅜	7	4	5⅜
8	10	5	8	5	1
9	11	8⅝	9	5	8⅝
10	13	0¼	10	6	4¼
11	14	3⅞	11	6	11⅞
12	15	7½	12	7	7½
13	16	11⅛	13	8	3⅛
14	18	2¾	14	8	10¾
15	19	6⅜	15	9	6⅜

Blocks sold as 8x16 are actually 7⅝ in. by 15⅝ in. to allow for the size of mortar joints in standard block construction. Remember that cement blocks are not uniform. Add ¼ in. to every 10 ft. of wall length to take this into account, and before beginning to build, make a trial stack to measure the precise height your wall will be.

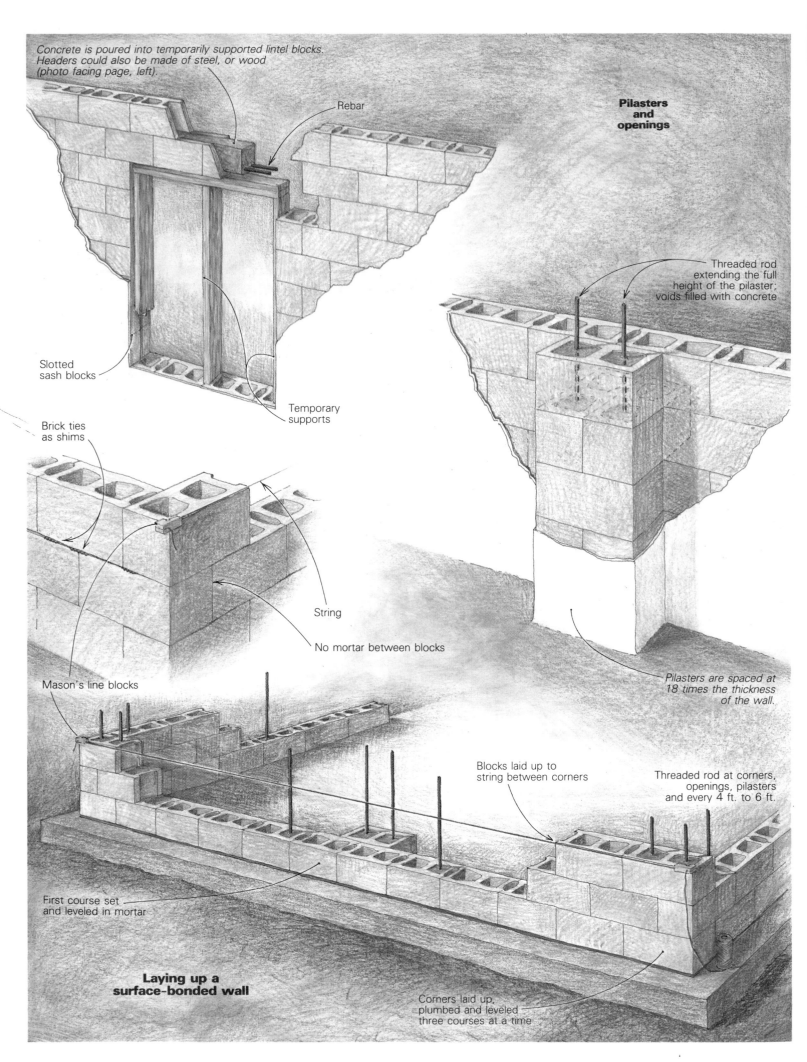

Concrete is poured into temporarily supported lintel blocks. Headers could also be made of steel, or wood (photo facing page, left).

Rebar

Pilasters and openings

Threaded rod extending the full height of the pilaster; voids filled with concrete

Slotted sash blocks

Temporary supports

Brick ties as shims

String

No mortar between blocks

Pilasters are spaced at 18 times the thickness of the wall.

Mason's line blocks

Blocks laid up to string between corners

Threaded rod at corners, openings, pilasters and every 4 ft. to 6 ft.

First course set and leveled in mortar

Laying up a surface-bonded wall

Corners laid up, plumbed and leveled three courses at a time

The first course of block is laid and leveled in mortar, as at right. The corners are built up three courses, then a level line is strung between corners. Dry block fills in up to it. Shims are used where necessary to keep blocks aligned. A mason's corner block, which holds a level line that is also the correct plane for the face of the blocks, is visible on the second course. Once the walls have been built up around openings, headers must be installed. These can be steel, concrete or wood, as above.

building up the corners three courses high. Check them for plumb with the 4-ft. level held vertically, and for level with a water tube. Then stretch a taut string between the top outside edges of each course of the built-up corners. Use mason's line blocks (available where you buy concrete blocks) to secure the line at each end. Fill in the length of the wall up to the string, and repeat the process every three courses, inserting metal shims as necessary to keep the wall level and plumb. If more than $\frac{1}{8}$ in. of shimming is required, use mortar instead. Check the wall for plumb at least every three courses. Connect new segments of threaded rod as you go.

Pilasters—These are engaged columns that reinforce the wall against lateral forces and keep it from buckling under heavy loading. For basement walls, pilasters should be on the inside to resist the pressure of the surrounding earth. The Kaynors put theirs outside to get them out of the living area. They are tied in by rotating the blocks of every other course 90° so that they become a part of the wall itself (drawing, facing page). Threaded rod or rebar should extend through the block cores the full height of all pilasters. After the wall is laid up, fill the voids of the pilasters with concrete.

For above-grade construction, pilasters are usually spaced along the wall at a distance equal to 18 times the thickness of the wall (for example, every 18 ft. for a 12-in. wall, or every 12 ft. for an 8-in. wall), or on a shorter wall, at midspan. The pilasters on the house shown here are on 8-ft. centers for earth-bermed walls, which is probably a good precaution for any below-grade construction.

Weight-carrying beams should also be supported by pilasters at each end. Be sure that the beam pockets extend into the wall at least

3 in. to get good bearing surfaces. Once the beam is in place, you can continue dry-stacking blocks in the usual manner.

Openings—For doors and windows, just omit blocks in the proper locations. This is where half-blocks come in handy. With these, you don't need to cut standard block down to size. The blocks at each side should be slotted sash-blocks, which accept a metal or wood spline that attaches to specially made framing.

Headers are required above openings, as in any other type of construction, and they should be properly sized for their span and load. Consult standard tables, codes, or an engineer if necessary. Headers can be made of wood (photo above left), steel angles or U-shaped bond-beam blocks. You support the blocks temporarily over the opening you want to span, then fill their cores with rebar and concrete. Once the concrete has cured, remove the supports, and you have a solid beam.

Coating the wall—Once all the blocks are stacked and the pilasters are filled with concrete, the walls are coated with surface-bonding mix (see sources of supply, below). Commercial mix consists primarily of mortar and strands of fiberglass chopped into $\frac{1}{2}$-in. lengths. Add water according to the instructions, and mix with either a garden cultivator or a mason's hoe (the kind with two large holes in the blade). The mix will cure in about an hour and a half, so don't whip up too much at one time. Hose down the block wall so that the mix won't dry out too quickly, then trowel on a $\frac{1}{16}$-in. to $\frac{1}{8}$-in. coat of the paste.

Both sides of the wall get surface-bonded. Use a hawk to hold a comfortable amount of the mix while you work, and press its edge against the wall to limit slop and spilling. Use a

plasterer's trowel, a steel trowel about 12 in. long, and work from the top of the wall down so you can moisten the block as you go if it begins to dry out in hot weather.

The USDA breaks the procedure down into four steps. First, with a series of sweeps of the trowel, spread the mix 2 ft. or 3 ft. upward from the hawk over a section about 5 ft. wide. Then even out the surface by going over the area lightly with your trowel slightly angled. Repeat these two steps over the area just below the block you've just covered, and cover as much area as you can in 15 or 20 minutes. Lastly, clean the trowel in water and retrowel the plastered area with long, firm, arced strokes to achieve a final, smooth surface.

The glass fibers bridge the joints between blocks, and the tensile strength of the wall increases as the concoction cures. Because the fibers are so short, the system won't work if you lay the block in mortar before coating the walls with the bonding mix. The fibers would be spanning almost their entire length, and this would destroy their effectiveness. The interior surfaces of walls can be textured with a light pass with a stiff brush. Mortar pigment can be added during mixing to color the wall. Surface-bonded walls can also be stuccoed or furred out, if you prefer something other than the bonding mix as a finished surface. □

Manufacturers of surface-bonding mix
Fiberbond Surface Bonding Cement: Stone Mountain Mfg. Co., Box 7320, Norfolk, Va. 23509.
Q-Bond: Q-Bond Corp. of America, 3323 Moline St., Aurora, Colo. 80010.
Stack & Bond: Conproco. Box 368, Hooksett, N.H. 03106.
Surewall: W.R. Bonsal Co., Box 241148, Charlotte, N.C. 28224.
Quick Wall: Quikcrete, 1790 Century Circle, Atlanta, Ga. 30345.

Site Layout

On a flat lot, footings can be oriented with precision using batter boards, string and a water level

by Tom Law

Driving that first stake is always exciting. It doesn't matter whether you've been designing and dreaming about the house for years, or you're beginning the first day of actual work on what you hope will be a profitable contract. Laying out the site in preparation for foundation work is your first chance to visualize the house full scale in its setting. The accuracy of your layout and the foundation it defines will also determine how much you will have to struggle to make your house tight and square.

Unless you are building on a sloped site, all you will need to do the job right is a 100-ft. tape measure, a water level, a ball of nylon string, enough lumber for batter boards, a helper and the application of some practical geometry.

Let's assume that the house is a simple rectangle, and that one of the long walls faces south. If precise solar orientation is important, use an accurate compass or one of the many siting devices available commercially. If such precision is not necessary, just stand facing the midday sun. Your outstretched left arm will point to the east and your right arm to the west. Unroll a ball of string along this axis for a distance a few feet greater than one of the long

walls of your building. I use braided nylon string because it will take an awesome amount of tension before it breaks, and because braided string doesn't unravel and can be used indefinitely. It comes in a highly visible yellow as well as in white.

Preliminary layout—Select one end of the layout as a starting corner and drive a small stake into the ground. You can use almost anything for a stake—a timber spike or a tent peg—as long as it holds the string off the ground. Now measure back down the string the length of the wall, drive another stake, stretch the string between the two stakes and tie it off. This lets you adjust the placement of the house on the site, and gives you an idea of where to locate the batter boards. Taut strings and accurate squaring are not necessary at this point, as long as the outside dimensions of the house are accurate.

To lay out the rectangle, pull another string from the corner you just established, at a right angle to the long wall. This is where your helper is needed. One person pulls the string while the other guesses at 90°. Measure along

this second string the width of the house and drive the third corner stake. The fourth corner can be found by measuring. Now step back and study the house placement on the site. Stop and think on this one a while—it's a permanent decision. If you are satisfied, you can begin preliminary squaring.

With a 100-ft. tape, measure the diagonals. If they measure the same, then you have created a rectangle. If not, you have a rhomboid, and you will have to adjust the two corner stakes opposite the long, south wall until the diagonals are about equal. Getting to within one or two inches at this point is close enough. I use a steel tape because cloth tapes stretch. A leather thong tied to the metal loop on the zero end of the tape will help you to pull hard and hold a dimension at the same time.

Crouch with your forearms braced against your thighs, and use your body weight to pull against the person on the other end of the tape. On the zero end, hold the leather strap, not the tape, and when you are squaring strings that are suspended from batter boards later on in the layout, keep the tape from lying on the string and deflecting it. Another method of get-

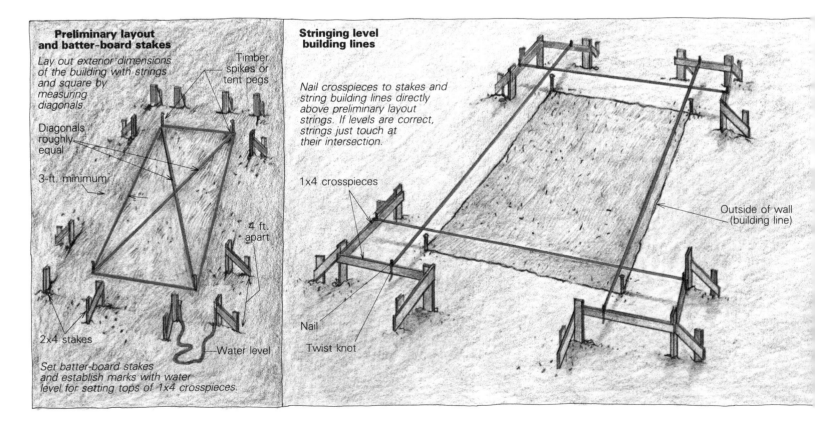

Preliminary layout and batter-board stakes

Lay out exterior dimensions of the building with strings and square by measuring diagonals.

Timber spikes or tent pegs

Diagonals roughly equal

3-ft. minimum

2x4 stakes

4 ft. apart

Water level

Set batter-board stakes and establish marks with water level for setting tops of 1x4 crosspieces.

Stringing level building lines

Nail crosspieces to stakes and string building lines directly above preliminary layout strings. If levels are correct, strings just touch at their intersection.

1x4 crosspieces

Nail

Twist knot

Outside of wall (building line)

From *Fine Homebuilding* magazine (October 1982) 11:26-28

ting an accurate reading on a tightly stretched tape is to hold the 1-ft. mark rather than the zero. This allows you to grasp the tape with both hands when holding it over a string intersection. With this method, remember to tell the person on the other end of the tape that you are "burning a foot," or "cutting a foot," so the measurement can be adjusted accordingly.

Once this preliminary layout is approximately square and located where you want it on the site, you can set up batter boards. Batter boards are fixed in the ground out beyond the excavation lines. They are temporary wooden corners used to tie the string that accurately defines the perimeter of the building at the outside-of-wall line, or *building line*, and the outside-of-footing line, or *excavation line*. If excavation is required within the perimeter of the structure, for an interior footing or a line of piers, you may want to establish batter boards to hold strings for these lines as well.

I use 2x4 stakes, 3 ft. to 4 ft. long, with 1x4 crosspieces. Usually this is lumber that has been used at least once before. Sharpen the stakes with a circular saw so they will drive easily with a sledgehammer. You'll need three stakes per corner, set about 4 ft. from each other, for a total of 12 for the rectangular house we're using as an example. Drive these stakes about 3 ft. outside the preliminary strings and parallel to them. This placement gives you enough room so that the excavation won't undermine the batter boards, and they can be used until the walls of the house are actually framed. It's a temptation to be exacting in placing your stakes, but you needn't take the time to be too fussy. A good foundation requires precision, but this comes from the strings the batter boards will eventually carry, not from the batter boards themselves. Be sure that the 2x4 stakes are rigid enough to withstand an occa-

sional bumping. If they aren't, nail 1x4 braces near the top of the 2x4 stakes, drive another stake where the brace touches the ground, and nail them together.

Stringing level building lines—I like to nail all the crosspieces at the same level whenever possible. When the strings are in place, this gives me a vertical reference anywhere on the perimeter, which is a real advantage in determining the depth of footings, or the height of foundation walls and concrete block. Since a lot of my foundations are block, I like to set my crosspieces (where the strings will eventually be tied) so that their top edges will be at the same height as the top course of block. To figure this, you must start at the bottom of the footings. In the colder parts of the country, the bottom of footings must always be at or below the frost line to prevent heaving during the winter. If the frost line is 32 in., for instance, and the depth of the footing itself is 8 in., then it will take 24 in., or three courses of 8-in. block to reach grade. Add another three courses, as a convenient and attractive foundation height, and you have a total of 56 in. from the bottom of the footing trench. Keeping this number in mind, measure 24 in. up from the ground on any of the batter-board stakes—the finished height of the block foundation—and make a level line.

I don't own a builder's level or transit, and I've never needed one in my 20 years in the trades; I use a water level (you can make your own). Whatever you use, mark level lines on all the batter boards at 24 in. Accuracy is very important here. Then nail the 1x4 crosspieces with their top edges even with the level marks.

When the crosspieces are up, pull a new string for the south side of the building over the crosspieces of batter boards on each end. Align

it directly above the preliminary layout string by sighting it from above or using a plumb bob. Tack a nail in the top of the cross member and tie one end securely. At the other end pull the string as tight as you can. This establishes a line of elevation, so you don't want it to sag. Use a twist knot to tie it off to another nail. The twist knot (drawing, below right) will keep a nylon string taut, while still allowing it to be released instantly for resetting. This knot doesn't work well with cotton line. Continue stringing until all the lines are up. If everything is level, the strings will just touch as they intersect a few feet in front of the batter boards.

Squaring the corners—The next step is to square the stringed corners, this time using the 3-4-5 check. These numbers refer to the sides and hypotenuse of a right triangle. Since 6-8-10 and 12-16-20 triangles are proportional to a 3-4-5, use the largest one you can for optimum accuracy. The intersection of the strings of each corner defines the 90° angle of the 3-4-5 triangle. You'll need a helper to measure and adjust the strings until the hypotenuse is exactly proportional. With one person holding zero (or 12 in., if you are "burning a foot") on the tape, the other person can mark the legs of the triangle on the string with a pencil, and then knot a short length of string loosely around the mark. Double-check the measurement, and then tighten this knot. Measure from knot to knot to get the hypotenuse, and adjust the strings on the batter boards if necessary. These adjustments will require driving new nails into the top edge of the crosspieces. Pull the previous nail as you correct the position of the string. If you don't, it can get very confusing when the strings come down temporarily for digging the footing trenches. When you take the string down and put it back up, check

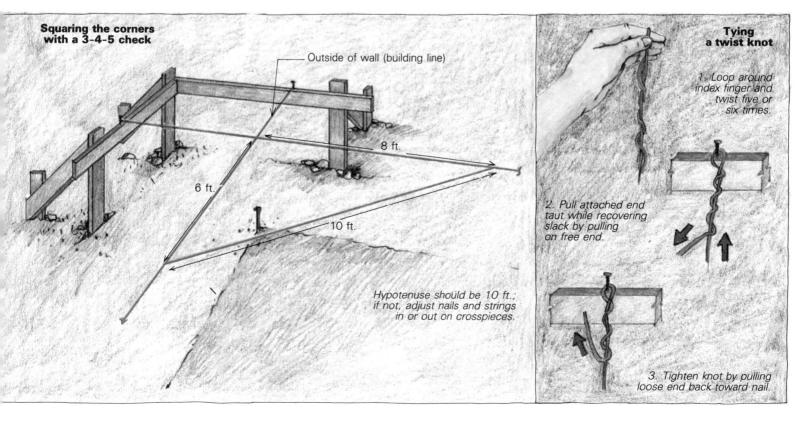

Squaring the corners with a 3-4-5 check

Outside of wall (building line)

8 ft.

6 ft.

10 ft.

Hypotenuse should be 10 ft.; if not, adjust nails and strings in or out on crosspieces.

Tying a twist knot

1. Loop around index finger and twist five or six times.

2. Pull attached end taut while recovering slack by pulling on free end.

3. Tighten knot by pulling loose end back toward nail.

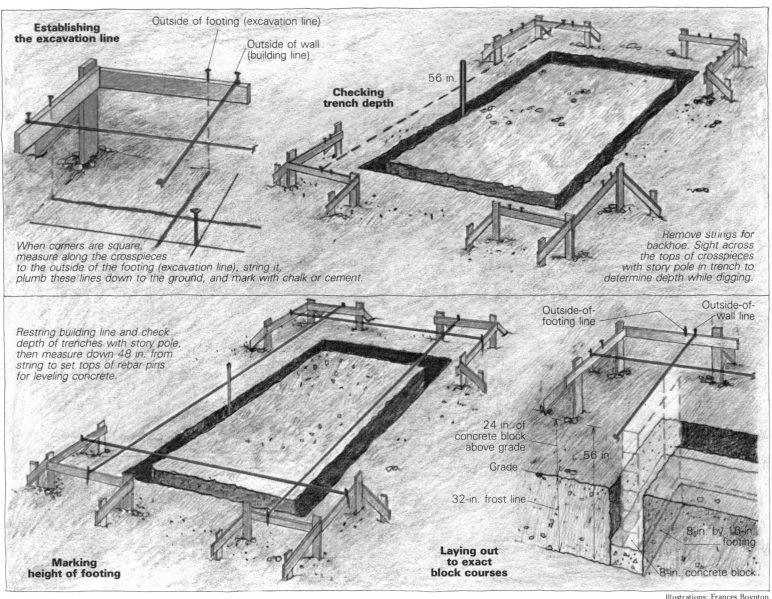

Establishing the excavation line

Outside of footing (excavation line)

Outside of wall (building line)

When corners are square, measure along the crosspieces to the outside of the footing (excavation line), string it, plumb these lines down to the ground, and mark with chalk or cement.

Checking trench depth

56 in.

Remove strings for backhoe. Sight across the tops of crosspieces with story pole in trench to determine depth while digging.

Restring building line and check depth of trenches with story pole, then measure down 48 in. from string to set tops of rebar pins for leveling concrete.

Marking height of footing

Laying out to exact block courses

Outside-of-footing line

Outside-of-wall line

24 in. of concrete block above grade

Grade

56 in.

32-in. frost line

8-in. by 16-in. footing

8-in. concrete block

Illustrations: Frances Boynton

the length to the knots, because nylon string stretches. To finish squaring up, use the 3-4-5 check on another corner, then check the diagonals again.

With the strings squared up to represent the eventual building lines (the outside-of-wall lines), and an elevation established, draw a plumb line down from the strings on the face of each cross-member, and write the wall thickness and the amount that the footing will project beyond the outside of the wall on the batter board. To reduce confusion, I drive nails and hang strings only on the outside-of-wall line and outside-of-footing line.

Excavation lines—Usually a footing is twice as wide as the foundation wall it supports, and as deep as the wall is thick. Footings are contained either by building a wooden form, or by digging a trench and using the undisturbed earth as formwork. I usually use the trench method. To show the backhoe operator where to dig, I plumb down to the ground from the outside-of-footing line and stretch a string at grade. I mark over this string with lime or cement dust as if I were marking out an athletic field. You can also use scouring powder with a shaker top. The backhoe operator should hold

the outside tooth of his bucket to the line, and dig to the inside.

Checking trench and footing depth—The batter boards give a quick vertical reference for determining how deep to trench. In our example, the bottom of the footing is 56 in. down from the top edge of the crosspiece. Instead of strings, which during excavation should be wound around a stick, use a story pole with a 56-in. mark on it. Stand the stick in the trench and sight from the top of one batter board to another. The mark on the story pole should line up with them.

When the machine work is finished, string the lines on the outside-of-wall line (the building line). Pull them very tight. Shape up the sides and corners of the trench with a shovel, maintaining the 56-in. depth you can now check by measuring from the string.

Next, set the depth of pour for the concrete. I use ⅜-in. or ½-in. steel reinforcing rods about 6 ft. to 8 ft. apart to indicate depth during the pour. Cut them about twice the depth of the footing so that you can drive them into the ground. Measuring down from the string to the top of the rebar, carefully tap them with a sledge until you read 48 in. on the tape. This

will give you an 8-in. footing at the 32-in. frost line, and six courses of block on top will bring you up to the string (drawing, above right). Then pour the concrete level with the top of the rebar. I use a garden rake to push the concrete around and for initial screeding. I hold the rake in a vertical position to smooth the top of the concrete and jitterbug the coarse aggregate down into the mix. You also might want to use a 2x4 screed short enough to fit between the rebar depth indicators, but it's not necessary to trowel the surface smooth. If you are pouring a foundation wall on top of the footing instead of laying block, the same techniques can be used, but remember to form a keyway in the footing to receive the next pour, and check with local codes to see if vertical rebar is required to tie the footing to the foundation wall.

The next day the concrete will be hardened sufficiently to begin working on the foundation walls. I usually drop a plumb bob down from the outside of wall lines and snap chalklines on the green footing. If the foundation is to be concrete block, then marking the corners will be enough since the mason will be pulling his own lines from corner to corner on each course. □

Tom Law is a builder in Davidsonville, Md.

Stepped Foundations

Using modular forms to build a foundation on a hillside

by Michael Spexarth

To the builder who's about to put in a foundation, building sites can pose a variety of problems. Assuming that soil conditions are stable, the flat lot with easy access is at the no-problem end of the scale, while the 10-in-10 (45°) "view lot" occupies the nightmare position. The standard spread-footing/stemwall foundation is the most common footing for the easy site. Here in the San Francisco Bay Area, the pier-and-grade-beam foundation is generally used on the steep ones (see pp. 76-79).

A lot of sites fall between these two extremes, and confronted with one of these, I like to build a modified spread-footing foundation. This is called a stepped foundation, and building one is standard procedure on a site that has stable, well-drained soil with a slope between 2 in 10 and about 5 in 10. I would seek an engineer's advice on any slope greater than 5 in 10, for soil conditions and steel requirements.

Stepped foundations take on the appearance of staircases as they stretch out across a lot, and their level changes make them look like the product of very complicated formwork. Well, they are complicated, but I avoid needless complexity by using modular form panels that can be used over and over again.

The panels—When I built the foundation in the photo below, I used 2-ft. by 8-ft. panels made from 2x4s and ½-in. plywood. The most durable forms are made with exterior-grade AC plywood, with the A side in contact with the concrete. The 2x4s are used on edge to reinforce the plywood panels around the perimeters, and on 16-in. centers in the field. Panels that are 8 ft. long are perfect for stepped foundations on hillsides that have a slope of about 2 in 10, while 4-ft. by 2-ft. panels are appropriate for a slope approaching 5 in 10 (top drawing, below). If your site is in the 2-in-10 range, the 8-ft. panels are especially handy if the outside perimeter of your foundation or retaining wall is divisible by 8, such as 24 ft. or 48 ft.

The inside panels are the same length as the outside ones, except at the corners. There an allowance has to be made for the thickness of the stemwall you are forming—8 in. for a two-story building, 6 in. for a one-story building, according to the Uniform Building Code. The foundation shown here supports a one-story house, so the inside-corner form panels had to be 7 ft. 6 in. in the north-south direction, and 7 ft. 2 in. in the east-west direction (bottom drawing).

Although it would be nice if foundation plans and hillside contours all came in 8-ft. increments, they don't, and a builder has to improvise when the form panels don't quite match the site. If you have an odd-sized foundation, a foundation with a lot of corners or a site with grade changes, you will have to cut or add small sections to panels, usually at the corners. Fortunately it's easy to add a smaller panel to act as a spacer or an extension when needed.

Excavation—When you plan the excavation, you have to decide how long the excavation cuts, or shelves, should be for the footings. For the 2-in-10 slope under this foundation, 8-ft. panels over 8-ft. shelves worked out fine. If a variation in the slope makes it necessary to cut shelves of different lengths, you can gang panels together or let a panel overlap the one below it for an adjustment.

Once we had our batter boards positioned, we stretched string lines about a foot off the ground to mark the centerline of the foundation. Then we shook handfuls of white lime over the string lines to mark their position on the ground. Most backhoe operators tend to wander right or left, depending on whether they are right or left-handed, and on sloped sites they have to con-

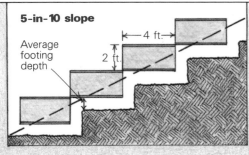

5-in-10 slope

Average footing depth

4 ft.

2 ft.

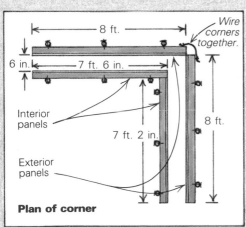

Plan of corner

8 ft.

Wire corners together.

6 in.

7 ft. 6 in.

Interior panels

Exterior panels

7 ft. 2 in.

8 ft.

Plywood panels *reinforced with 2x4s are used by the author to form stepped foundations on sloping sites. Opposing pairs of panels hold concrete for the stemwalls—the footings below them are formed in excavated trenches. At the downhill end of each step, a plywood dam tacked to the panels keeps the concrete on the right level.*

sider the potential for rolling their equipment. The limed line is their reference point, and allows them time to consider their position on the hill. They get safer use of their equipment, and we get a straight excavation, which cuts down on the hand-digging.

On this job, we had the operator use a 24-in. bucket, and every 8 ft. we had him drop down 2 ft., leaving 24-in. wide level steps 8 ft. long in the ground. The shallowest point in each trench should be equal to the depth of the footing required by the building and the climate. Don't expect that the shelves will be exact in length or that they will be spot-on level. Such adjustments are made as the form panels are placed.

After the steps had been dug by the backhoe, we ran string lines from our batter boards to mark the outside edges of the stemwall. Then we set the outside panels to the line marked by the string. Where the slope or the excavation made large gaps at the lower edge of the panels, we nailed a plywood skirt to hold in the concrete, as shown in the drawings below. We also closed up the open ends of the forms with plywood dams.

The panels are held in place by nails driven through steel stakes into the 2x4s along the panel tops and bottoms (photo facing page). Three stakes per panel and two nails per stake are enough to secure them. I like to use steel stakes for most foundation work because they are easier to drive into hard soil, loose rock, sandstone or shale. They are also easier to remove than wooden stakes because they exert less friction against the soil.

After checking to make sure that the outside forms were square and plumb, we added 2x4 kickers (or braces) to stiffen them. Most panels had three kickers, and we made sure that the highest points of the forms, near the steps, were well-braced. Once the outside panels were secure, we set the inside forms. We started in the middle of a run, and worked toward the corners. The inside panels matched the length and height of the outside ones, and we nailed plywood ties across their tops every 3 ft. or so to keep the distance between the panels at a steady 6 in. To set the distance between the panels along their middles, we cut 6-in. lengths of 2x4 to act as spacers. With these in place, we ran tie-wire loops between the two panels and twisted them tight. Then we removed the spacers. The loops keep the forms from spreading too far as they are filled.

Steelwork—Our plan called for two pieces of #4 rebar at the top and at the bottom of the foundation. In a stepped foundation, the steel is also stepped. Horizontal bars become vertical at the step, then horizontal again. Given the number of bars embedded in a 6-in. wall, things can get crowded at the step. I space the bars so they will be evenly distributed at this point, 2 in. from the face of the concrete and 2 in. from each other. If the panels are overlapping by 12 in. to 18 in., I add one or two vertical bars. The step is the weakest point in this type of foundation, so it's important to have plenty of steel there.

Since a stepped foundation means stepped

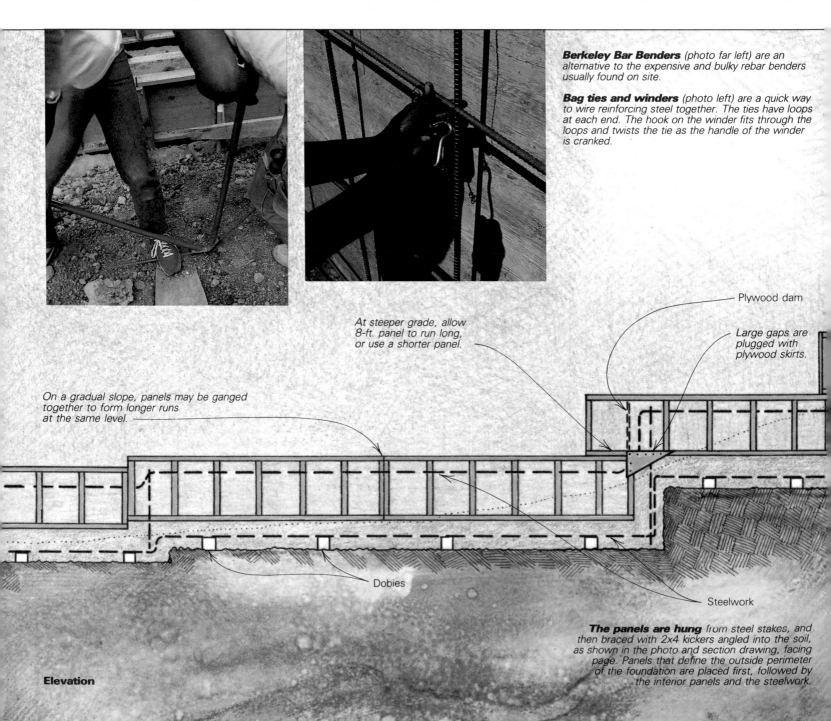

Berkeley Bar Benders (photo far left) are an alternative to the expensive and bulky rebar benders usually found on site.

Bag ties and winders (photo left) are a quick way to wire reinforcing steel together. The ties have loops at each end. The hook on the winder fits through the loops and twists the tie as the handle of the winder is cranked.

Plywood dam

Large gaps are plugged with plywood skirts.

At steeper grade, allow 8-ft. panel to run long, or use a shorter panel.

On a gradual slope, panels may be ganged together to form longer runs at the same level.

Dobies

Steelwork

The panels are hung from steel stakes, and then braced with 2x4 kickers angled into the soil, as shown in the photo and section drawing, facing page. Panels that define the outside perimeter of the foundation are placed first, followed by the interior panels and the steelwork.

Elevation

steel, there is a lot of rebar bending in a job like this. We used Berkeley Bar Benders to put the steps in the steel (photo facing page, far left). They are two 3-ft. lengths of pipe that have a 3-in. length of pipe welded onto their ends at a 60° angle. These 3-in. lengths are slotted to accept a piece of #4 rebar. One worker uses a bender to hold the rebar steady while another levers the bar to the desired angle. If you don't have to fold rebar week in and week out, the benders, manufactured by Berkeley Bar Bender (1215 Shattuck Ave., Berkeley, Calif. 94709), are an inexpensive (about $40) alternative to the heavy-duty cutter-benders that are used by concrete specialists.

We stepped the lower pair of bars to match the stepped panels, and tied them to dobies, which are precast 3-in. cubes of concrete. The dobies are attached with tie-wire to the rebar, and they keep the rebar 3 in. above the ground, as required by code.

Since we had plenty of rebar that needed tie-wire on this job, we used bag ties and a winder to speed things along. Bag ties are short cutoffs of foundation tie-wire (6 in., 8 in. or 10 in. long) with a loop on both ends. They come strung together in rolls of 1,000. To use one, you wrap it around the steel with one hand, and insert a winder through the two loops. A winder is a steel hook that's free to spin in a wooden handle. You crank on the handle, and the hook winds the wire and the steel together (photo facing page, center). No more tangled rolls of tie-wire or awkward angles for pliers.

We pump-poured this foundation because it was well off the street, but the forms would take any mix. I start a pour with one pass around the perimeter to fill up the footings. Then I fill the forms halfway and check them for any distortion, separation or split wood. If I have a problem at this point, I know I'm going to have a disaster later, so I stop the pour while I sort things out. This accomplishes two things: I can fix the questionable formwork, and the concrete has time to set up and carry its own weight, which takes some of the burden off the forms. If I have several workers on the job, they continue the pour at other areas of the foundation.

Once the forms are topped off, we rough-screed the stemwalls using the top of the forms as a guide. Then I set the bolts. As the mix sets up, we remove the plywood ties across the forms. The stemwall top can then be finished as desired. This is a good time to pull out the steel stakes—you've got about 24 hours before they become a permanent part of the foundation.

If you carefully scrape the panels and coat them with form-release after every job, they will eventually pay for themselves. Form-release is an oil-base coating manufactured by companies that sell concrete accessories, such as the Burke Company (2655 Campus Drive, San Mateo, Calif. 94403). Most small contractors just use old motor oil. Around here, many builders rent out their form panels to help each other out, and to defray costs. □

Michael Spexarth is a contractor and part-time building inspector in El Cerrito, Calif.

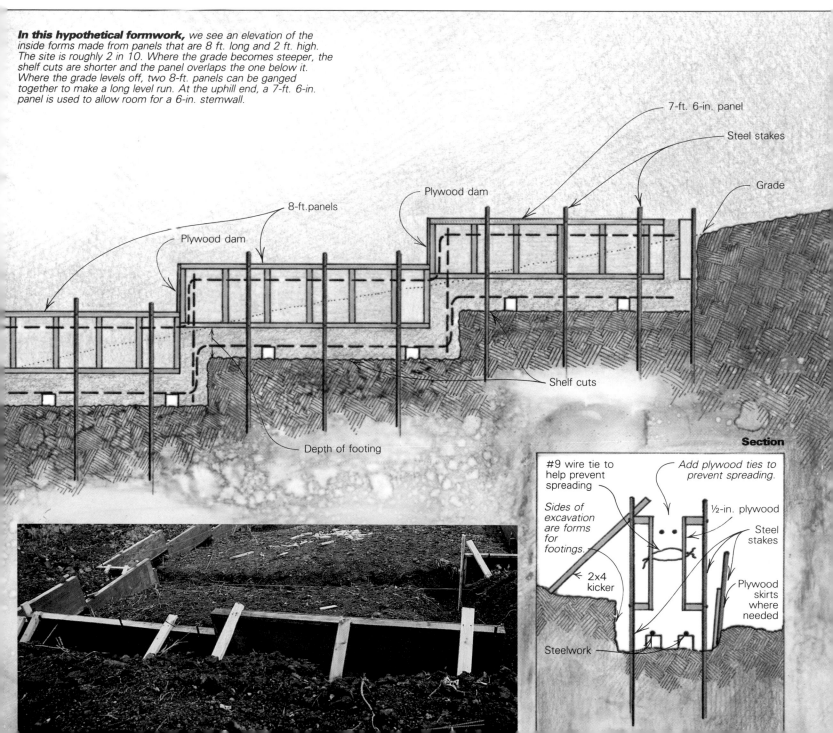

In this hypothetical formwork, *we see an elevation of the inside forms made from panels that are 8 ft. long and 2 ft. high. The site is roughly 2 in 10. Where the grade becomes steeper, the shelf cuts are shorter and the panel overlaps the one below it. Where the grade levels off, two 8-ft. panels can be ganged together to make a long level run. At the uphill end, a 7-ft. 6-in. panel is used to allow room for a 6-in. stemwall.*

7-ft. 6-in. panel

Steel stakes

Grade

Plywood dam

8-ft. panels

Plywood dam

Shelf cuts

Depth of footing

Section

#9 wire tie to help prevent spreading

Add plywood ties to prevent spreading.

Sides of excavation are forms for footings.

½-in. plywood

Steel stakes

2x4 kicker

Plywood skirts where needed

Steelwork

Facing a Block Wall with Stone

A good rock supply, tight joints and hidden mortar are the secret to a solid, structural look

by Tim Snyder

Building with fieldstone and building with concrete block represent two extremes in masonry construction. Concrete blocks aren't especially interesting to look at, but they go up fast, and it's easy to build a sturdy wall with them. Stone construction demands patience, skill, and above all, lots of rocks. Even with these ingredients, the different shapes and sizes of the material make it tough to keep a wall of stone plumb and strong. Given these considerations, it's easy to appreciate a construction technique that combines the beauty of stone with the strength and practicality of concrete block.

Larry Neufeld laid up his first stone face to cover a block chimney in a house that he and his brother were building. He had never worked with stone before, but as a general contractor he knew enough about masonry to take on the project. By the time work began on the solar addition shown here, he had developed a technique and style that take the best from both building materials. The finished wall—20 tons of mass facing the windows and skylights on the south side of the addition—shows little mortar at all, and unless you examine the joints carefully, they seem to be dry-fit.

A flexible system—Neufeld's method uses found stone, and thanks to the New England countryside, he can usually gather what he needs from the fields and stone walls on the owner's property. Working against a 6-in. thick block wall, he lays up a face 8 in. thick, using odd-sized stones from 2 in. to 7½ in. thick. The void behind the stone is filled with mortar, which sets up around the masonry ties set in the block's joints.

Neufeld's system can work just as well with a poured wall or a bearing wood-frame wall, as long as the footing is beefed up to hold the extra load, and there is a mechanical connection between wall and face.

It's good to begin the job with plenty of

Hiding the block. With a depth of 8 in., the stone face that covers this block wall doesn't require rocks of uniform thickness. Careful fitting is still important, though. At left, Neufeld works against temporary grounds that frame an opening in the wall. These boards were later replaced by the oak casing shown on the facing page. The finished wall faces a bank of windows in a solar addition, and looks like a solid stone wall.

stones. As you look for rocks, pick out natural corners, base-course stones with especially flat, broad faces and pieces with unusual colors or mineral formations. Toss these in separate piles before you start building, and each time you sort through your rock, take stock of the sizes and shapes you've got. Cataloging like this can make the job go a lot more quickly. But even with a good collection of stone, you can expect to be missing a few key pieces. In the middle of a job, Neufeld often finds himself driving more slowly past stone walls after work, seeking out an elusive corner or curved face.

Before laying up the face, be sure that any wall or ceiling surfaces that will be adjacent to the stones are finished. This means drywalling, paneling, plastering and painting earlier than you normally would, but it's far easier to do this work before the face goes up.

Laying up the wall—For bonding stone to block, Neufeld uses a mix of three parts sand to one part portland cement. Working his mix in a wheelbarrow, he adds just enough water to make a very stiff mortar. Then at the front of the barrow he adds a bit more water and trowels up a small section of wet mix. The stiff mix is used between the stones so that no mortar will flow out of the joints onto the exposed rock face. The wet mix fills voids closer

to the wall, and bonds the back sides of the stones solidly to the block.

Neufeld uses a tape to check the thickness of the face as he lays it up. A level isn't much help because of the irregularities in the stone, so he uses it only for rough checking. Working against a plumb block wall is pretty good insurance that the stone face will be plumb, but with many rough facets to account for, Neufeld does plenty of adjusting by eye. Fortunately on this job, a temporary post had to be nailed up to support the second-floor overhang where the circular stair would go. By plumbing the post both ways, Neufeld was able to use it to align the face of the wall as he laid it up. He also measured against the post to check the arc of the curved wall section.

The secret to achieving the dry-stack look is to test-fit all stones carefully before laying them in place, and then to keep the mortar away from the face edge of the joint. Test-fitting the stones is the first step in building the face, and it's like working on a big jigsaw puzzle. Working horizontally across the wall, you have to find stones that fit well together. A good fit means not only that the joints are tight, but also that they are staggered vertically, just as they would be in a solid, structural wall (see Neufeld's finished wall, above).

In many instances, you have to do some coaxing to get a joint right. And it's always a

good idea to flatten the top edges of your stones slightly before casting them in the wall. This ensures a stable surface for the next course of stones to rest on.

Neufeld uses a mason's hammer to knock off leading edges, and a cold chisel to fracture thick stones. Sometimes you can split a sedimentary rock along its bedding lines, but more often than not you'll end up with random fragments. This is one reason why Neufeld prefers to trim off as little as possible, using smaller pieces to fill gaps rather than trying for an ideal fit between two stones. He doesn't like to chip into the exposed face of a stone if it can be avoided, explaining that a split face has a harsh look that will never be lost inside a house.

Your test-fit stones should be able to rest on the previous course without falling off. They don't have to be exactly plumb at this stage, but you're looking for a gravity fit. Once you're satisfied that a group of stones fits well into the wall section, memorize their relative positions and remove them from the wall. Then prepare a bed of mortar by packing some stiff mix on top of the previous course. Work out from the block wall and leave the inch of joint area closest to the outside face bare of mortar. Lay down just enough mortar so that each stone will seat securely in its preassigned position. After pressing the

Hiding the mortar. The key to the dry-fit appearance is to keep mortar away from the face edge of each joint. Neufeld packs the mortar close to the block, left, then presses the stone in place and seats it in its mortar bed with several hammer blows, as shown above.

Balance and alignment. At right, stone chips inserted along joint lines to serve as temporary wedges prevent tall, thin stones from leaning out of plumb. They're removed after the mortar has set. The facing is thick enough to conceal a heating duct in the wall. Below, corner and curve construction depends on a good selection of shapes and sizes. A mixture of small and large stones also makes it easier to stagger the joint lines.

stones into their mortar bed, Neufeld sometimes uses a hammer to help seat them.

As you seat the stones, check to make sure they're plumb. Broad, narrow stones that don't extend the full depth of the face tend to lean out farther than they should. To make minute adjustments in orienting these stones, Neufeld inserts small rock fragments that serve as temporary wedges. They hold the stones in alignment until another course is laid up and the mortar sets; then Neufeld removes them.

It's best not to pack mortar behind a course until the mortar between the stones has set. Then trowel in the wetter mix to fill the space behind the stone, and you're ready to test-fit another course. This way the wetter mix can't ooze out of the joints and dribble down the face of the stone.

At the end of the day when you're using up

the last of your mortar, don't fill all the way up behind the last course of stones. It's better to leave a slight depression because this forms a keyway for the mix you trowel in the next day. Another important practice at the end of the day is cleaning the stone you've laid up. Go over the joints with a pointing chisel or a sharp piece of wood, and rake them back so that there's little or no mortar showing. Then use a broom to sweep down the face of the wall so that any drops of mortar are removed before they adhere.

Curves, corners and openings—To build the curved section of the face, Neufeld traced the clearance arc for the circular stair onto the concrete floor. Then he fit and laid up the stones as if he were working on a straight section. The only differences were that he had to use smaller stones to get smoothly around the arc, and that he could no longer sight off the temporary post to check for plumb. He used a level instead.

Successful corners are mostly a matter of having a good variety of cornerstones to choose from. Your first inclination may be to overlook small, right-angled rocks in favor of large, squarish stones. But what you actually want is a mixture of large and small; this creates overlapping joints and integrates the corner with the rest of the wall, as shown in the photo, bottom left.

You don't need an exact 90° angle to make a cornerstone. The secret is to aim for a right-angle average over several courses. Stones that come within 10° of 90° should work in a corner, as long as you get a good combination of large and small, acute and obtuse.

At door or window openings, both the stone face and the block wall are exposed. On this job, Neufeld hid this joint with trim. Before constructing the face, he erected temporary grounds from 2x stock along the trim lines. Once the face had been built into these plumb and square housings, they were removed and replaced by finish trim.

Fine points—Neufeld admits that it takes time to develop technique and style, a consistent choice of stones that will look nice in the finished wall. He likes to play one shape off against another, but stresses that the joint lines should give an impression of horizontality. Using a mixture of small and large stones is important to the overall composition, and also makes it easier to stagger the joints. But with this method of laying up stone, you've got the flexibility to try out your own ideas. Neufeld says that on his next job, he'd like to do a rock pattern in relief.

Building this 32-ft. wall took Neufeld about 400 hours. Since he was the general contractor for the entire solar addition, working on the wall kept him at the site through the arrival and departure of most of the subcontractors. Because the structural part of the wall—the concrete block—was finished at an early stage, there was no need to rush in laying up the stone. Having the time to find and fit the rock is an important advantage. □

Foundations on Hillside Sites

An engineer tells about pier and grade-beam foundations

by Ronald J. Barr

Most houses built on conventional sites sit atop a spread-footing foundation. It has a T-shaped cross section (small drawing, below center), and supports the house by transferring its own weight and the loading from above directly to the ground below. Spread footings can be used on sloping sites by stepping them up the hill (small drawing, below right), but this usually requires complicated formwork and expensive excavation. For slopes greater than 25°, the structurally superior pier and grade-beam foundation (large drawing, below) may be less costly. Apart from making steep sites buildable, pier and grade-beam foundations make it possible to build on level sites where soils are so expan-

sive that they could crack spread footings like breadsticks as the earth heaves and subsides.

How they work—The pier and grade-beam foundation supports a structure in one of two ways. First, the piers can bear directly on rock or soil that has been found to be competent. This means it's stable enough to act as a bearing surface for the base of the pier. Second, the piers rely on friction for support. This is what sets pier and grade-beam foundations apart from other systems. The sides of the concrete piers develop tremendous friction against the irregular walls of their holes. This resistance is enough to hold up a building. A structural engineer can look at the soil report

for a given site, study the friction-bearing characteristics of the earth and then specify the number, length, diameter and spacing of the piers that will be necessary to carry the proposed structure.

The rigs used to drill pier holes for residential foundations have to be able to bore holes 20 ft. deep or more. Foundations have to extend through unstable surface soil, such as uncompacted fill, topsoil with low bearing values and layers of slide-prone earth. Anchored in stable subterranean strata of earth or rock, deep piers don't depend on unstable surface soil for support.

The slope of the lot and the quality of the surface soil influence the size and placement

A pier and grade-beam foundation. Deep piers extend through layers of loose topsoil to lock into the stable soil below, which can support the weight of a house. The grade beams follow the irregular contour of the site, and transfer the building's loads to the piers.

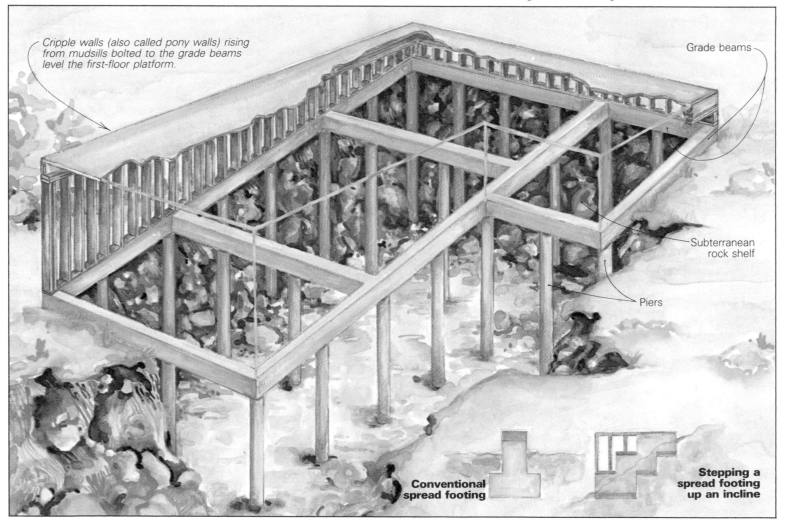

Cripple walls (also called pony walls) rising from mudsills bolted to the grade beams level the first-floor platform.

Grade beams

Subterranean rock shelf

Piers

Conventional spread footing

Stepping a spread footing up an incline

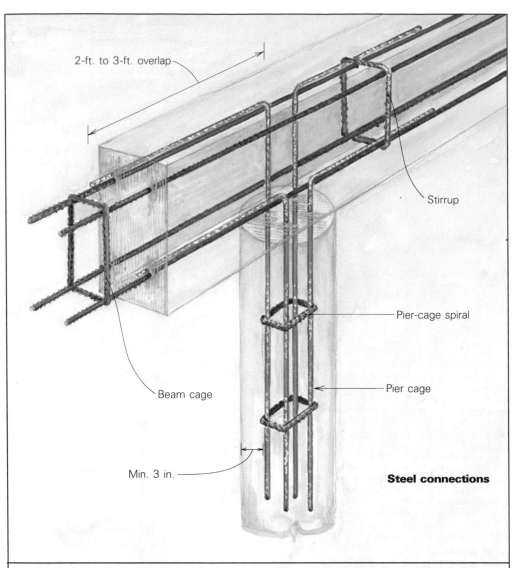

2-ft. to 3-ft. overlap

Stirrup

Beam cage

Pier-cage spiral

Pier cage

Min. 3 in.

Steel connections

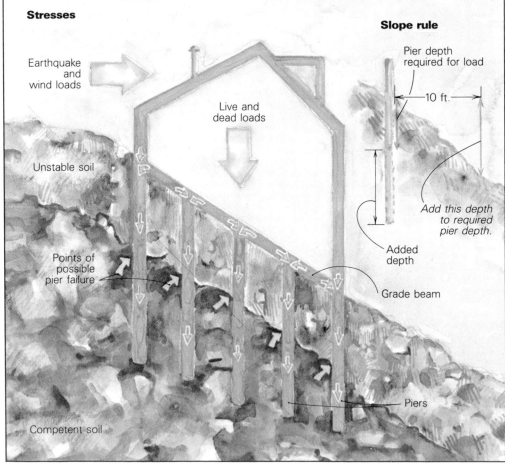

Stresses

Slope rule

Earthquake
and
wind loads

Live and
dead loads

Pier depth
required for load

10 ft.

Unstable soil

*Add this depth
to required
pier depth.*

Added
depth

Points of
possible
pier failure

Grade beam

Piers

Competent soil

of reinforcing steel in each pier (drawing, left). The steeper the lot and the more suspect the soil's stability, the bigger the steel. This is because the pier doesn't just transfer the compression loads from the house to the ground. It also resists the lateral loads induced by winds, earthquakes or by the movement of surface soils, which result in a kind of cantilever beam-action on the piers, as depicted in the drawing below left. They have to be strong enough to resist these forces—pier cages made from 1-in. rebar are not uncommon in extreme cases.

Grade beams—As the name implies, a grade beam is a concrete beam that conforms to the contour of the ground at grade level. Grade beams require at least two lengths of rebar, one at the top and one at the bottom. Foundation plans often call for two lengths at the top and two at the bottom, tied into cages with rectangular supports called stirrups. The bars inside the beams work like the chords in a truss, spreading the tension and compression forces induced by the live and dead loads of the structure above.

The sectional size of the grade beam depends on the load placed on it, and on the spacing of the piers. Pier spacing is governed by the allowable end-bearing value of the piers, or by the allowable skin friction generated by the pier walls and the soil under the site. Pier depth ranges from 5 ft. to 20 ft.; pier spacing, from 5 ft. to 12 ft. o.c. Pier diameters are normally 10 in. to 12 in., and can go up to 30 in. Beam sections start at about 6 in. by 12 in. and go up.

Because of the number of variables involved in designing a pier and grade-beam foundation, the engineering is best left to a professional. It's not happenstance that there are lots of lawsuits over foundation failures—many of them on slopes.

What the engineer needs—An engineer designing a pier and grade-beam foundation has to know what lies beneath the surface of the site. A soils engineer's report is usually required for this information (see p. 77). Sometimes the city or county building department will have records of soil characteristics in your area. The engineer also needs a copy of your house plans to calculate the loading on the foundation. If you can, tell the engineer what pier size your excavator can drill. If you're planning on making the steel cages yourself, ask your engineer about rebar diameter requirements. Although he may specify #5 or #6 rebar, which is impossible to bend in the field, it could be that more #4s (½-in. rebar) would do the job just as well.

Make sure you find out the spacing and size

Engineering considerations. **The steel embedded in each pier must resist the lateral loads from wind or earthquakes, which are transferred to the piers at the junction of the unstable and competent soils. If piers are embedded in a slope, the amount of slope rise in a 10-ft. run must be added to the pier depth required for load.**

Pier and grade-beam in expansive soils. A well-engineered system has a 2-in. gap between the grade beam and the soil, to allow for expansion. The pier should be straight, not bell-shaped at the top, so loads are transferred down to the stable soil.

of the anchor bolts your foundation will need for fastening sills to the grade beams. For some downhill slopes, the cripple walls (drawing, p. 73) will need additional blocking, and anchor bolts placed 2 ft. o.c.

Piers on a slope must be longer to compensate for the lack of soil on the downhill side (inset drawing, facing page, bottom). Figure the additional length as follows: Tie one end of a 10-ft. long string to the stake marking the pier location, and pull it taut downhill and hold it level. The distance from the end of the string to the ground directly below has to be added to the pier hole to compensate for its location on a slope.

Void boxes—Ask your engineer whether you can cast (pour) the grade beams directly on the soil, or whether they should have void boxes under them. A void box (drawing, right) is a gap, usually about 2 in., between a grade beam and expansive soils. Some builders put a 2-in. thick piece of Styrofoam at the bottom of the grade-beam forms before the pour. Once the forms have been stripped, they pour a solvent along the base of the grade beam to dissolve the Styrofoam. Another method is to fold up cardboard boxes until you've got a stack 2 in. thick, and then lay them at the bottom of the forms. You don't have to remove them—they'll rot away.

When I worked as a building inspector, I saw what can happen to a house when void boxes aren't included in grade-beam pours over expansive soils. A couple of home owners asked me to find out why their foundation had settled in the center of the house. There was a large fireplace there, and they thought it was the culprit. But it wasn't. The fireplace showed no signs of movement. However, outside the perimeter grade-beam foundation, recent rains had caused the clay to expand, raising the foundation (and the exterior walls) as much as 2 in. in places. The center of the house had stayed put, and everything else had risen around it. None of the doors or windows worked, and every wall in the house was cracked. Voids under the grade beams would have prevented all this.

Problems—Once the foundation design has been determined and your driller is on site, it's almost inevitable that something unexpected will happen. For instance, your excavator may not be able to drill to the depth specified on the plans. Sometimes you can get away with this; other times you have to provide for another pier nearby. If you're drilling friction-bearing piers, you may have to change their diameter and spacing, or change their end-bearing condition. Check with your engineer for the correct course of action.

Drill alignment sometimes slips during drilling, and the auger can break into an adja-

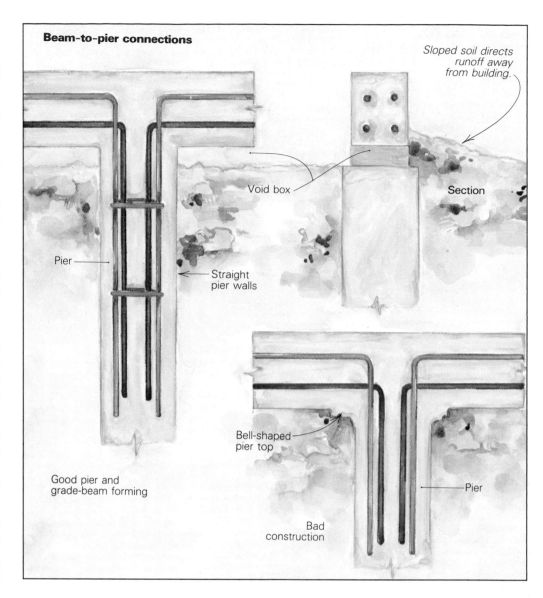

Beam-to-pier connections

Sloped soil directs runoff away from building.

Void box

Section

Pier

Straight pier walls

Good pier and grade-beam forming

Bell-shaped pier top

Pier

Bad construction

cent pier hole. If this happens, check with your structural engineer. The remedy will depend on the load-bearing requirements at that particular part of the foundation.

Watch the soil as it's being removed from the pier holes. Its appearance should change in the upper half of the hole, as the drill leaves the topsoil layer and passes into the more stable soil. If you notice an abrupt change in the color or texture of the soil at the bottom of the hole, get in touch with your soils engineer right away, because a change in soil type may mean you'll need to drill deeper piers to make sure you're into solid earth. Shallow piers are the cause of most pier and grade-beam failures, so it's best to err on the side of caution and drill those holes deep.

A little while ago I drew up some plans for remedial foundation work for a house that was suffering from short piers. Actually, some of the piers were long enough to extend into solid material, and this uphill portion of the house hadn't moved. But the downhill half rested on shallow piers that weren't engaged with the competent soil layer. The awesome power of the surface layers moving downhill had pulled the house's grade beams apart and stretched their reinforcing rods like taffy. Above these cracks in the beams, the house was slowly being torn in half. Just getting a

drill rig to the site required dismantling part of the house, and the necessary foundation repairs ended up costing $60,000.

After the drilling—Grade beams should be centered over the piers. If one is out of alignment by more than a few inches, the eccentric load placed on the pier could eventually crack it in half. Rebar splices should overlap by 3 ft. And make sure that the tops of the pier holes haven't been widened to create a bell shape (drawing, above). Such a bell can be a bearing surface for expansive soils, and can cause uplift problems during wet weather.

A pier and grade-beam foundation usually doesn't need a special drainage system, but it's important to direct runoff away from the building. Banking the soil a few inches up the exterior side of the grade beam is generally all that's needed, and won't appreciably intrude into the void under the beam (drawing, top). The important thing is to make sure that running water doesn't wash away the soil around the piers. This erosion might not be critical if the piers are end-bearing, but if they are friction piers, the washing action could severely limit their load-carrying capabilities. □

_____

Ronald J. Barr is a civil engineer. He works in Walnut Creek, Calif.

A Pier and Grade-Beam Foundation

Advice from a contractor on one way to build on a steep slope

by Michael Spexarth

As good home sites get harder to find, many builders are looking at hillside lots that used to be considered unbuildable. A builder who can cope with the special foundations required on steep sites can often buy a lot for a modest sum, and build a house that enjoys an attractive view of the scenery below.

Here in the San Francisco Bay area, chances are good that you'll find a pier and grade-beam foundation under just about every new hillside home. This type of foundation links a poured concrete perimeter footing to the ground with a matrix of grade beams and concrete pilings, some up to 20 ft. deep. The resulting grid grips the hillside like the roots of a giant tree (see pp. 73-75). Slopes in excess of 45° can be built on with this kind of foundation. And in areas where there are landslides, expansive soils or earthquakes, a pier and grade-beam foundation may be required by local building codes.

As a rule pier and grade-beam foundations need more reinforcing steel than conventional foundations, and require special concrete mixes. On the other hand, they usually call for less formwork than perimeter foundations do. What's more, the footings for grade-beam foundations don't have to be level.

Soil survey—A soil engineer's report is usually required by local codes, and even if it isn't, it's smart to get one. Because the tendency on steep sites is to overbuild, the survey's recommendations may reduce the number and size of piers, which designers or engineers with less knowledge of geology will spec to err on the side of caution. This kind of needless overbuilding wastes money.

To make the survey, the soil engineer will make test drillings or trenches at various spots around the site with a drill rig that typically bores a 6-in. dia. hole. Or he will bring in a backhoe. At various depths, core samples are taken with a cylinder that is lowered into the test hole. Although they will vary from site to site, these holes are usually 20 ft. deep or less because that's the limit of most drilling rigs used to bore the finished holes. The engineer will take the soil samples back to the lab and analyze them to determine their bearing and expansion characteristics.

These tests, taken together, give a picture of the soil types and conditions at various depths, and of the natural water courses. Armed with this data, the soil engineer makes recommendations about the number and the diameter of the piers needed, and their depth. He specifies the steel-reinforcing requirements and the minimum distance between the piers. In northern California, the typical soil survey costs between $1,000 and $1,500.

Site preparation—The first order of business is to get rid of unwanted plants, loose topsoil, logs and other debris that isn't considered a permanent part of the landscape. Once the lot is cleared, mark stumps, rocks and areas of soft, wet soil with stakes and flagging. The safety of your drill operator could depend on his knowing what's where.

Upslope lots are usually more difficult to deal with than downslope lots. Here's the rule of thumb around here. A foundation on a downslope lot will probably cost twice as much as a foundation on a level site; a foundation on an upslope lot will be three times as expensive. In addition to moving the weight of the entire house uphill, the builder has to deal with the problem of hauling off tons of dirt.

Most soils are compacted in nature, and when drilled into or excavated, they can expand to two or three times their undisturbed volume. A 16-in. dia. by 18-ft. deep pier hole will contain about 1 cu. yd. of compacted soil, which will become as many as 3 cu. yd. of loose soil. So if 15 pier holes are drilled, you might have 45 cu. yd. of loose soil covering the site, or a volume 20 ft. by 30 ft. by 2 ft. deep. This soil can change the contour lines in the site plan, the elevation of the house, the loading on the foundation, and the drainage of the hill. So you have to move it out of there.

In the Bay area, some building departments allow loose soil to be evenly distributed up to 3 ft. deep if the slope is less than 18°. Such soil-dispersal zones must be away from natural drainage channels. On steeper slopes, the loose soil has to be removed or buttressed with a retaining wall to keep it from creeping downhill. Other building departments ask the soil engineer to recommend soil removal or dispersal, then the county inspectors check the project to see if they agree. Each jurisdiction has its own policies governing drilling spoils. You should find out what they are to avoid trouble later.

Most builders stockpile the soil during construction, and then use it for landscaping once the building is finished. If the soil is unusable, or there's just too much of it, it has to be hauled off. We usually rent a dump truck and a tractor with a front-end loader to move the stuff, and take it to the nearest fill site.

Layout—The standard batter board used to locate the corners of perimeter foundations isn't the best solution for layout on slopes. Batter boards don't work well because most pier holes are drilled by a tractor-mounted auger, and as the operator maneuvers the rig around the site, chances are good that the batter boards and some of the survey stakes will be crunched into little splinters. In addition, 12-in. survey pins marking the exact location of piers can shift around in loose topsoils. A D6 Cat weighing around 10 tons may push loose topsoil downhill 1 ft. to 3 ft. as it lumbers around the site. So although the pins may appear to be in the correct relationship to one another, they might have moved downhill with the topsoil. This can cause a lot of problems when you start building your forms.

In order to ensure the correct siting of the house, and to guarantee exact layout points for drilling each pier, I arrange to get my surveyor on the site as the holes are being drilled. The surveyor centers the transit over a survey hub at a fixed elevation away from the drilling area, and then uses a chain (surveyor's tape measure) to locate the corners of the foundation and piers as calculated degree angles are turned. This ensures that no matter what soil movement has affected the preliminary stake positions, each pier hole is located precisely, right at the time it's drilled. The extra cost involved in having the surveyor on site is usually negligible compared to the additional labor and material necessary to do the whole job over again if your pier holes are drilled in the wrong places. If you can't get your surveyor on the site during drilling, the best alternative is to place reference points outside the drilling area to double-check stake location as the pier holes are being bored.

Drill rigs—There are two basic types of drill rigs: tractor or crawler mounted, and truck-mounted. Within these two categories you'll

Dirt flies as a tractor-mounted auger pulls soil from a 20-ft. deep pier hole. Rigs like this can work on slopes as steep as 45° or a little more. The ever-present helper with a shovel positions the auger, plumbs the drill shaft and directs dirt away from the hole.

Rebar for grade-beam and pier cages is assembled on site with jigs nailed to the top of sawhorses. The slots in the jigs position the #4 steel as #3 rebar stirrups are tied in place. Finished grade-beam cages await installation atop the woodpile. A rebar offcut used as a stake holds the woodpile in place.

find rigs that are as different as the people who operate them. The truck-mounted rigs have an auger attached to a boom that can reach several drill targets from one setup point. Such trucks do less damage to a site than a tractor, but they can't negotiate slopes greater than about 30°.

Tractor-mounted augers can drill in terrain that is inaccessible to trucks, and they can work on slopes up to 45° or slightly more. Sometimes an operator will anchor his tractor's winch cable to a stout tree or to a dead-man driven into the ground uphill. Some operators carry a power pole on their cat. They auger a hole at the top of the slope for the pole and hook the cable to it. This lets them hang on the side of a steep slope as the holes are drilled. It's a chilling sight.

Drill operators generally charge by the hour, by the footage (the cumulative depths of the holes they drill) or by the job. They base their fees on the capabilities of their equipment, on the difficulty of the job and on the kind of soil they're working in. The rig that's pictured in this article cost $130 per hour, and it took 12 hours to drill 16 pier holes. And with a crawler-type rig, there is an additional delivery charge (often called travel time) tacked onto the hourly rate—$150 for this job. In rocky locations, you can expect a surcharge that will cover the cost of broken equipment, such as auger teeth.

Although these rigs can drill holes up to 20 ft. deep, soil conditions can change every few feet. One hole can be 6 ft. deep and hit rock, and 8 ft. away the next hole might be 18 ft. deep before the same layer of rock is engaged. The variables are surprising, so un-

less you're good at poker (or bridge), do this type of bidding liberally.

During drilling, soil lifted out by the auger is deflected away from the hole by a helper using a shovel (photo, p. 76). The helper will also need a hose to water down the auger bit. Water lubricates the cutting action, reducing the time needed to bore the holes, and lengthens the life of the equipment.

The loose dirt is left at the bottom of the hole because it's too far down to clean it out. Because pier and grade-beam foundations work by friction rather than direct bearing, this loose soil isn't usually a problem. But several years of drought can cause the soil around the piers to shrink and reduce the friction that holds the house on the hill. Enough shrinkage, and the house begins to settle.

You can limit the possibility of soil shrinkage by pouring a few gallons of water and a third to a half-bag of portland cement into each hole after the final depth has been reached and the auger has been removed. Then have the operator re-insert the auger for a minute's worth of mixing. The auger works like a giant milkshake machine to blend cement, water and tailings. When it's lifted out again, the mix should be thin enough to ooze off the drill bit. Once it hardens, this slurry forms a pad that will take direct bearing, and so lessen the chances for foundation settlement during dry years.

Once the holes are drilled, we cover them with plywood to keep out loose topsoil and debris. If there's even a remote chance that small children might wander onto the site, we stake the plywood to the ground and cover it with loose topsoil to make the area uninter-

esting. We also notify the neighbors that there are deep holes covered with plywood on a lot nearby, and we post the area.

Concrete and steel—A pier and grade-beam foundation has a lot of steel in it, and an equally large volume of concrete—100 cu. yd. is not uncommon. These foundations are a lot like icebergs, with their bulk concealed below the surface. With this much steelwork, the builder who sets up an efficient way to fabricate the beam and pier cages can save money.

On a recent job we assembled #4 rebar cages on sawhorses using #3 rebar stirrups and spirals. We easily bent the #3 steel with a rebar bender, and wired the straight pieces in place with standard foundation tie wire (photo above). Some foundations call for heavier steel, and anything over #4 (½-in. dia.) is difficult to bend. If your specs show a lot of #5s, #6s, #7s, consider having the stirrups and spirals fabricated at a metal shop and then delivered to your site for cage assembly.

The foundation can be monolithic, or it can happen in two pours. This latter way produces a cold joint between piers and grade beams. For this job we poured the piers first, and let the cages run long for ties to the grade beams. We suspended each pier cage in its hole with a rebar offcut placed under a stirrup, as shown in the top left photo on the facing page. After the piers were poured, we bent the ends over and tied them into the beam cages (photo facing page, top right).

Forms—When we build our forms depends on the configuration of the rebar. If the grade-beam steel is a series of interconnected cages, we tie them together and then build the forms around them. This allows for adjustment if any piers are out of alignment. The outside of the forms defines the shape of the house, and registers the placement of the mudsills.

If unconnected individual bars are used for the grade-beam steel, it's easier to build the forms first and then place the rebar. Since the forms are built according to the contours of the site, nothing has to be level. We pre-drill the mudsills, insert the foundation bolts and then place them in the wet concrete (photo bottom right). Leveling the house occurs when the foundation walls, sometimes called pony walls, are built.

Pumped concrete—The standard pumping mix in this area contains ⅜-in. pea gravel in a six-sack mix. This aggregate size allows you to use the lower-cost grout pumps, which have a 2 or 3-in. dia. hose. These pumps are pulled behind a pickup, and can pump the concrete 150 ft. up a 45° slope. The six-sack mix has one bag more portland cement than the usual mix. The richer mix is needed to achieve a higher compression rating than you get with the smaller-size aggregate.

You should have water available for lubricating the pump hose, washing loose soil out of the forms and, if pouring the beams separately from the piers, for removing any dirt that may have accumulated on top of them. For cold-joint connections, an epoxy bonder applied to the top of the piers just before the grade beams are poured will help ensure a good bond between the two pours.

Pumpers usually know about how much water they want in the mix, and they will check the mud in the transit mixer for water content. A stiffer mix can clog a dirty hose, and it makes the pump work harder. Since adding more water will lower the psi rating of the concrete, try to hire a pumper who keeps his equipment in good condition. Tell the batch-plant dispatcher you are using a pumper on a steep slope. Adding a water-reducing agent will keep the mix soupy enough to pump and still achieve the required psi compression rating when it cures. Around here this admixture costs only about $2 per cu. yd., and it adds about 1½ in. of slump to a load.

But I still prefer a stiffer mix because it won't run downhill as much as a wet batch. We make an initial pass around the forms, filling them half full. With a stiff mix, the concrete is able to carry some of its own weight by the time we pour in the rest.

If you happen to hear that glorious sound of bursting forms and splitting stakes, halt the pumping immediately. If you wait for 30 minutes to an hour, the mix should set up enough to allow you to continue, despite a weak spot in the formwork. In the meantime, fill other areas of the foundation. Even in mild weather, stripping forms the next day won't keep your engineer happy, but it allows you to get your form lumber back in usable shape. □

Michael Spexarth is a general contractor in El Cerrito, Calif.

A pier cage hangs on a rebar beam (above left) as concrete is pumped into the hole. The protruding bars will be bent over to tie into the grade-beam steel. Once the concrete piers have set up, the pier-cage steel is bent over to tie into the grade-beam cages, as shown at top right. Because there are so many splices involved in the steelwork, Spexarth first installs the cages, then constructs the forms around them. Redwood mudsills are cut to length, and placed near the appropriate footing.

As the wet concrete for the grade beam is screeded flush with the top of the form, the sills are embedded in the wet concrete and checked for side-to-side level. The J-bolt protruding from the sill will be worked down into the concrete, and the nut will be tightened after the beam cures.

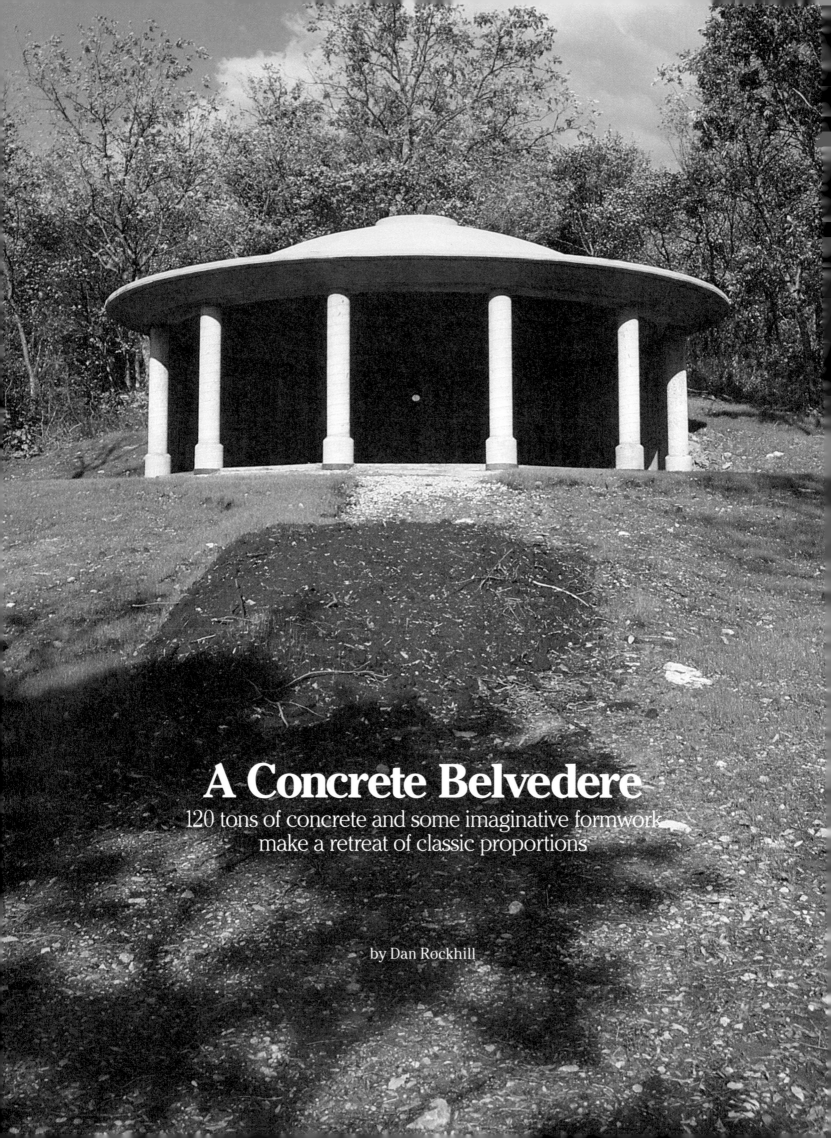

A Concrete Belvedere

120 tons of concrete and some imaginative formwork
make a retreat of classic proportions

by Dan Rockhill

Although it may look like the temple of Vesta in Rome or perhaps a Civil War monument, this concrete belvedere (photo facing page) is actually a country retreat for a city couple who live in eastern Kansas. It is used for family outings and meditation.

When our clients, Roy and Marilyn Gridley, first came to us and described the kind of building they wanted, they had two requirements. First, it had to be round. Round buildings are reminiscent of structures built by Pawnee Mandan Indians, a subject in which our clients are keenly interested. Second, the structure had to be indestructible. After having had a wood-frame building on the site vandalized, occasionally shotgunned and finally burned to the ground, they were wary of conventional building materials. They also told the architect, John Lee, that they wanted this shelter to be a distinct contrast to their traditional home in town. The only amenity would be a cast-iron woodstove.

Weighing the alternatives—While Lee began design studies in proportion and window placement for different seasons, I began to research construction techniques and costs. Stone was our first choice, but we had to eliminate it because stonework inevitably means very high labor costs. Concrete block looked too commercial, brick too institutional and ferro-cement a little too sculptural. We wanted something that would not be prohibitively expensive. This disqualified high-tech solutions like inflatable balloon forms covered with ferro-cement or molded plastic forms made to our specifications. An additional drawback to all of these methods, with the exception of ferro-cement, was that none of them allowed you to get a sense of craftsmanship—a quality we felt was necessary for the success of this project.

These considerations lead us early in the project to decide on concrete for both the walls and the roof. We knew that we could use concrete to support the weight and thrust of a clear-span roof. Concrete also met the requirement of being almost indestructible. The problem, then, became choosing the right shape for the roof.

Our clients had brought Lee their own model of the retreat, built out of a Quaker Oats box. The model didn't have a roof on it because they weren't sure of what we could do. Neither were

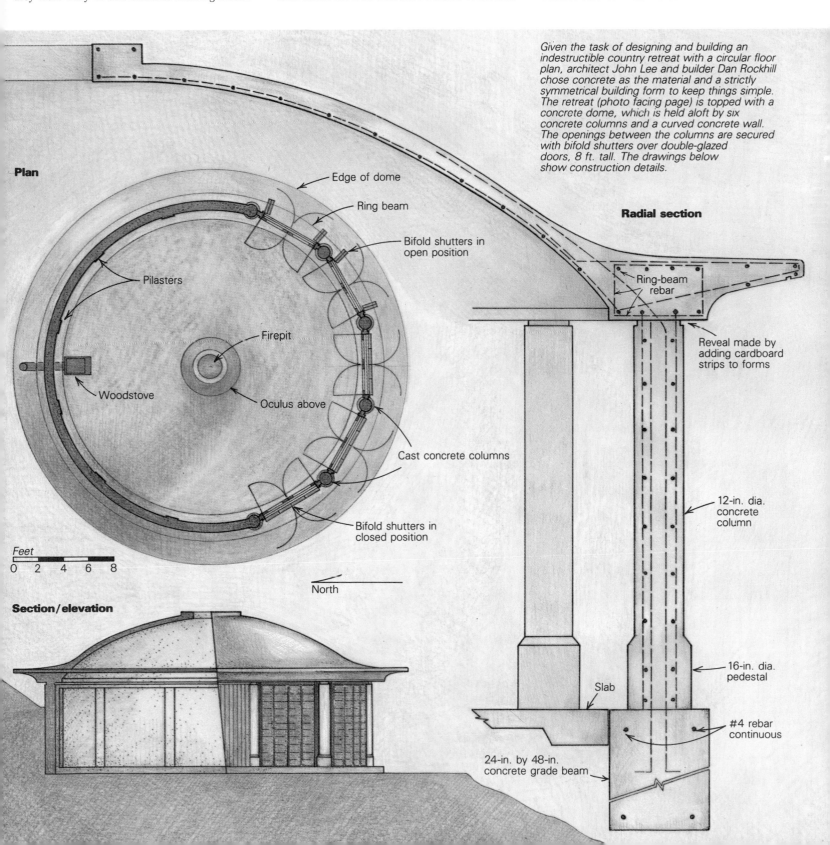

Plan

Edge of dome

Ring beam

Bifold shutters in open position

Pilasters

Firepit

Woodstove

Oculus above

Cast concrete columns

Bifold shutters in closed position

Feet
0 2 4 6 8

North

Section/elevation

Given the task of designing and building an indestructible country retreat with a circular floor plan, architect John Lee and builder Dan Rockhill chose concrete as the material and a strictly symmetrical building form to keep things simple. The retreat (photo facing page) is topped with a concrete dome, which is held aloft by six concrete columns and a curved concrete wall. The openings between the columns are secured with bifold shutters over double-glazed doors, 8 ft. tall. The drawings below show construction details.

Radial section

Ring-beam rebar

Reveal made by adding cardboard strips to forms

12-in. dia. concrete column

16-in. dia. pedestal

Slab

#4 rebar continuous

24-in. by 48-in. concrete grade beam

northern section of the circle. Each bar protruded 2 ft., and would eventually be tied to the steel in the curved wall.

A curved wall—More than half of the building is enclosed by a solid arc of concrete, 10 in. thick and 8 ft. tall. This wall holds back the earth to the north, and supports a concrete ring beam that encircles the building, tying the structure together. Along the southern section, the weight of the ring beam is borne by the cast concrete columns (section drawing, previous page).

We made the wall forms in 8-ft. by 8-ft. sections, with ¾-in. plywood plates and 2x4 studs. After laying out the curved plates on the plywood, we cut them out with a jigsaw As shown in the drawing at left, there is an inner form and an outer form. We lined the inside of both forms with ½-in. plywood, which bent to the shape without any kerfs on the concave side. Once we had the forms in place, we applied another layer of ½-in. plywood, which staggered the joints to tie the forms into a single unit. We then faced that with 1x4 #3 pine placed vertically to give a board-form impression (photo facing page, top). We left out a few rows of the 1x4s at the points directly opposite the columns. These would form shallow pilasters in the wall, continuing the rhythm of the columns around the wall.

The only problem we had with the curved forms was that the plywood kept remembering that it wanted to be flat. This caused us to lose the constant arc of the circle at the joints between form sections. We gradually smoothed out these kinks when we set the joists out from the center. The joists would eventually support the dome forms, and they also served as a platform for placing the concrete during the wall pour. Once we had the forms in the right place, we fastened their bottom plates to the grade beam with a powder-driven fastener in the center of each stud bay. Then we braced the entire assembly with tiebacks (angled 2x4 braces) at every other stud.

We used snap ties to hold the 10-in. wall both together and apart (photos this page). Snap ties are steel rods that pass through concrete forms. They have ridges on them that separate a pair of plastic cones by a consistent distance (drawing, left). The cones bear against the inside of the forms, while the rods continue through the forms and past the studs and walers (the horizontal bracing) to be held fast by wedges. The flange on the end of the snap-tie rod is 8¼ in. from the cone, which allows enough room for a ¾-in. plywood form, a 2x4 stud, a 2x4 waler and the wedge. Snap ties come in various lengths so that you can build walls of different thicknesses. After the pour you snap off the rods flush with the wall, and the cones pop out. The ties are available from most concrete-product distributors, who will usually rent the wedges too. It's the simplest system available and easily modified for out-of-the-ordinary formwork. In our case, we used 1x4s for the walers, and made up the difference with wood spacers.

The spacing of the snap ties is determined by the pressure of the concrete on the forms. It's better to over-design than to risk a form breaking during the pour. Our snap ties were on 16-in.

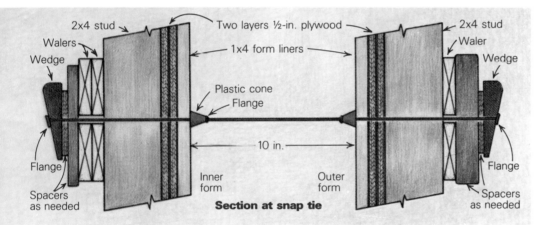

Snap ties are steel rods that pass through concrete forms and secure the forms at a consistent spacing. As shown in the drawing, plastic cones ride on the rods, and bear against a flared lip on the rod and the inside surface of the forms. In the photo top right, the form on the right is in position, the one on the left still needs to be drilled for the rods to pass through it. On the outside of the forms (photo top left), steel wedges that bear on the walers draw the snap ties together. This form will be complete when a Sonotube column form is added atop the pedestal. A 10-in. wide section will be removed on the wall side of the tube to create an integral connection with the wall.

Drawing labels: 2x4 stud · Walers · Wedge · Two layers ½-in. plywood · 1x4 form liners · 2x4 stud · Waler · Wedge · Plastic cone · Flange · 10 in. · Flange · Inner form · Outer form · Flange · Spacers as needed · Spacers as needed · **Section at snap tie**

we. Most roof solutions over drum-shaped bases are conical in section, but a cone-shaped roof didn't ring true for this building. A dome seemed like the right shape—its gentle rise culminating at the top in a round opening called an oculus (the Latin word for "eye"). I thought we could find a way to build one, and with encouragement from our clients and our engineer, Mike Quinlan, I went ahead and projected some costs. All along I felt we could benefit by the regular geometry of the formwork and save labor costs as a result. This became even more apparent after running an estimate. Repetition became the key to making the project affordable. If we could figure out how to make one section of the formwork, we could make jigs to speed the construction of the other sections.

On the side of a hill—The site, in rural Douglas County, Kan., is a heavily wooded hillside with a dramatic view to the south. We decided to put the building near the top of the ridge, with its north wall nestled into the hillside and its south wall open to the view and the sun. As shown in the plan drawing on the previous page, the southern rim of the dome is supported by six columns, spaced about 5 ft. apart. After laying out a few test circles on the ground, we settled on a 26-ft. dia. space. Once our soils expert approved the site for our plans, we cut a road in, dug a doughnut-shaped trench for the grade beam and poured 25 yd. of concrete to support the walls and the dome. Along with the grade-beam steelwork, we included ½-in. vertical rebar stubs on 16-in. centers along the centers, both horizontally and vertically. It was plenty strong for a 10-in. thick, 8-ft. tall wall.

We set the steel in place on a 16-in. grid in both directions, and used ½-in. rebar (#4) for the wall. Next, we placed the grid on the snap-tie centers so that we could wire the whole thing together. It is very important for steelwork to be absolutely rigid because the concrete will push it around during the pour if it's the least bit flimsy.

Once we had the steel in place, we oiled the inside face of the form with a mixture of 5 gal. of diesel fuel and 4 qt. of used motor oil. This is an inexpensive homemade form-release agent we used in lieu of costly commercial blends. Then we set the outside sections. After we had them in place, we ran our 1x4 walers, drilled holes for the snap-tie rods and pulled the whole thing together with the wedges.

Columns—Six 12-in. dia. columns carry the domed roof along the south-facing part of the plan. Both ends of the curved wall terminate at a column (photo facing page, left). The other four columns stand about 5 ft. apart, creating five openings that frame the view. To give the columns a wider base, we first cast a 16-in. dia. pedestal for each one. The pedestals are 16 in. tall, and beveled to meet the slimmer columns at a pleasing angle. To form the columns we used Sonotubes, which are cardboard forms that are torn off after the concrete has set. The tubes are simple to set up, and cost about $25 apiece.

At the tops of the column and the wall forms we inserted a strip of cardboard to reduce the width of the forms. This makes a reveal at the top of both the wall and the columns, which makes the dome look as if it's floating atop its supports. We continued the interior formwork of 2x4 studs and plywood around the front of the building to help stabilize the columns (photo bottom right) and to serve as support for the subsequent construction of the dome.

The walls and columns were pumped with 17 yd. of 3,500-psi concrete. We used a pump with a boom and a 5-in. nozzle. A rig like this is expensive—$350 per half-day in our area, but it is easily directed to any point on the site. We found the pump to be well worth the money.

We added a super-plasticizer, available from concrete suppliers, to the concrete to make it flow more easily. This is a liquid that is added on site to the mix. Most walls require a high-slump mix, and this is especially important when you want the surface to be imprinted with the form-board wood grain. Because concrete begins to lose its strength when you add too much water, it's not good practice to "soup it up." The Portland Cement Association cautions that an extra gallon of water per yard weakens the mix by 5%, and increases the shrinkage by 10%. The plasticizer increases the slump from about 4 in. to 8 in., and at the same time it actually increases the strength of the concrete. In a compression test, our 3,500-lb. concrete broke out at 4,000 psi in 28 days.

While the concrete was still wet, we sank steel dowels into the wall and columns on 12-in. centers to tie the dome to its supports. We gave the wet concrete a light sting with a vibrator to

The inside wall form is in place, and sections of the exterior forms are being assembled. Joists radiating from the center of the building help to align the forms, and provide a handy work platform during the wall pour. Note the gap in the vertical 1x4s on the inside wall form. It will result in a thin pilaster, reflecting the column on the opposite side of the building.

Rockhill continued the circular formwork to encompass the south side of the building. There it served as support for the dome formwork, and bracing for the column tubes.

help consolidate it, and we rapped both sides of the forms with rubber mallets to bring the cement paste to the surface. This gave the walls a great wood-grain impression.

Forming the dome—After the walls were pumped, we immediately started preparing the the dome formwork. We had already placed 64 joists atop the upper waler, which acted as a header. Each joist was supported at the center of the building by a framed core of studs, which we called the king post. The 64 joists divided the circumference into 16-in. wide segments. These joists carried the weight of the trusses, which in turn supported the dome form.

We had prefabricated the trusses in the shop even before we put in the foundation. We made

64 of them, of three different lengths. We staggered these lengths around the outside, from a long to an intermediate with short lengths in between (top photo, next page). With a simple jig, two of us made all the trusses in 22 hours. The top of the bowstring truss was made by bending 1x4 fir in two layers. The joints at the three corners were beefed up with plywood gussets.

Although we had most of our construction details worked out before we broke ground, there was one nagging problem that we weren't able to resolve until the last minute. We didn't know how to cover the dome trusses with a skin of form material. Our solution was to use the centerlines of the columns and pilasters as origins for what I called the 12 great circles. These 12 meet at the center, or oculus, in a swirling

Bowstring trusses of three different lengths were used to form the dome. Each truss was centered over its own joist, which in turn was braced to the ground. The lip at the edge of the dome was formed with 2x10 braces cut to the contour of the lip and covered with ½-in. plywood.

Pine 1x4s sheathe the dome trusses. In plan, twelve 1x4s radiate out from the edge of the oculus to the center of each column and pilaster. The gaps between them were then filled in with shorter lengths of 1x4.

pattern that looks much like the aperture of a camera lens (photo middle left). We infilled between these shapes with more 1x4s bent over the trusses.

Another detail we didn't have a clue on was the dome lip. Lee did many design studies while I considered the formwork options. The engineers wanted some overhang at the edge of the dome for structural purposes, and we wanted it for a place to sit. The big question was the severity of the compound curve and whether or not we could use plywood to form it. We ended up cutting 2x10s into the finished profile, and affixing them to the outside wall studs (photo top left). Once we had it right, we made 64 of them. These were sheathed with ½-in. plywood segments, radiused on their outer edges and kerfed on their backs for easier bending.

Dome steel—The dome's oculus is open and outfitted with a cover and a combination storm and screen window. I scared up some pretty high-priced lids in my search for a good metal hatch for the opening. Everything designed for the purpose was much too expensive, so we bought an old tractor-wheel hub and welded some hinges onto it for the cover. The wheel is contained by an 8-in. wide by 12-in. tall concrete compression ring.

Steel is the backbone of concrete, and that was certainly evident as we tied piece after piece in place to reinforce the dome. The steel virtually obscured the forms below. To elevate the lengths of rebar into the concrete, we placed them on wire bolsters, which rested on the form boards. All told, the dome contains over 6,000 linear feet of rebar. Most of it went into the ring beam and the dome lip (photo bottom left). When tapped, the lip makes a distinct "ping"— a sound unlike any I've heard from concrete.

Pumping the dome—Because this was an open form, we had to use a low-slump concrete or it would have run off the slope. We started at the top (the dome is only 3 in. thick here) and worked our way around the roof, floating the concrete with darbies. We had one man underneath rapping the forms with a mallet to bring the cement paste to the surface and to minimize blemishes and bubbles. Eventually we had to curtail this rapping because the vibrations caused the concrete to lose its set and run down the slope.

Our most difficult task was estimating how much concrete we needed to finish the dome. We held our loads to five yards per truck—otherwise it would spill out the back as the truck climbed the steep grade to the site. We started with 15 yards and came up short. And this was after two pros had told us, with great conviction, that it would be 15 yards for the whole job. So we had a callback for three more yards and

The dome, left, has over 6,000 linear feet of rebar. Most of it is in the dome lip and the ring beam that bears on the columns and the wall. Rockhill and his crew worked the concrete from the top of the dome downward, which required an exceptionally stiff batch of concrete to keep it from running off the form.

were still short. Another call for four more just made it. We now had 44 tons of concrete on the roof formwork, and it didn't budge—much to the relief of the guy inside tapping the forms. We sprayed the fresh concrete with curing compound, and left it alone for 28 days to give it a chance to reach its full strength.

A month later we climbed back up the hill with crowbars in hand. By this time we were very anxious. After all our work we hadn't seen even a glimpse of our product. We pried the entire shell off in a few hours, yelling and tossing splintered forms every which way. The only thing that slowed our pace was the black widow spiders that had moved into the dark and damp spaces behind our forms.

Stripping the forms off the domed ceiling was surprisingly difficult. We joked about pulling the king post out and watching all the formwork fall down. Well, we pulled out the post, but nothing happened. The trusses just hung there, making it impossible to enter the building and work. So we put the king post back in place and pried at the trusses. We ended up taking two-thirds of them apart piece by piece from the underside. Then we couldn't get the plywood off the inside walls. We finally had to devise a way of pulling it away with a come-along, which the plywood would resist until it literally burst off the walls and flew across the space at us. We actually had to put up a plywood deflector shield.

The low-slump concrete we had to use on the dome produced a honeycomb effect in some areas, not unlike popcorn in appearance. It was a bit of a disappointment, but there wasn't anything we could have done to prevent it. It's not a structural defect, and it was easy to patch.

After the forms were stripped, we poured a slab and colored it with dark red pigment. We left a 3-ft. circle at the center for a firepit. We also started our 12 radial expansion joints out to the columns and pilasters from this center.

Because concrete is not impermeable to water, we had to spend a few days putting a cementitious coating on the outside to waterproof it. We used Thoroseal (Thoro System Products, Standard Drywall Products, 7800 N. W. 38th St., Miami, Fla. 33166).

Doors—The spaces between the columns are filled with two layers of doors. The exterior doors are heavy shutters that protect the building from vandals. The inside doors have screens and glazing. Having 20 custom doors made up by a manufacturer was out of our budget so we built them ourselves. We decided to make them out of redwood because of its workability, resistance to decay and low maintenance.

We started with the shutters. They had to be thick but needed to hold tight to the building when open. A folding shutter seemed best in meeting this objective. We needed four shutter panels in each opening. The panels have a 2x4 Z-truss core, which is in turn covered with 1x12 redwood. We banded each shutter with 1x4s all around. Because each shutter panel weighs close to 100 lb., we attached them to the jambs with four 4½-in. by 4½-in. butt hinges.

We also built the combination doors out of redwood but instead used a triple layer of ¾-in.

stock. We glued and screwed these layers together, alternating the layers of rails and stiles to create mortise-and-tenon-like connections. This was important because the doors are close to 8 ft. tall and, with their tempered insulating-glass inserts, weigh over 120 lb. each. This layering also yields a thick door (2¼ in.), which is in keeping with the heavy concrete building.

We set the 2x6 redwood trimmers in place with seven drop-in anchor bolts per column (for more on fasteners, see article on pp. 38-43). Each trimmer had been dadoed seven times across its grain (about 16 in. o. c.) to receive a glued-in-place plywood insert that had been cut with a jigsaw to conform to the circumference of the concrete column (drawing, right). This gave us a good firm mount and a flat surface which we could square across the opening and set the sub-jamb to—another 2x6 which we could square and plumb accurately. This went all the way to the floor over the column pedestal. We made our doors and shutters narrow enough to fit inside these rough openings, with a little room left over to shim our prefinished 1x8 jambs precisely where we wanted them.

Without a doubt, this was the most enjoyable building project I've ever worked on. Three college students and I did the entire concrete portion of the project, from form building to dome floating, in a mere six weeks.   □

Dan Rockhill is a contractor and associate professor of architecture at the University of Kansas at Lawrence. Photos by the author.

Each column bay is closed off by a pair of doors on the inside that swing inward, and a pair of bifold shutters on the exterior side that swing outward. The interior doors accommodate either a double-wall tempered-glass insert during cold weather or an insect-screen panel for hot summer months when the bugs begin to fly. The door jambs are affixed to 2x6 trimmers, which are anchor-bolted to the concrete columns, as shown in the drawing below.

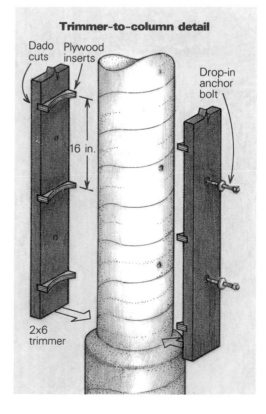

Trimmer-to-column detail

Dado cuts
Plywood inserts
Drop-in anchor bolt
16 in.
2x6 trimmer

Radiant-Floor Heating

An overview of the updated version of an ancient system

by Dennis Adelman

I first discovered radiant-floor heating in southern France, where I subsequently designed and installed a number of systems. Since then, just about every person I've spoken with who lives in a house with modern radiant-floor heating thinks it's the greatest thing since sliced bread. It's economical, invisible and technologically elegant. And it makes for comfortable heat, which, after all, is the most important measurement.

These systems consist of loops of plastic tubing embedded in a concrete slab, which you can finish like any other. Warm water flowing through the tubing heats the concrete, which becomes a wall-to-wall radiator, warming your living space from the floor on up.

Radiant-floor heating isn't a new idea. The hot springs that supplied the Roman baths were run through clay pipes that first passed under the palace floors before arriving at the bathhouses. Two thousand years later, French heating engineers in the 1950s made several unhappy attempts at using floor heating in low-cost housing projects. The assumptions were that since hot-water central heating was by far the most common form of heating in west-central Europe, and since masonry was by far the most common building material, it would be brilliance itself to combine the two. Copper or black steel pipes were simply embedded in concrete-slab floors and hooked up to conventional boilers. A promising idea,

n'est-ce pas? But it had some disastrous short comings. The designers used the boiler water at boiler temperature: 194°F. They also put their pipes too deep in the slab and much too far apart to produce comfortable, uniform heating. Tenants complained about burning their feet in some places and freezing them in others. Sore ankles, varicose veins and headaches were common afflictions, and radiant-floor heating fell into disrepute.

Radiant-floor installations are not unheard of in the U. S. (Frank Lloyd Wright championed them for years), and there has never been a single disastrous mistake here that alienated the public. But even when it was properly installed and spaced, the copper, galvanized or black steel pipe used to carry hot water through the slab limited the longevity and effectiveness of early RFH systems. All types of metal pipe are subject to corrosion, and if the slab ever shifted or cracked, the pipe would be likely to crack and leak. The expense and inconvenience of repairing pipes embedded in a floor slab kept RFH from developing a reliable reputation.

By the late 1960s, European researchers had come up with a better idea of how RFH systems should be engineered. The advent of flexible plastic tubing meant that their findings could be put to use.

In Europe, the tubing of choice is either polypropylene or high-molecular density reticulated polyethylene. Suppliers guarantee these materials for 30 years. In the U. S. and Canada, polybutylene is more readily available. These materials make up the big three of RFH tubing. They are all flexible, cheaper than copper, remarkably resistant to abrasion, puncture or freezing damage, and even capable of transporting 194°F water in domestic plumbing systems. They can't be glued or welded, so to join them you use special compression fittings.

When I began building in France, I invariably came up against the memory of The Mistake of the 1950s. Many of my clients were skeptical of the technical breakthroughs that I was telling them had made RFH the best system around. Since then, plastic tubing has proven itself, and the number of installed RFH systems on the continent is increasing at an astounding rate.

Why radiant heat?—The idea of using part of a building's structural system (in this case, the floor) as a radiator is quite sensible from a thermal point of view. To measure human comfort, scientists don't use the ambient air temperature (which a thermometer or thermostat reads) but rather, the *mean radiant temperature* (MRT), which is defined as the average of the room air temperature and the room wall temperatures.

Here's why mean radiant temperature is a better indication of comfort. If you are sitting in a well-heated room with your back up to a single-glazed window, you'll probably feel cold, in spite of the fact that the room isn't. The single pane of glass is a cold wall, and it pulls the MRT below the comfort level at that particular spot in the room.

You don't have to stop with eliminating the cold wall. If you create a warm wall, the wall-temperature part of our formula goes up, so the air temperature can be reduced by the same amount and the level of comfort will be the same. You can lower your thermostat setting a few degrees without feeling colder. This can save 10% to 20% on energy consumption.

A manifold should be installed before the tubing for the RFH system is unrolled. This one was fabricated on site. Once the tubing has been laid out, it should be filled with water under a pressure of at least 60 psi. This tests the tubing for leaks, and lets it assume its final position before the concrete pour.

From *Fine Homebuilding* magazine (August 1984) 22:68-71

Design guidelines—The RFH systems discussed in this article are all *hydronic* systems. Heat is distributed by circulating hot water through plastic tubing in a concrete-slab floor. Hydronic RFH systems require a small electrical pump to move the heated water from whatever heats it (technically called a *calorifere*—usually, a boiler) to the floor.

Although materials and techniques are constantly being refined, there are some basic design requirements that are common to all residential RFH systems. Comfort limits the maximum temperature of the surface of the floor to 79°F. The floor should also be heated as evenly as possible. The best way to do this is to run 80°F to 125°F water through the tubing, and to space the tubing runs between 8 in. and 16 in. apart, depending on the temperature of the water and the requirements of the zone to be heated. Spacing is much more important than the depth at which the tubing is buried (usually 3 in. in a 4-in. slab). Depth affects the speed at which the slab heats up, but not the eventual temperature of the floor.

The 80°F to 125°F temperature range is quite a bit lower than typical boiler-heated water, as we've seen. But it is precisely the temperature range at which alternative heating sources like heat pumps, active solar and geothermal operate best. Boilers work too because there are ways, which I'll explain, to bring their output temperatures down.

Where the RFH meets the house exterior walls, perimeter insulation must be carefully installed (drawing, above right). This serves two purposes. The first is to prevent heat losses from the slab to the outside. In almost every case of slab-on-grade construction, over 97% of the heat loss from the slab will occur at the perimeter walls. The second purpose is to provide an expansion joint between the house walls and the slab.

There is a fiery debate over whether you should insulate under the slab itself. I don't think it's worth it. At 6 ft. below grade, the annual average temperature of the earth is equal to the annual average temperature of the building site. Six feet of earth have an R-value equal to about 3 in. or 4 in. of insulation (depending on the relative humidity of the earth). So for my money (and yours), it's worth insulating below the slab only if the water table rises up into the 6-ft. zone under the slab, or if you don't want all the thermal mass that 6 ft. of earth represents—if, for example, your house already has a lot of thermal mass. (Too much mass would take to long to heat).

An average RFH system in an average house requires several hundred yards of tubing. To cut down on electrical pumping costs and general awkwardness, this total length should be divided into several loops. This also gives you the chance to regulate each loop individually. It's a good idea to make the loops the same length (200 ft. is about the maximum for ½-in. tubing; 300 ft. for ⅝-in. tubing).

Controlling the temperature—RFH systems have one troublesome characteristic. Even if you ask very nicely for more heat or less heat,

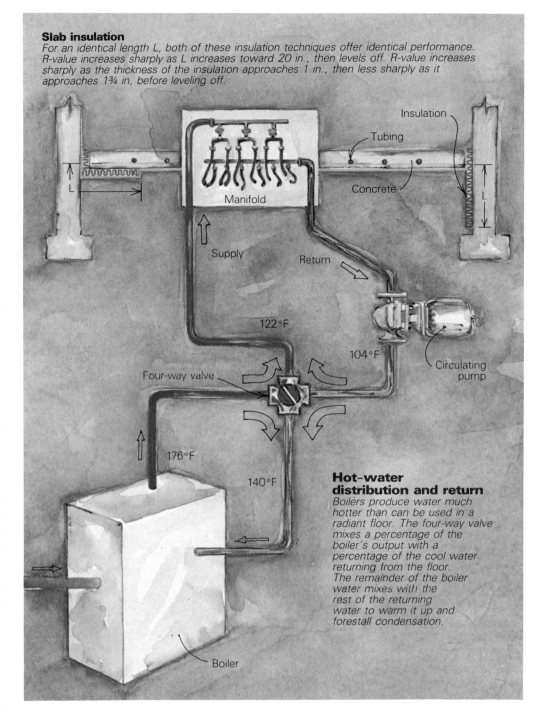

Slab insulation
For an identical length L, both of these insulation techniques offer identical performance. R-value increases sharply as L increases toward 20 in., then levels off. R-value increases sharply as the thickness of the insulation approaches 1 in., then less sharply as it approaches 1¾ in, before leveling off.

Insulation
Tubing
Concrete
L
L
Manifold
Supply
Return
122°F
Four-way valve
104°F
Circulating pump
176°F
140°F
Boiler

Hot-water distribution and return
Boilers produce water much hotter than can be used in a radiant floor. The four-way valve mixes a percentage of the boiler's output with a percentage of the cool water returning from the floor. The remainder of the boiler water mixes with the rest of the returning water to warm it up and forestall condensation.

they are very slow to respond. I know of three ways of dealing with this difficulty: hi-tech, low-tech and bad-tech.

I've recently seen some radiant-floor installations that have an ordinary wall thermostat that tells the calorifere what to do. This is bad tech. Using an indoor thermostat on an RFH is like trying to use spurs to control the speed of your car.

Let's say that you've set your thermostat to 68°F. The floor starts to receive heat, and the room starts to heat up. When the air temperature reaches 68°F, the thermostat shuts off the calorifere. But the slab has absorbed lots of heat, and this heat will be transferred to the room no matter what the setting on the thermostat. So by the time the slab has given up its stored heat, the room temperature reaches a steamy 77°F. Then the room starts to cool down. When it reaches 68°F, the thermostat again kicks in to ask for more heat. By the

time the slab can respond to the demand, the room temperature has slid down to a chilly 58°F or 59°F.

The high-tech solution uses—what else?—microchip technology. It is called Progressive Anticipatory Regulation (PAR). The brain of the PAR gathers information from an aquastat (which monitors the water temperature flowing into the slab) and an outside temperature probe. The PAR unit is pre-programmed with the anticipated thermal response times of the house, and the time the RFH system requires to answer a heating demand. By regulating the heat output of the calorifere itself or a motorized mixing valve that regulates the temperature of the water flowing into the slab, the PAR is capable of assuring a constant temperature in the house.

Once adjusted and fine-tuned, the PAR works like a charm. You can even program it to set back the temperature for nighttime, or to turn the heat back up just before you return from a trip.

The complete system (controller, exterior probe, motorized three-way valve and so forth) cost upwards of $750 at the time of this writing. It saves 10% to 15% in fuel consumption, so its use in larger installations is justified, especially where the other components of the system are equally sophisticated.

The low-tech approaches are more complicated and are the ones that interest me the most. They demand the participation of the home owners. While boilers that use liquid fossil fuels (oil, gas) or electricity can be controlled automatically with thermostats, aquastats or PAR, you can also adjust them yourself if you want to. Boilers that burn solid fuels like coal and wood work the same way as the others, but they are not automatic.

Most of the installations I have done have been for self-reliant, conservation-oriented clients who are eager to understand the inner workings of their homes, and who want to use wood as their heat source. Although they all concede the necessity of the small electric pump that runs the water through the RFH, they are loathe to increase their electrical dependence (and their cash outlay) by resorting to motorized valves and such.

We have found that RFH systems can run very nicely on wood, if the home owners provide the fine tuning that the system demands. Different installations require different approaches, but in every case, the users have been happy with their systems.

Powering the RFH system—If you use active solar or a heat pump as a heat source, your water temperature will be 85°F to 105°F. Perfect. But most people use a boiler and step its 194°F water down to 105°F to 125°F.

Non-electric boilers have two major components—a firebox where fuel is burned to release heat, and a heat exchanger, where heat from the firebox is transferred to the fluid (water) that flows to the distribution system.

If you try to regulate a boiler so that it delivers water at a temperature lower than 140°F, you will encounter two problems. The first is that modern boilers—except those specially made to power radiant systems—are inefficient at lower temperatures. The other problem is that if the water returning from the distribution system enters the boiler at less than 140°F, it will cool down the metal of the heat-exchanger coil sufficiently to provoke condensation of the water vapor present as a by-product of the combustion process. This will eventually lead to corrosion and shorten the life of the boiler.

In the case of oil-fueled (and to a lesser extent coal-fueled) boilers, there is the additional presence of sulfur dioxide gas in the exhaust fumes. If these fumes mix with the condensed water vapor, sulfuric acid will form on the coil, and damage it even more quickly.

You can prevent these problems by installing a three or four-way mixing valve between the calorifere and the RFH. The valve regulates water temperature by mixing cool water coming from the slab with hot water coming from the boiler. I usually use the simpler four-way, which looks like a Maltese Cross. There are openings to allow the connection of four water lines, and an interior adjustable damper that allows the flow relationships between the four ports to be altered at will (drawing, previous page). The water coming out of the boiler at 175°F is diverted in two directions in the valve. The water returning from the RFH is diverted in a similar way. The damper regulates the temperatures at the other two ports.

This valve is the key to the regulation of a boiler-powered RFH. It can be motorized to allow hookup to a PAR, or it can be controlled manually. I've found that there are all sorts of tricks you can do with RFH systems.

For the installation I talk about below, we used a wood boiler. The heating needs of the house are met by adjusting the four-way valve manually. But we also wanted to make heat generation independent of heat distribution, so instead of hooking the boiler directly to the RFH, we interposed a hot-water storage tank.

This resolved several problems inherent in a wood-powered RFH. First, it let us get about 12 hours of heating out of an efficient three or four-hour burn in the boiler. The heat released is stored in the water in the storage tank for later use.

The tank also serves as a heat sink that can be put to use when the electric power that drives the pump fails. The boiler-tank circuit works on natural gravity circulation (thermosiphon) and does not depend on electricity to function. If the power goes, the RFH ceases to function, but the storage continues to absorb heat. We hinged the top and one side of the insulated box that houses the storage tank so that it can be opened a bit to let the stored heat radiate directly into the living space.

If the owners fail to stoke the boiler, an aquastat on the storage tank shuts down the pump at 80°F so cold water doesn't circulate.

Installation—The system we installed for John and Mary Henderson, who live on one of the Gulf Islands of British Columbia, was typical. They wanted to jack up an A-frame so a new first story could be built under it. Once the roof was raised, the new walls were up and the area where the floor was to be poured was cleared, John and I started preparations for the radiant-floor heating system. First we smoothed and leveled the ground. We had the advantage of having a fine clayey subsoil as a base. If it had been gravelly, we would have spread builder's sand on the ground to cover anything sharp that could puncture or tear the insulation, the vapor barrier, or the tubes.

Inside the perimeter walls we dug trenches 2 in. deep and 2 ft. wide to accommodate sheets of polystyrene beadboard insulation. We wanted their upper surface flush with the ground on which the slab was to be poured.

Before laying the beadboard in place, we spread 6-mil poly over the ground and extended it 4 in. up the exterior walls. This was our vapor barrier.

Next, we laid 6-in. by 6-in. mason's mesh over the vapor barrier and the beadboard, stopping 3 in. or 4 in. from the wall. As is ordinarily done, we tied together the overlapping pieces of mesh with baling wire. Aside from consolidating the slab once the concrete was poured, the mesh also provided us with a nice grid that we could tie our tubing to. The space between the edge of the mesh and the wall was left to allow space for strips of beadboard that we placed vertically against the exterior of the slab-to-be. Our insulation and our expansion joint were ready.

We started laying out the tubes by attaching one end to the in manifold and then uncoiling the tube, and tying it to the mesh with string every 10 ft. or so as we went along (see the sidebar at right). I don't use wire, because it might bite into the tubing. You can make pretty tight turns in polybutylene tubing. The recommendation is five times the diameter of the tube—about 4 in. for our ¾-in. stuff. This part of the job needs at least two people. Three is definitely not a crowd.

Before the concrete pour, we filled the system with water at a pressure of at least 60 psi (use a pump if your on-site pressure isn't this high). This let us check for leaks, put the tubing under pressure during the pour and cure, and helped the tubing find its place in the floor before the concrete imprisoned it forever. (If you watch your garden hose in the instant after turning on the tap you'll have noticed that it jumps and bounces a bit as the pressurized water flows through it. The jumpy-bouncy position is the one we wanted our tubing to find. It meant a little less stress on our plastic once the concrete hardened.)

There isn't very much difference between pouring a radiant slab and one that doesn't radiate. Just be careful not to put holes in the tubing with track shoes or wheelbarrows. Fortunately, polybutylene and other plastics are all incredibly rugged materials, and there aren't many risks if you're reasonably careful.

Starting at the side of the slab where the concrete is arriving, I usually slosh down just enough concrete to cover the mesh and tubing. As the work progresses, we put down boards so the wheelbarrow can be moved to the far reaches of the space without having to roll right over the tubing. At the same time that this is all happening, one or two people are setting up wet screeds (thin snakes of concrete that are leveled off on top at the finished slab height). When this first stage of the pour is completed, the wet screeds are set up enough to be used as guides for screeding the rest of the slab.

At the Henderson's house, when we all came back several hours after the pour to finish up the job, one person set himself to cutting through the part of the vertical peripheral insulation that protruded above the finished slab. With the excess mortar pulled off by the troweling, we covered the beadboard with a thin coat to hide it and give the floor slab a nice, finished look. □

Dennis Adelman is a hydronic consultant and systems designer in Durango, Colo.

Planning the layout

This article is meant to be an overview, so I'm not going to get into the technical nitty-gritty, but here's a simplified example of sizing one loop in a larger RFH system.

You begin by calculating each room's heat loss at the 99% ASHRAE design temperature of the site (see the ASHRAE handbook). Next, you convert these results to heat loss per sq. ft. of floor area, and make corrections for the eventual floor covering over the bare slab, if any, and any difference between the 68°F base temperature of the tables and the temperature you wish the RFH to supply.

Using this figure, you can choose a tube spacing based on the average water temperature you hope to use. This will also give you the proper tubing density (linear feet per square foot of slab area), so you can figure out how long the loop needs to be.

The most readily available tubing material in British Columbia was 100-ft. coils of ½-in. ID polybutylene. This was what we used on the Henderson's house. In the Pacific Northwest, where we were working with 130°F water, a 100-ft. coil of polybutylene will heat a 10x13 room. Based on this, I calculated that we'd need 600 ft. of tubing for the slab. This meant that we would have six loops.

I tried once to write a computer program that would draw each loop in a projected installation and tell how many feet of tubing should be used. In the time it took me to write the program, I could have fitted up the whole state of Utah with RFH. And of course the program didn't work. Now I'm back to using scaled graph paper for my planning and although it gets messy, the results are reliable. The configurations we used are shown in the drawing at right.

You don't under any circumstances want to place joints, unions or splices under the concrete. The layout that looked best on paper was to position the in and out manifolds near the center of the house and to let the loops radiate outward like the petals of a flower.

It's essential to have your manifolds ready before you actually lay out the tubing. They give you a sturdy starting and finishing point for each loop, and they make it easy to fill the completed loops before the pour. In Europe, it's easy to buy rather elegant extruded brass units, but here I've been making my own out of black iron or copper tubing. I support them on a temporary framework of rebar until we frame up, when I cut the rebar away and fasten the manifold to studs. During construction on this job, they stuck out in the middle of the room, but when the framing was finished they found a discreet hiding place in a service box built into a partition.

As a general rule, the first portion of each loop, where the warmest water will be flowing, should head toward the coldest part of the room, which will usually be the exterior walls. But it shouldn't come closer than 12 in. to the exterior wall, regardless of the spacing used elsewhere in the floor. This distance optimizes the delicate balance between warming up the coldest part of the floor and minimizing heat loss at the perimeter.

Plan things so that you will arrive back at the return manifold at the end of each loop with about the same small amount of excess tubing in your hands. This will keep the pressure drop in each loop (a function of length) about the same and ensure that each loop in the system will carry equal amounts of warm water. —D. A.

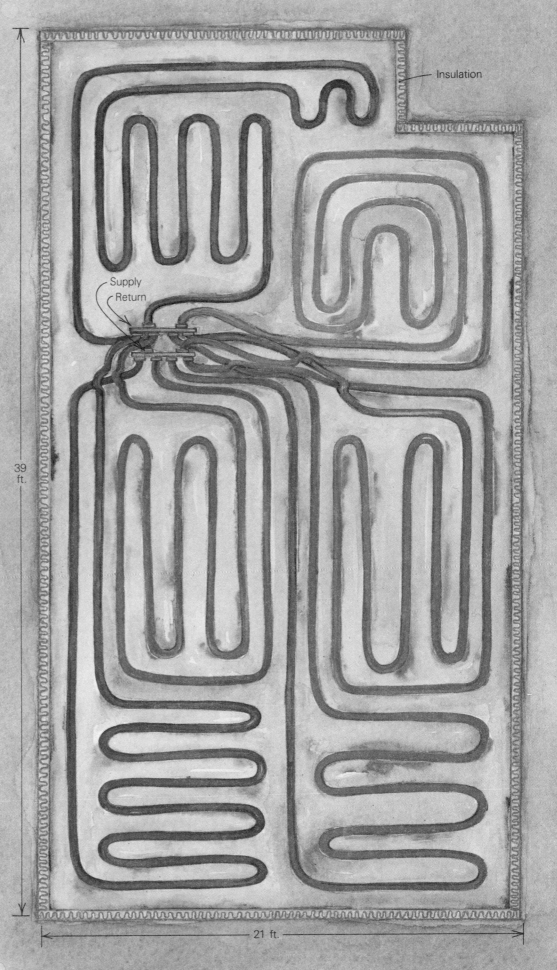

A typical radiant-floor layout
Six hundred feet of polybutylene tubing in six separate loops heat this house in British Columbia. Having short loops cuts down on the size of the electric pump required and gives the owners the flexibility to regulate each loop separately. Loops should all be about the same length to keep the pressure drops equal, and at least 12 in. from the walls to minimize heat loss at the room's perimeter.

Warm Floors

A well-designed, radiant-floor system can deliver heat where and when you want it

by Michael Luttrell

My company's interest in radiant floors originated in our work on solar space-heating systems. We learned that in winter it was not cost-effective to generate, from an active-solar system, the 120°F to 150°F required for a fan coil or the 160°F to 180°F required for baseboard systems. We discovered that radiant floors, on the other hand, require water temperatures of only 80°F to 100°F. This lower demand temperature can double the percentage of the winter heating load carried by the sun (the solar fraction), and it makes active-solar space-

heating systems feasible. We soon realized that radiant-floor heating systems have a large number of other virtues. Radiant heat is silent, clean, invisible, healthy (no dust or pollen blown around), flexible in concept and design, adaptable to any fuel source, affordable (see the sidebar on the facing page), energy efficient and extremely comfortable.

Radiant heating has a long and fascinating history, reaching back over 4,000 years (for some recent history, see Dennis Adelman's article on pp. 86-89). Rather than delve into the past here, I

From *Fine Homebuilding* magazine (June 1985) 27:68-71

Modern radiant-floor heating systems begin as lengths of polybutylene tubing laid in a pre-arranged pattern over gravel (facing page, top) or on the wood of a subfloor. The tubing is then embedded in concrete, so that the slab floor can be heated by pumping warm water through the tubing. Supply and return lines meet at the manifold (facing page, bottom). Supply lines, marked by the red arrows, will transport the heated water to the slab. Return lines, marked in blue, will bring the cooler water back to the heater. The pressure gauge is the best indicator of leaks during the concrete pour.

intend to focus on the materials, equipment, controls and design considerations that we use in our radiant floors.

Problems with copper—Radiant floors are constructed by embedding tubing in concrete—either slab on grade, thin slabs poured over wood floors, or 1½-in. thick mortar beds. Warm water, at 80°F to 100°F, from any source, is pumped slowly through the tubing. The concrete draws the heat out of the tubing, and radiates it into the living space. Most of our work has been in the custom residential market, and the majority of our floors have been thin slabs poured over wood (for more on thin-slab floors, see pp. 99-103).

Radiant floors with copper tubing are common. In researching our first few projects, we talked with everyone we could find with experience in installing or living with radiant heated floors. One of our best sources of information was a contractor who has been installing copper radiant floors for nearly 40 years. About half of his current business is repairing older radiant floors. We found that even though the copper-based radiant floors provide impressive comfort, their operating lifespans are limited. Typically, they last between 20 and 30 years before the first leak, and by 40 to 50 years, most of these floors are no longer usable.

Copper tubing usually comes in 60-ft. rolls (or rarely, 100-ft. rolls), and thus there are several soldered or brazed connections in the concrete in a typical radiant-floor system. This combination of dissimilar metals in the presence of lime and residual moisture is corrosive, and most leaks in copper radiant floors happen at the solder joints. The second most common location of leaks is where the copper crosses the steel reinforcement and creates dielectric corrosion points. The third cause of leaks is cracks in the concrete that rupture the copper.

We set out to discover a way to retain the many virtues of radiant-floor systems and to eliminate the disadvantages inherent to copper-tubing systems. What we found was polybutylene. The combination of polybutylene and several recent technological developments, such as slab insulation and electronic control systems, promises a renaissance of the most comfortable heating system ever known.

Design procedures—Radiant floors require an engineering commitment that is set, as it were, in concrete. There is no possibility of adjusting the tubing once it is installed. In this sense, there is no tolerance for error. On the other

Dollars and sense

I don't like to compare the cost of radiant floors to costs for other heating systems, because radiant floors aren't really comparable in terms of comfort, efficiency, noise level, or design freedom. Nonetheless, such comparisons have to be made. Around here, the standard alternative to radiant-floor heating is a forced-air system with a gas furnace or a heat pump. In my experience, radiant-floor systems need cost no more than quality systems of this type.

*Installation costs—*Slab-on-grade radiant floor designs are the least expensive since there's no incremental cost for the concrete. And customers of ours who have coupled radiant heating with an active-solar system for hot water have further reduced installation costs through federal and state tax credits.

With very little experience, workers can install tubing at a rate of 100 sq. ft. per hour. Adding several hours for manifolding, pressure testing, and standing by during the pour costs out at about 40¢ to 70¢ per sq. ft. Travel time, organization, overhead, profit and other expenses result in a sale price that averages between $1.00 and $1.25 per sq. ft. For concrete over wood-frame floors, there is an additional cost of $1.00 to $1.50 per sq. ft.

The selection of the best heat source for radiant floors is totally independent of the floor system itself. We have used solar systems, gas boilers, water-heating woodstoves, heat pumps, and our current favorite, the domestic water heater. The ability to eliminate a boiler is a tremendous first-cost benefit. For loads or use patterns greater than the capacity of the water heater, we use boilers to provide both hot water and space heat.

Let's look at the installation costs for two radiant-floor heating systems,

excluding permit and inspection fees, business licenses and other local variables:

2,000-sq. ft. house, slab-on-grade construction, four heating zones: Radiant-floor installation: $3000; Hydronics package (cost of tubing, valves, manifolds and related system hardware): $1,000; Nautilus water heater: $400; Total: $4,400.

3,000 sq.-ft. house, concrete over wood, six zones, high-efficiency boiler heat: Radiant-floor installation: $4,200; Hydronics package: $1,200; High-efficiency boiler (for floor and domestic hot water): $1,700; Concrete floor poured over wood subfloor: $3,300; Total: $10,500.

*Operating costs—*Here's where real savings can come into play. Computer simulations of radiant-floor homes vs. conventionally heated homes indicate potential savings of 15% to 30% in fuel costs. That analysis did not model multiple zone control, which could increase savings to as high as 60%. Our experience confirms utility bills that are half what you'd pay to heat the same space with other systems.

Radiant-floor systems are only part of a design sequence that strives to optimize every energy-related design factor, from insulation to appliances. Nonetheless, study after study has shown that for energy-efficient households, lifestyle is more important than technology. The major adjustment factor in lifestyle that makes this possible is comfort level. Almost everyone would like to have an energy-efficient home; no one likes large utility bills. But many people are unwilling to make sacrifices for efficiency. What we are trying to develop is a technology for energy-efficient lifestyles without any loss of comfort or convenience, in a home of conventional appearance. —M. L.

hand, the technology itself is most forgiving. It is hard to install a radiant floor that will not work if a few simple guidelines are followed.

The earthquake potential is an important concern in California, where we do most of our work. So we always recommend that an engineer go over plans for radiant floors in new construction or retrofits.

For most floors poured over tubing, we use standard concrete with pea-gravel aggregate. We usually pump the concrete for convenience and to protect the tubing. The thin slabs poured over wood-frame floors need be only 1½ in. thick, so the dead load added to the floor is only 15 lb. per sq. ft. Lightweight concrete, or Gypcrete (900 Hamel Rd., Hamel, Minn. 55340), a poured-in-place gypsum mix, can also be used for even lighter weight.

In a well-built modern house, most heat is lost through the windows and doors in exterior walls. So in laying out the tubing runs, we try to establish a heating pattern to cancel out this loss. Our current procedure is to make several

passes in front of windows and doors at 6 in. o. c., then several at 9 in. o. c. before spiraling in and out of the center of the room. We also space the tubing more closely in bathrooms to provide some extra heat.

We consider the floor coverings, appliance locations and probable furniture placement. For example, we do not want to heat under the refrigerator, which would make it work harder, or under permanent counters or any place a fixture or wall might be fastened down in the future. Tubing runs under a threshold or wall plate can be protected with a small piece of stainless-steel plate. Pipe runs through unheated spaces should be insulated with ⅜-in. or thicker foam-rubber pipe insulation.

We use a lot of tubing. Even in our mild (3,000 degree-day) climate, we use about 1½ ft. of tubing per sq. ft. of floor. This is as much as a mile of pipe in a 3,500 sq. ft. home. The tubing is inexpensive, and this quantity allows our systems to operate under less favorable conditions like low water temperature, high heat loss, and thick carpets installed over the floor. Abundant tubing provides a faster thermal response to overcome the temperature lag, which in the past has often been mentioned as a disadvantage of radiant-floor heating systems.

All this tubing is divided into individual loops between 150 ft. and 300 ft. long. Loop length shouldn't vary by more than 10%. By keeping all the loops the same length, balancing is automatic. On some installations, we include balancing valves to permit fine thermal tuning, although we seldom have to use them.

Zones and controls—We want to heat only those spaces in a home that are, or are about to be, in use. Radiant hydronic floor systems are easy to zone, schedule and control. In our installations, we use a combination of zoning and scheduling to maximize convenience and comfort while minimizing operating expense. (Compare this to a central-air system with a single thermostat for the entire house.)

Generally, one or more runs of tubing are laid out to provide heat for a particular zone of floor area. A zone might be a single room, a set of rooms (a study and a bathroom, for example) or even part of a room (the eating area in a combination living and dining room). Each zone can be controlled independently by thermostats, timers or even home computers. A single circulating pump can serve from one to twelve zones.

When we design radiant-floor heating systems, we consider three types of zoning: functional, structural and modulation.

Functional zoning is appropriate with open floor plans that have no clear physical boundaries. For example, zoning the area under the dining table, or around the TV, or a conversation pit. Radiant techniques heat objects and people, not air, so discrimination is possible.

Structural zoning refers to the separation of heating functions to those rooms that can be physically isolated and may be seldom used or are preferred cool, such as guest bedrooms, or formal entertaining rooms. These can be set at minimum temperatures when not in use.

Modulation refers to varied tube spacing, which modulates the heat-flux density. We try to vary the spacing based on design criteria. For example, we usually provide closer spacing in the bathroom, where people are usually lightly dressed, or in front of sliding-glass doors, where heat loss is quite high.

Control strategy is a critical part of design decisions. In the simplest control, a setback thermostat can overcome the thermal lag often mentioned as an inconvenience. This type of thermostat can be programmed to turn on the heat several hours before the space will be used, depending on the lag time, and turn the heat off (or down) a few hours before the end of use. Most setback thermostats have manual overrides, weekend features, and two setback periods, for day and night.

There are different opinions about thermostats for radiant heating systems. There is great virtue in using simple thermostats, which clients are familiar with and know how to operate. Although thermostat sensors are shielded from radiant heat rays, and are thus more sensitive to convective (air) temperature, there is a positive and constant relationship between air and radiant temperatures in a stable environment. Thermostats like the Honeywell T-87 are common, inexpensive, uncomplicated and easy to service. We have perhaps 200 such thermostats in service, and they seem to work well.

The second level of control can be provided by a BSR Home Control system (available from Sears and Radio Shack outlets), which can allow flexible control of the heating system on a timer, and permit override switching from any location in the house, or, through the use of a telephone modem and controller, from any location in the world. This is a practical approach for clients with unpredictable schedules, or for weekend or vacation homes.

Computer aficionados who are looking eagerly for some practical justification for all that expensive equipment will be glad to learn that zoning and scheduling are ideal for computerization. Although the development of hardware and software is still in its infancy, more is appearing day by day, and for the adept experimenter, the possibilities are numerous.

Materials and equipment—The polybutylene appropriate for radiant floors is stamped "ASTM D-3309" and "100 PSI at 180°F." It is available from plumbing-supply houses at about 20¢ per foot in 100-ft. or 1,000-ft. rolls. Shell Oil Co. (One Shell Plaza, Houston Tex. 77002) manu-

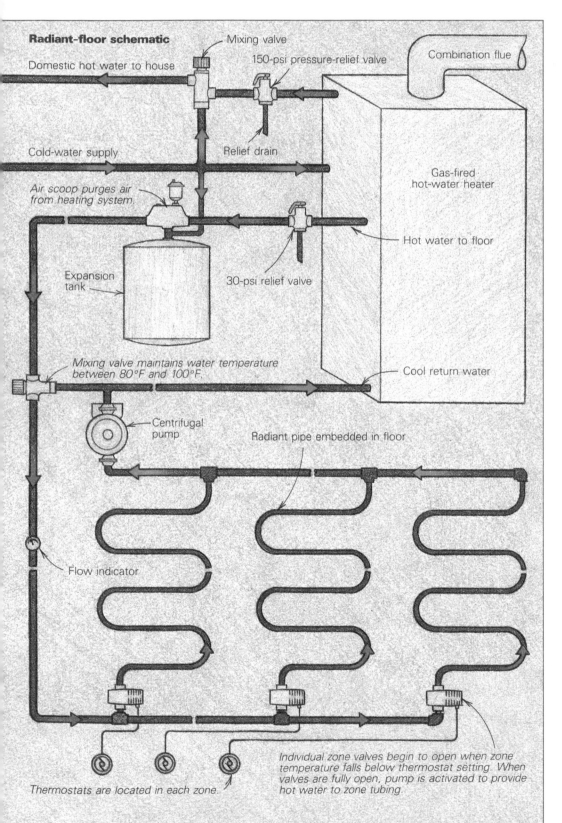

Radiant-floor schematic

Mixing valve

Domestic hot water to house

150-psi pressure-relief valve

Combination flue

Cold-water supply

Relief drain

Gas-fired hot-water heater

Air scoop purges air from heating system.

Hot water to floor

Expansion tank

30-psi relief valve

Mixing valve maintains water temperature between 80°F and 100°F.

Cool return water

Centrifugal pump

Radiant pipe embedded in floor

Flow indicator

Individual zone valves begin to open when zone temperature falls below thermostat setting. When valves are fully open, pump is activated to provide hot water to zone tubing.

Thermostats are located in each zone.

factures only the raw material, but they do have helpful literature on polybutylene available. Polybutylene pipe is extruded by a number of companies, with each one using its own brand name. For radiant floors, we use ½-in. ID pipe, except for long runs to remote manifolds or sub-systems, where ¾-in. ID pipe is used. The tubing is easily cut with a pocketknife or tubing cutters. It can be hand-formed into a radius 10 times the outer diameter, or 6 in. for ½-in. ID.

Lengths of polybutylene can be joined with heat-fusion welding, compression fittings, or crimp fittings. The easiest of these in the field is the crimp fitting, which uses an aluminum or copper alloy ring precision-crimped around a barbed insert fitting. Except in rare cases, we never plan pipe joints in the concrete. Ideally, the only joints in the system are at the manifold and at valves.

For temporary use during pressure testing and pouring of the concrete, where things tend to get banged around a bit, we use inexpensive acetal plastic fittings and aluminum rings, which are easier to crimp. Because the acetal may become brittle and crack under certain conditions, we use only brass fittings and copper-alloy rings when we're putting the final system together (The Failsafe System by Marshall Brass, 450 Leggitt Rd., Marshall, Mich. 49068).

The tubing is springy and will float, so it is important to anchor it firmly before the concrete pour. For fastening tubing in a slab-on-grade project, we use iron tie wire, strips of duct tape, or plastic ties—whatever we have on hand. The only purpose of the ties is to hold the pipe in place while the concrete is poured. For concrete over wood, we prefer a nail-down plastic clip without sharp edges, such as Sioux Chief Tube Talons, #552L (Sioux Chief Manufacturing Co., Box 397, Peculiar, Missouri 64078). In either case, we fasten the tube to the reinforcing steel or the wood floor at about every 3 ft. to 4 ft., and about every 1 ft. at the curves.

Water flow in the floor is controlled by different types of valves. *Balancing valves* are rotating flapper-type valves, not intended for positive shutoff, but rather for reducing the flow of water. We install them on the supply manifold, near the pump and heating equipment. They are adjusted by a screwdriver slot in the stem, and secured with a locknut. Thermal balancing is a complex procedure because every time a valve is adjusted, it affects the flow in all the loops. The easiest way around such complications is to design a system of equal loop lengths.

Zone valves control the flow of heated water to each zone. Each zone valve is controlled by its own 24-volt thermostat. We use the TACO 571 zone valve (1160 Cranston, Cranston, R. I. 02920) because it has few moving parts and an integral switch that starts the pump when the valve is fully open.

Isolation valves are used to isolate part or all of the system for repair or service. For emergency shutoff valves, such as the main feed water, we prefer a quick-acting quarter-turn full-port ball or disc valve such as the Milwaukee Butterball (Milwaukee Valve Co., 2375 S. Burrell St., Milwaukee, Wis. 53207). For occasional isolation of subsystems, we use less expensive gate

Radiant floors can be installed over wood subfloors. The tubing is anchored to the subfloor, and either lightweight or standard concrete is poured over it to a thickness of about 1½ in.

valves. For any valves or hose bibbs (used for flushing, purging and draining) that should not be operated by non-service people, we remove the handle and fasten it nearby. These valves and hose connections should be placed in such a way that they allow the entire system and each major subsystem to be isolated and drained so that debris and air can be bled from the system. The pump also needs isolation valves on either side of it.

Traditionally, cast-iron pumps have been used for radiant floors. Because we are trying to use the smallest centrifugal pump we can to minimize energy costs, we prefer a Grundfos (Grundfos Pumps, P.O. Box 549, Clovis, Calif. 93613) stainless steel or bronze ½-hp pump, for up to 3,000 sq. ft. These pumps have ceramic bearings that require no maintenance, and are exceptionally quiet.

The source of heat is entirely independent of the radiant-floor system. We have used cast-iron boilers, instantaneous water heaters, Heatmaker combination boilers by BGP Systems (141 California St., Newton, Mass. 02158), solar systems and water-jacket woodstoves. Our favorite system uses the same water heater that is used for domestic hot water. For this approach, we like the Nautilus, by American MorFlo (MorFlo Industries Inc., 18450 S. Miles Rd., Cleveland, Ohio 44128) or PGCS by A. O. Smith (P.O. Box 28, Kankakee, Ill. 60901) with 85% efficiency.

Installation procedures—If we're installing a radiant system over wood-frame construction, the tubing is secured to the plywood along snapped chalklines. When all the manifold con-

nections are made, we fill the tubing with water and pressure-test the system at 100 psi. A drop on the pressure gauge or a high-pressure spray of water means there's a leak.

Leaks or ruptures in the tubing can be repaired even during the concrete pour with the connectors and splicing equipment we keep on hand for such emergencies. Fortunately, this doesn't happen often. All splices should be carefully wrapped with electrician's tape and foam rubber; and the location of the splice should be noted for future reference.

During the pour, we maintain 60-psi pressure in the tubing. We also make sure that one of our crew is around to watch the pressure gauge and make sure that the concrete crew treats the tubing gently.

After the pour, unless freezing weather is anticipated, we reduce the pressure to between 30 psi and 40 psi as a visual indicator that the system is intact. In freezing weather, we either use antifreeze in the tubing or we remove the water from the tubing by draining or blowing after the pour. Polybutylene can withstand freeze-thaw cycling, but the concrete may crack as a result of the slight expansion of the tubing. As construction proceeds, we leave a gauge in the system that the carpenters can see. If they shoot a plate nail into the tubing, we hope they'll notice the pressure drop and give us a call. It is usually several months later when we return to cut away the temporary manifold and hook the floor up to its heat source.　□

Michael Luttrell is a general contractor in Napa, Calif. Photos by the author.

An AirCore Floor

Warm air from a sunspace and a floor of concrete blocks heat a bedroom addition

by Bill Phelps

Two thousand years ago, wealthy Romans were enjoying the comfort of radiant-floor heating. They circulated hot air from a central furnace through hollow bricks under the floors of their villas. The systems were called *hypocausts*, meaning fire from below, and they worked very well. The only major drawback was the cost of operating them. During the peak of the heating season they could consume more than two cords of firewood per day.

A modern version of this ancient technique turned out to be an ideal solution for a passive-solar addition I recently completed in Jackson, Wyo. We substituted solar energy collected in a sunroom for the firewood, and concrete masonry blocks for the hollow bricks below the floor.

This modern hypocaust, called an AirCore system by the National Concrete Masonry Association, uses the hollow cores of concrete blocks for air ducts. The blocks can be stacked in walls with their cores aligned vertically or laid on their sides below a concrete-slab floor with the cores aligned horizontally. In either case, as the warmed air circulates through the block cores, it gives up its heat to the wall or floor mass, which in turn radiates the heat into the living space.

The solar addition we built for Sophie Echeverria included a new master bedroom and a small sunroom (photo right), which was to be used as a back porch and as a sitting room on sunny days. She wanted the sunroom to help heat the addition, but there was no need to store the solar energy it collected in the sunroom itself. Consequently, we needed a form of remote heat storage that could be placed inside the bedroom. An AirCore floor was the answer.

To provide ducting for the air, the concrete blocks are laid on their sides in rows and columns so that the cores line up and the rows and columns of blocks form a rectangular blockbed. Across the width of the blockbed at one end is a supply plenum, from which the heated air can be drawn into the cores of the blocks by means of a fan. The warm air heats the blocks and the concrete slab above them. This mass will soak up heat from the air as long as the air is warmer than the blocks. After the air has traveled through the entire length of the blockbed, it is collected in a return plenum and routed back to the sunroom to be heated again.

A concrete slab heated from below with an AirCore blockbed can be 13°F to 15°F warmer

---

Bill Phelps lives in Jackson, Wyo.

The sunroom for the bedroom addition faces south, and with its quarry-tile floor stores considerable heat. Hot air from the sunroom collects near the ceiling and is ducted into the AirCore floor of the adjacent bedroom. After giving up its heat to the floor mass, the cooled air is returned to the sunroom to be warmed again.

than a conventional slab floor. This makes the floor more comfortable to walk on, and it also helps to heat the room above it. It does so in two ways. The first way is by radiant heat transfer. The surface of the heated slab radiates heat directly to any object or surface nearby that is cooler than it is. Heat is also delivered to the room by natural convection currents. This happens when the layer of room air that touches the floor absorbs heat from the concrete, is warmed and rises. Cooler air then moves in to replace it and so completes a convection loop.

Designing the blockbed—When the time came to fit an AirCore floor into this addition, there were plenty of constraints to work around. Architect Bob Gordon managed to design the addition so that it worked well with a peculiar house that had only one square corner, two prows, and a roof that pitched in two directions

at once, with roof beams running at obtuse angles. As if this weren't enough, property-line setbacks restricted the size of the addition, and none of the walls of the house faced south.

The addition fits comfortably on the original structure, as shown in the top drawing, facing page. It includes a small south-facing sunroom, and a new entry tucked between the house and the addition. The bedroom is compact, but feels open and roomy thanks to the light from the sunroom and a window on the east wall. We placed the AirCore floor under the bedroom.

The size and shape of the bedroom was just right for a rectangular blockbed 8 ft. by 20 ft. Twenty feet is the maximum length that's recommended for blockbeds. If the bed is longer, greater fan power is needed to force the air through it, and the air coming out of the bed will be cooled too much to heat the far end of the blockbed. A length of about 16 ft. is considered to be ideal.

To reduce the extra air pressure required by the longer bed, we used 8-in. blocks, even though 6-in. blocks are preferred in most cases. The 6-in. blocks add slightly to the air-flow resistance, but they take up less vertical space, cost less and have a higher percentage of mass per unit than 8-in. blocks. In either case, it is important to use high-density sand-and-gravel blocks to increase the mass of the floor.

As shown in the bottom drawing, facing page, the plenums we put at each end of the blockbed were rectangular. They extend the full 8-ft. width of the blockbed and measure 8 in. by 12 in. in cross section. That is actually a little larger than they need to be. If the air duct that connects to the plenum is at the end of the plenum, as it is here, the cross-sectional area of the plenum need be only as large as the cross-sectional area of the duct that enters it. If the air duct enters at the center of the plenum, the cross-sectional area of the plenum need be only one-half the area of the duct.

A rectangular plenum worked well in our case since the blockbed is only 8 ft. wide. But if the blockbed had been much wider, a tapered plenum would have helped to distribute the air flow evenly across the entire width of the bed. Mike Nicklas, an architect with Innovative Design in Raleigh, N. C., has designed many AirCore systems and prefers to taper the supply and return plenums for this reason. The tapered plenums are reversed at each end of the blockbed so that the bed geometry, in plan, is a rectangle or square enclosed by a parallelogram. The supply

From *Fine Homebuilding* magazine (December 1985) 30:50-54

and return air ducts enter the plenums at their widest ends, which are positioned at opposite corners of the blockbed.

We located the fan in a closet (photo p. 97) that was conveniently positioned over the supply plenum. The fan could have been placed at either end of the blockbed, although the return side is generally better. We framed a separate compartment at the end of the closet to isolate the fan and the air duct. The compartment was insulated, and we lined it with acoustical board. Then the fan was connected to flex duct and mounted with vibration isolation brackets to minimize fan noise. We used a squirrel-cage fan with the motor mounted externally to avoid overheating the motor.

Air leaks and improperly insulated ductwork in unheated spaces are two causes of poor performance in AirCore systems. Sealing the ductwork, plenums and blockbed against air leaks was an easy job in our case. The blockbed and the plenums were completely covered by a concrete slab, and we sealed the flex-duct connections with duct tape.

A backdraft damper—Another important component that we installed is a backdraft damper. It works like a check valve to ensure that air can flow in one direction only through the ductwork, and to prevent what is called reverse thermosiphoning when the fan is off. Warm air from the blockbed can rise by natural convection back out through the supply duct and into the sunroom as temperatures in the sunroom begin to drop. Cool air from the sunroom would then be drawn into the blockbed through the return duct to complete a convection loop in the reverse direction of the fan-forced air flow. As a result, some of the heat that is stored during the day would be lost back to the sunroom at night.

Commercially available backdraft dampers must be placed in a level section of the ductwork to function properly. But our duct chase was next to the sunroom wall directly above the fan, and we didn't have a section of level duct. So we substituted an inexpensive homemade backdraft damper that works well. We made it from a piece of plywood with a hole sawn out of the middle. We cut a piece of 2-mil polyethylene (4-mil would also work) to fit over the hole and stapled the poly to the plywood along the top edge to act as a hinge, as shown in the drawing on p. 97. The plywood is positioned slightly out of plumb so that the plastic is held closed against the inside of the plywood by gravity when the fan isn't running. When the fan kicks on, the plastic is easily pulled open by the air pressure.

We installed the damper in a sheet-metal box built to serve as an adaptor between the rectangular wall register and the round flex-duct coming from the fan. One side of the box is completely open and covered by the wall register. The opposite side of the box is fitted with a round duct connection and an elbow to which we attached the flex duct. The distance between these two sides was kept as long as the chase would allow to provide enough room for the plastic flap to swing freely.

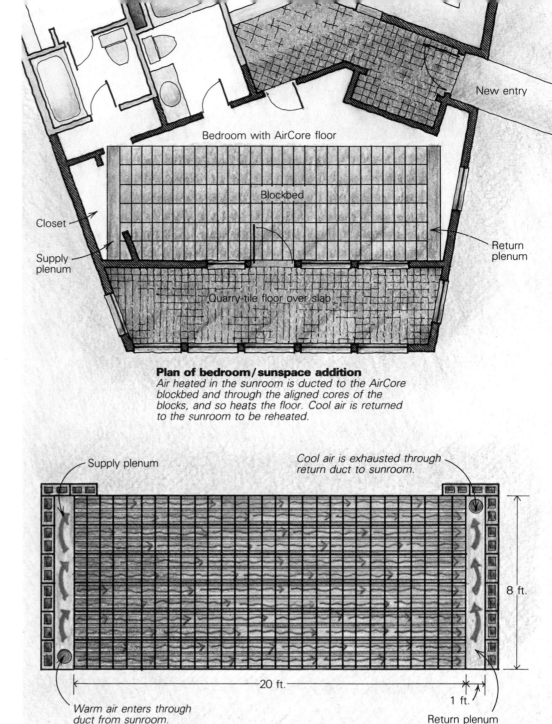

Plan of bedroom / sunspace addition
Air heated in the sunroom is ducted to the AirCore blockbed and through the aligned cores of the blocks, and so heats the floor. Cool air is returned to the sunroom to be reheated.

Plan of blockbed

System controls—The controls we used to operate the fan are simple and inexpensive. An in-line cooling thermostat was positioned out of the sun in the sunroom. This thermostat kicks on the fan when the temperature in the sunroom rises above 85°F. A standard switch wired in-line between the thermostat and the fan acts as a manual override. We used a rheostat instead of a switch to provide variable-speed control for the fan, thereby giving the system some flexibility to adjust to different seasonal conditions.

More complicated controls are required if an AirCore system is used for both cooling and heating or if maximum efficiency is a concern. In these cases, a differential thermostat is substituted for the cooling thermostat. A differential thermostat measures the difference between temperatures at two locations rather than the temperature at one location. Temperature sensors are installed with the differential thermostat

so that it can keep track of the temperatures of the blockbed, outside air and inside air, along with the temperature of the heat source.

As soon as the air in the sunroom is at least 8°F to 10°F warmer than the blockbed, the fan is turned on. It continues to run until the difference in those temperatures drops to only about 2°F to 4°F, at which point the fan is turned off. This means that the fan runs only when the conditions are best for transferring heat into the blocks. If the AirCore system is also used for cooling, the differential thermostat is programmed roughly in reverse of the heating mode. Cool night air is drawn from outside through the blockbed when the outside air is cooler than the blocks.

Insulation—We insulated the concrete-slab floor of the sunroom with extruded polystyrene. It has good compressive strength and low water-vapor permeability, but polyisocyanurate foam boards can also be used. We placed a 4-in. layer around the perimeter and a 2-in. layer below the sunroom slab and bedroom blocks.

The need for insulation below the slab is a controversial subject. Some people think that almost all of the heat loss from a floor slab is at the perimeter. But much depends on the climate and the average yearly ground temperature. Although deep-earth temperatures are moderate compared to outside temperatures, they are still much cooler than comfortable indoor temperatures. This fact, combined with the large volume of earth below the slab and the large surface area of contact between the two, can cause a lot of heat to be lost below the slab, especially in colder climates. When the temperature of the slab is increased for radiant-floor heating, the rate of heat loss below the slab is even greater.

The National Concrete Masonry Association (P.O. Box 781, Herndon, Va. 22070) recommends that R-5 insulation be placed below an AirCore floor when the climate averages 4,500 heating degree days (HDD). The insulation should be increased to R-10 at 6,000 HDD; above 8,000 HDD, insulation should be at least R-15 to R-20. As for perimeter insulation, NCMA recommends that R-5 be installed when heating degree days average at least 2,000. It should be increased to R-10 at 3,500 HDD, and to R-20 at 6,000 HDD. Above 8,500 HDD, at least R-25 to R-30 should be provided. NCMA also recommends that all ductwork be insulated, along with the supply and return plenums. If any insulation is left exposed in the plenum, local fire codes may require that it be covered with gypsum board or other fire-retardant material.

We placed a 6-mil polyethylene vapor barrier below the insulation. It prevents ground moisture from wicking into the blockbed, where it would be picked up and carried into the sunroom when the fan is on. Enough moisture in the blockbed would create ideal conditions for mold or mildew to sprout and fill the sunroom with unwanted odors.

Selecting the floor covering that will go over the AirCore slab is often the last step in the design process. Ceramic tile is the best choice because it conducts heat as efficiently as the concrete slab. Carpet or even wood flooring can be used over a radiant heating slab, but both of them act as insulation between the room and the heated floor.

Construction—The simplicity of an AirCore floor is most apparent when the time arrives to install it. The first step is to set the perimeter insulation into place and then backfill as you would for a conventional concrete slab. The backfill material is leveled and compacted in layers until the sub-grade elevation is reached. The sub-grade elevation must be low enough to allow for the finish flooring, a concrete slab (usually 4 in.), the concrete blocks, insulation, and a screeded layer of sand between the insulation and the compacted backfill (drawing, facing page). The layer of sand is necessary to smooth over the rough backfill and provide a level surface for the insulation and the blocks.

The blockbed we installed covered only a portion of the total area of the floor. This presented us with two options. We could compact the backfill at one elevation and then fill in the areas not taken up by the blockbed with washed rock, which does not require compaction. Or we could compact the backfill at two elevations,

one for the blockbed and one 8 in. higher where there was no blockbed. We chose the second option in order to save the cost of the rock. Once the blockbed was in place, a smaller amount of washed rock filled in the transition area between the two elevations.

The vapor barrier can go over or under the layer of sand. We placed it below. We did that in response to a phenomenon that occurs at building sites. It seems that whenever you unroll a large piece of plastic, the wind immediately begins to blow. Placing the poly under the sand gets it out of harm's way in a hurry.

We needed a layer of sand about 3 in. thick because our backfill was full of rocks and ended up pretty uneven after it was compacted. Crusher sand was the least expensive, but any clean sand will work. The sand must be level and at the right elevation so that the concrete slab will be the correct thickness and end up at the proper elevation. The best way to ensure that is to set level screed boards for an elevation guide. A long, straight 2x4 can then be pulled over the screeds, working it back and forth as you make your way down the length of the screeds (top photo, p. 98). The excess sand that builds up in front of the 2x4 is removed and used to fill in the low pockets. Since this blockbed was only 8 ft. wide, we were able to get it close enough by shooting in four grade stakes at the corners and then eyeballing as we worked the 2x4 across the sand.

The insulation board is set directly on the leveled sand. The 2-in. extruded polystyrene that we used was firm enough to walk on once in place. After all the insulation was down, we marked the exact location of the blockbed and plenums on the insulation by measuring off the

Design guidelines for an AirCore system

The first designers of AirCore systems, in the 1970s, worked on their own. Without much help, they developed design procedures and installed many successful systems. The National Concrete Masonry Association (P.O. Box 781, Herndon, Va. 22070) has since stepped in and is actively working to combine the experience of those early pioneers with current research to provide comprehensive design guidelines. If you need assistance, NCMA is the best place to begin looking. In fact, if your masonry-block supplier is a member of NCMA, the association will help you design your system. NCMA also has two technical publications (50¢ each) on the subject: "AirCore Design Guidelines" (TEK 140) and "Second Generation Passive Solar Systems" (TEK 120).

Two factors need to be determined when designing an AirCore system. These are the size of the blockbed and the rate of air flow. The size of the blockbed determines its thermal-storage capacity. The air-flow rate, which is measured in cubic feet per minute (cfm), is the amount of air that travels through the blocks.

The thermal-storage capacity of the AirCore mass needed depends on the amount of solar energy that the building will collect and on how much thermal storage is already supplied. The first step in the design process, then, is to decide what the solar contribution (amount of south-facing glass) should be. Volume III of the *Passive Solar Design Handbook* (American Solar Energy Society, 1230 Grandview Ave., Boulder, Colo. 80302) shows how to find a cost-effective balance between solar contribution and energy conservation for every region of the United States. Once the optimum solar contribution is selected, this handbook can also be used to size the number of square feet of south glazing needed to get that contribution for 94 different passive-solar system types.

The next step is to calculate the total heat-storage capacity of the building without the AirCore mass. A simplified method that takes into account the type, thickness and location of each mass element is presented in Chapter F of the *Passive Solar Design Handbook*, Vol. II. It determines the diurnal heat capacity (the amount of heat that can be stored and then released on a daily basis). The calculation will yield the total diurnal heat capacity, or the number of Btus that the building can hold for each degree Fahrenheit of temperature (Btus/°F). In order to relate the building's ability to store

heat to its ability to collect solar heat, the total diurnal heat capacity is divided by the total area of south glazing. The resulting quotient represents how much heat can be stored in the building per square foot of south glazing and is expressed thus: Btus/°F/sq. ft. of south glazing.

A general rule of thumb for solar buildings recommends that the total amount of thermal-storage mass should be between 60 and 70 Btus/°F/sq. ft. of south glazing. The exact value chosen depends on the climate. Areas that get a lot of sunshine will need to use a higher value for total thermal-storage capacity. Subtracting the total diurnal heat capacity of the building without the AirCore mass from the recommended total from the rule of thumb will determine how much additional thermal storage should be added with the AirCore system.

The procedure can be carried one step further with a computer-aided thermal network simulation. A mathematical model of the building created for the computer keeps track of heat-flow rates and heat storage for every element of the building. Bion Howard, research engineer at NCMA, uses this technique to fine-tune a system for optimum performance. He begins by using the rule-of-thumb approach to estimate the

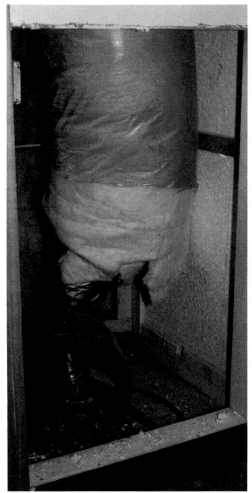

Seen through the access door, the squirrel-cage air-circulation fan for this project is conveniently located in a closet above the supply plenum, and directly below the intake register that's placed high on the sunroom wall above.

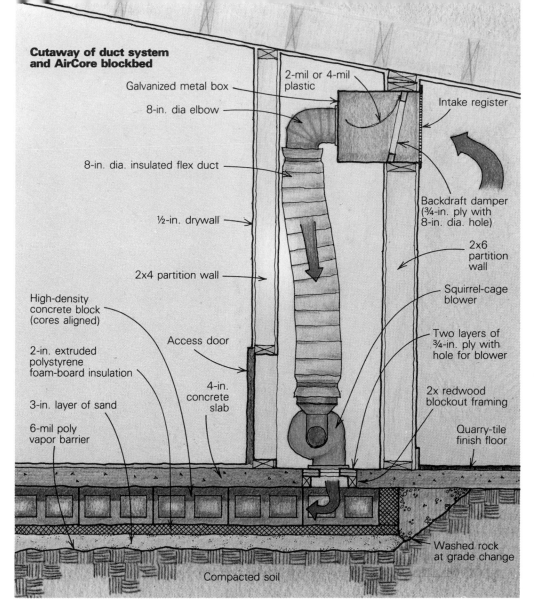

Cutaway of duct system and AirCore blockbed

Galvanized metal box

8-in. dia elbow

8-in. dia. insulated flex duct

½-in. drywall

2x4 partition wall

High-density concrete block (cores aligned)

2-in. extruded polystyrene foam-board insulation

3-in. layer of sand

6-mil poly vapor barrier

Access door

4-in. concrete slab

Compacted soil

2-mil or 4-mil plastic

Intake register

Backdraft damper (¾-in. ply with 8-in. dia. hole)

2x6 partition wall

Squirrel-cage blower

Two layers of ¾-in. ply with hole for blower

2x redwood blockout framing

Quarry-tile finish floor

Washed rock at grade change

---

amount of mass that will be needed. He then varies the amount of mass and the air-flow rate in successive computer runs until he is able to pinpoint which combination will deliver the best temperatures for transferring heat to the living space at the right time.

To set up a thermal network simulation that contains an AirCore system, it is important to know how well heat is transferred into the blocks from the air that is circulated through them. This rate of heat flow is measured by a relationship called the heat transfer coefficient. It is the number of Btus transferred each hour for each degree Fahrenheit difference in temperature between the air and the blocks and for each square foot of block-core area. A typical average design value, based on measured performance, is .48 Btus/hr./°F/sq. ft. of block-core area.

This lengthy design process is necessary whenever an AirCore system is going to supply a large percentage of the thermal storage in a building. But when a small AirCore system is to be used as additional or backup thermal storage for an attached sunspace, detailed design calculations seem unnecessary. In these cases, you may take your chances with the following rule-of-thumb procedure. Like any rule of thumb, its accuracy depends in part on the ability of

the user to adjust it to a particular situation and to respect its limitations.

First calculate the area of *solar glass*. Solar glass differs from south glazing. It is calculated by subtracting from the total area of south glazing an area equal to 7% or 8% of the floor area of the heated living space. For example, if a sunspace has 200 sq. ft. of south glazing and is attached to a home with 2,000 sq. ft. of heated living space, the area of solar glass would be 40 sq. ft. when 8% is used. For each square foot of solar glass, provide between 40 and 60 Btus/°F of thermal storage. If the weather is usually sunny or if very little thermal storage is provided in the sunspace, use the high end. If the weather is usually cloudy or if considerable thermal storage is provided in the sunspace, use a lower value.

How many square feet of AirCore blockbed will be needed to supply the thermal-storage capacity called for depends on the density and size of the blocks used and on the density of the concrete. The specific heat for each is also a factor. It is the amount of heat in Btus that a pound of a given material will hold per degree Fahrenheit (this information is available from NCMA). But if you are building an AirCore floor with high-density concrete blocks (125 lb./cu. ft.) and a 4-in. concrete slab, the total thermal-storage capacity for

one square foot of floor with 6-in. blocks is 15.31 Btus/°F. For 1 sq. ft. of floor with 8-in. blocks, it is 16.9 Btus/°F.

The final step in the design is to determine an air-flow rate. NCMA recommends that between 2 and 5 cfm of air flow be supplied for each square foot of solar glass in an AirCore system used for heating. Increasing the air-flow rate acts to increase the heat flow into the blocks but at the expense of greater fan power and higher operating cost. An air-flow speed of about 100 ft./min. through the block cores is satisfactory.

The air ducts used to transport the air to and from the blockbed are designed like any conventional forced-air system and require the experience of someone who is familiar with mechanical-systems design. The static-pressure drop or air-flow resistance in the blockbed itself should be about .1 in. of water. Values from .1 in. of water to .24 in. of water have been measured in existing AirCore blockbeds. The higher values indicate that there are excessive flow restrictions or that the fan is improperly sized. It is also a good idea to design a higher air-flow resistance in the ductwork (by changing its size or configuration) leading to and from the blockbed than the air-flow resistance of the blockbed itself. This will help to ensure an even distribution of air flow through the blockbed. —*B. P.*

foundation walls. We double-checked to be sure that the inlet and outlet to the plenums were correct, then we snapped chalklines on the surface of the insulation to outline the blockbed, and just set the blocks in place (middle photo).

The core holes of concrete blocks are actually tapered slightly so that they will fall off the mold when they are cast. There is some question about which direction that taper should face in the blockbed since a slightly different air turbulence pattern will result for each option. Since there is no definite answer about which way is best, we solved the problem by paying no attention at all to how we set them and ended up with a random mix.

We formed the sides of the supply and return plenums with a single row of blocks (bottom photo). Since this blockbed was built with 8-in. by 8-in. by 16-in. common blocks, we simply set the blocks with the cores facing up. This row of blocks also provides a ledge to support the cover we placed over the plenum.

Corrugated steel decking is recommended to cover the plenum and support the slab above it. A small quantity of steel decking, however, was going to be too expensive to have shipped in, so we decided to use a double layer of corrugated galvanized-steel roofing. To make up for the weaker metal cover, we wired a grid of #4 rebar and positioned it over the plenum when we poured the slab. We cut one hole in the corrugated metal at each plenum for an inlet and an outlet to connect to the air ducts. Over each hole we formed a blockout with 2x redwood to create the inlet and outlet holes in the concrete and secured them by nailing the steel roofing to the bottom of the blockout. We covered the blockout with scrap 2x material to keep concrete out of the plenum during the pour, and held the top of the 2x cover flush with the top of the slab so that it would not be in the way when the concrete was finished. It was later replaced by a double layer of ¾-in. plywood, which made a mounting plate for the fan. We cut the plywood to fit into the square hole left in the concrete when the temporary cover was removed and then cut a hole in the middle of the plywood to match the outlet hole of the fan.

On this job, the ductwork and the registers above the slab were exceptionally simple to hook up. The intake register was located at the west end of the sunroom near the ceiling. It was directly above the fan, so it required only a short piece of insulated flex duct to connect it with the fan. The register for the air returning to the sunroom from the blockbed was located at the east end of the sunroom near the floor. This means that the cool air returning from the bedroom is circulated over the quarry-tile floor of the sunroom, where it can be warmed again. It then rises and is drawn into the intake register to complete the cycle.

The only thing left to do was to wire the fan and controls and wait for the sun to come out. Our solar hypocaust works quietly thanks to all the steps we took to muffle the fan noise. It also manages to keep the sunroom from overheating, and the bedroom stays warm and comfortable. The nice thing is that it does it all without taking up any room in the addition. □

Laying an AirCore blockbed. **After the earth fill has been compacted, a poly vapor barrier is put down. Next, a 3-in. to 4-in. layer of sand is screeded level to make a smooth base for the blockbed (top photo). This is followed by 2 in. of extruded polystyrene foam-board insulation. The blocks, with their cores aligned, are laid directly on top of the insulation (middle photo). The completed blockbed (bottom photo) will be covered with a 4-in. concrete slab. A sheet-metal cover will be placed over the plenums (seen in the foreground) to keep the concrete out. Grading and insulation for the sunroom floor can be seen in the background.**

Max Jacobson

The Thin-Mass House

Concrete floors and interior stucco walls improve
heat storage in passive solar designs

by Max Jacobson

As designers of passive solar homes, our firm has been searching for an elegant yet economical solution to the problem of thermal storage. Instead of concentrating thermal storage in Trombe walls, rockbeds, water pools or eutectic salt arrays, we've now worked out techniques to use thin layers of stucco and concrete both as heat storage and as finished interior surfaces. We're not talking about standard masonry construction. Essentially, our approach takes the typical wood-frame stucco house and turns it inside out: Stucco layers become the finished interior walls, and thin slabs of concrete are used as floors, even in two-story wood-frame houses.

Masonry surfaces warm or cool us by means of radiation and conduction, and a thin masonry envelope can be a very comfortable environment if it's oriented to the sun and properly insulated. Used as finish surfaces, the mass becomes a part of the building's fabric, like the stonework in a whitewashed Mediterranean house or the earthen bricks in our own southwestern adobes. Surrounded in such homes by

solid, durable materials, the occupant feels a comfortable sense of permanence.

In addition to these characteristics shared with traditional masonry structures, the thin-mass house also has several advantages.

Lazy mass—Only the first few inches of the interior mass of a masonry structure effectively store heat. The mass in the center of the wall isn't doing much, and in fact, can act as a thermal bridge, conducting the higher indoor temperatures outside. Most of the mass in the center of a concrete wall doesn't have a structural purpose (that's why concrete blocks are hollow), yet it adds to the building's foundation requirements and increases loading during an earthquake. In our thin-mass house, we replace these lazy inches of mass with lighter, well-insulated materials—conventional wood framing and fiberglass insulation.

A little history—Concrete as a finish material has been around for a long time. In 1894, the southern California architect Irving Gill used

concrete to duplicate the glossy, hard-packed earthen floors he had found in Mexican adobe houses. He added pigments to the concrete mix to produce neutral, earthy colors.

By 1930, the staining and painting of concrete surfaces was widely practiced, although the concrete was made to look like stone or antiqued wood rather than packed earth. Typically, the surface was painted with a thin wash, followed by darker designs (photo, next page).

Integral coloring and surface coloring are both still used, but we can also use a new, less expensive method that restricts the color to a 1-in. or 2-in. layer over an existing slab.

Topping over on-grade slabs—When the site permits, an on-grade slab can serve as both the foundation and the ground floor. But clients' first reactions to the suggestion of on-grade slab floors are typically negative. They assume that the floor will look like the inside of a garage and that it will always be cold. But if the slab's perimeter is carefully insulated to prevent thermal bridging, the floor will be comfortable, its

The Lee-Carmichael house (previous page) is a thin-mass building in northern California. It has a heat load of only 205 therms annually, because of its 245 sq. ft. of south-facing glass and the thermal storage capacity of its stuccoed interior walls and concrete floors. Without these features, its yearly heat load would be about 630 therms (1 therm = 100,000 Btus).

temperature fluctuating narrowly around the mean internal temperature.

You can get a gorgeous finish floor by using integrally colored concrete for the entire slab. The pigments are just dumped into the transit mix truck as the load is readied for delivery. The pigment adds about $20 per yard to the cost, so it's important to use it only where it will be visible. In some cases, we prefer to use a two-step system for on-grade slabs, which allows more decorative effects and separates the finish work from foundation frenzy. We first pour a conventional slab, and then finish it with a 1½-in. thick layer of integrally colored concrete.

Because it's nearly impossible to reinforce such a thin layer of concrete (the wire mesh can come to the surface in places) the topping has to be bonded to the reinforced slab below to prevent shrinkage cracks. To do this, you must be sure that the slab has a rough surface. Don't trowel it smooth after screeding. The surface must be absolutely clean. Sweep (or, better yet, vacuum) it free of all dust and debris. Any areas that have had oils spilled on them must be thoroughly washed with detergent and flushed with a hydrochloric-acid wash.

Just before pouring the topping layer, spread a bonding grout of one part portland cement, one part sand and one-half part water over the moist prepared surface. Use a stiff broom to spread it into a 1/16-in. to 1/8-in. layer, then pour the topping layer promptly. If the grout dries out, no bond will be formed, and the topping will develop a network of cracks as it dries. The grout spreads easily, so wait until the transit truck is in sight to begin. Any contraction joints in the underlying slab must be repeated in the top layer.

Concrete on wood framing—We often pour thin slabs on wood-framed floors. Joists, of course, must be beefed up to carry the extra load. For a 3-in. slab, we are adding 37 lb. per sq. ft. to the normal value of 50 lb. per sq. ft. of combined dead and live load. This roughly doubles the floor load. You could double the number of floor joists, but we have found that it's more efficient to increase the depth, as shown in the chart below. The chart is for #2 Douglas fir with joists 16 in. o.c.

Span (ft.)	Joists (normal construction)	Joists with 3 in. concrete added
8	2x6	2x8
10	2x8	2x8
12	2x8	2x10
14	2x10	2x12
16	2x12	2x14

Assumptions: live load, 40 psf; dead load, 10 psf; + 37 psf for 3-in. concrete.

When a slab is poured on wood framing, we don't attempt to join the two physically—they remain separate, independent structures. Be-

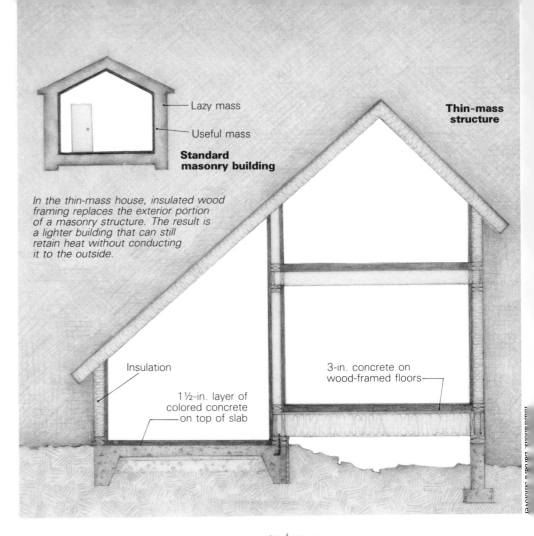

Lazy mass
Useful mass

Standard masonry building

Thin-mass structure

In the thin-mass house, insulated wood framing replaces the exterior portion of a masonry structure. The result is a lighter building that can still retain heat without conducting it to the outside.

Insulation

1½-in. layer of colored concrete on top of slab

3-in. concrete on wood-framed floors

The 1½-in. topping layer of colored concrete is poured onto an underlying slab prepared with a bonding grout. Foam board insulation at foundation perimeters works well in moist locations. Transite panels (cement/asbestos board) placed over the insulation protect it from impact damage.

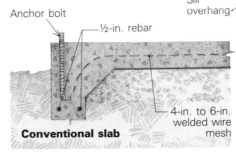

Anchor bolt

½-in. rebar

4-in. to 6-in. welded wire mesh

Conventional slab

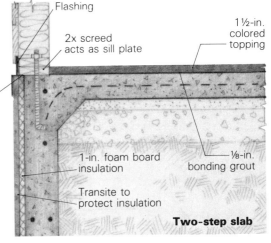

Flashing

2x screed acts as sill plate

Sill overhang

1½-in. colored topping

1-in. foam board insulation

Transite to protect insulation

1/8-in. bonding grout

Two-step slab

Painted concrete as a form of ornamentation was much in vogue in the 1930s. The geometric designs on this arch in the Los Angeles Public Library resemble mosaics.

Photos: Bob Lee

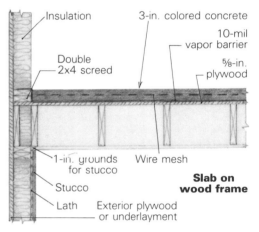

A *slab on wood framing* requires deep joists to handle the extra load (drawing, below). The plastic membrane keeps concrete from seeping through the plywood substrate. Sheeting wasn't used on the floor above because the underside of this floor is not a finished ceiling. The workers are screeding integrally colored concrete, using the doubled wall plates as guides. Left, the slab has been vibrated into place and is now being floated. Finish troweling will follow.

Insulation · 3-in. colored concrete
· 10-mil vapor barrier
Double 2x4 screed · ⅝-in. plywood
1-in. grounds for stucco · Wire mesh
Stucco · **Slab on wood frame**
Lath · Exterior plywood or underlayment

cause the concrete needs to be reinforced against shrinkage cracking, we use a 3-in. layer with welded wire mesh at its center.

After nailing down the plywood subfloor, we set double 2x screed plates where the walls will go, and lay a 10-mil polyethylene moisture barrier to prevent any moisture in the concrete from seeping through the plywood and discoloring the underlying wood. This step can be omitted if the framing won't be visible from below, but including it will retard the rate at which the concrete slab will dry, and minimize shrinkage cracking.

Spread welded mesh over the floor, and brace the walls with exterior plywood siding or shingle underlayment before pouring the concrete (photos above). To minimize damage to the floor during construction, it's possible to postpone pouring the slab until the structure is nearly complete, but it's difficult to screed the pour level with walls in the way.

Just as the wood floor frames receive a thin coat of concrete, the wall studs are covered on the inside with stucco lath and a 1-in. thick layer of integrally colored cement stucco.

Consistency and pattern—The colored topping layers are thin, so we use small aggregate. As a general rule, the size should not exceed one-third of the slab's thickness. For our 1½-in. slabs, we use ⅜-in. pea gravel.

The recommended proportions of a topping mix are one part portland cement to one part sand to two parts aggregate. Water should be limited to 45 lb. for each 100 lb. of cement. This blend produces a stiff mix with a 1-in. to 2-in. slump. You'll need to increase the water ratio if you have to pump the concrete.

Don't trowel until the sheen has left the surface. This will prevent water and fines (powder and small particles in the aggregate) from rising and causing the floor to "dust" with use.

After troweling, strike expansion joints to control the contraction that occurs during drying. This is especially important in 1½-in. slabs, where we place joints at a maximum of 4 ft. on center, in each direction. Jointing gives you a good opportunity to introduce scale and variety into the floor surface. Using a hand joint edger, you can cut freeform joints or a regular grid using chalklines as guides. Another option is to

use pattern-stamping tools that are pressed into the freshly troweled surface to create shapes that resemble bricks, stones or tiles. Ranging from small handheld devices to large rolling cages, these tools create a surface of relentless regularity—concrete that looks like it's trying to be something else. We prefer designs that preserve the hand-worked appearance of the jointing. In one example, we began by snapping chalklines in a 3-ft. by 3-ft. grid on the underlying slab. At each intersection, a glazed ceramic tile was placed on a small bed of grout, and tamped down to a 1½-in. height. These tiles became a dispersed screed for the later leveling of the colored concrete. After troweling, joints were struck between adjacent tiles (see next page). We have found, however, that 3-in. reinforced slabs can cover room-sized areas (15 ft. square) without any interior joints and without any appreciable shrinkage cracking.

Curing the finished concrete properly is important because a thin pour can easily dry out too fast. Premature drying will cause hairline cracking throughout the surface. In extreme cases, fissures extend through the pour and cause the slab to curl slightly at the cracks. Keep the slab wet and covered with burlap, plastic or waterproof paper for seven days. You can walk on it the second day, but be careful not to tear the protective material. Covering work areas with flattened cardboard boxes is a good hedge against damage.

Coloring options—Masonry suppliers have color charts of a wide variety of pigments. While integral coloring produces the most uniform results, dusting coloring powder over a freshly poured slab and working it into the surface also works well. The dust-on mixture consists of one part portland cement to one part sand and a variable amount of pigment. It is scattered like grass seed at a rate of 60 lb. to 125 lb. per 100 sq. ft. of floor, and it's pressed into the fresh concrete with a wooden float. More pigment gives a more intense color, and spreading the mixture in two separate casts (two-thirds first, then the rest) will result in more uniform coloration.

Painting and staining are options to consider if you don't plan on integral or dust-on color. Many paints and stains have been developed specifically to be used on concrete. Traffic will gradually wear through these finishes, and you'll have to renew them periodically, but protective coatings of clear varnish or shellac will help them last longer and produce a shiny glazed surface. Another trick is to mop the painted floor with a protective coating of starch. When the floor is washed, this coating will come up along with accumulated surface dirt. Mop on a fresh coat of starch to begin the process again.

Thermal performance—Because we use these thin layers of concrete as thermal storage in passively heated buildings, it's crucial that the building be oriented correctly to the sun, insulated well and weatherstripped. Even without any additional mass, such a building will store and release heat better than the average structure, assuming that it can be successfully vented during periods of overheating.

Exactly what are the thermal benefits of adding mass to a building that is already oriented correctly, insulated and weatherstripped? To answer this question, we analyzed a recent design of ours, the Lee-Carmichael residence, in Glen Ellen, Calif. (photo p. 99).

This area has a mild climate, only 2,900 degree days annually. We designed for a low temperature of 29°F. Any benefit we can demonstrate for the thermal mass in this house will be greatly magnified in more severe climates.

The conclusions drawn from our analysis show clearly that it pays to design a house to be passively heated, regardless of whether thermal mass is used. Adding thermal mass substan-

Above right, a 1½-in. thick topping layer of integrally colored concrete poured onto an underlying slab is being worked into a finished surface. Tiles on 3-ft. centers set in mortar before the topping was poured act as a dispersed screed. The finished walls are protected by plastic. After cleaning, the finished floor was sealed (right); a coat of tinted wax will follow.

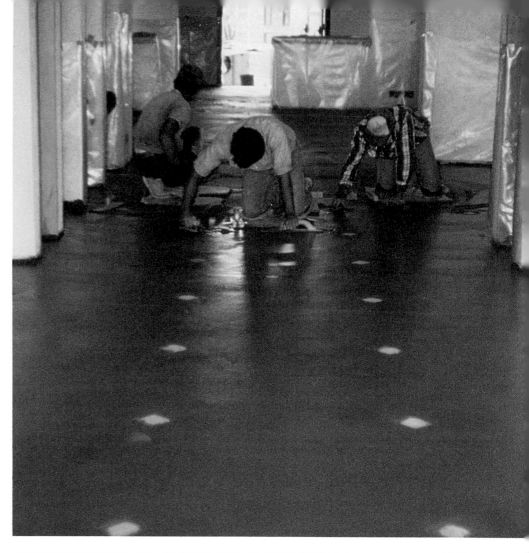

Solar orientation and thermal mass

Type of building	Heat load (therms per year)	Fuel cost (first year)	Value of future heat savings (current dollars)
Non-solar Average orientation, insulation, infiltration	630	$397	-
Solar without mass Proper orientation, insulation, weatherstripping	434	273	$2,416
Solar with wall mass only 1 in. stucco over 2,258 ft.²	334	210	3,648
Solar with floor mass only 3 in. concrete over 1,500 ft.²	248	156	4,712
Solar with floor and wall mass (Lee-Carmichael house)	205	129	5,238

This chart compares savings achieved with solar orientation and added thermal mass in a house heated by a natural gas-fired furnace working at 70% efficiency, burning fuel that costs 44¢ per therm (100,000 Btus) input. The Lee-Carmichael house, with both floor and wall mass, has a heat load of 205 therms per year. With solar orientation minus the added mass, the heat load would have been 434 therms. Projected over 30 years (adjusted for inflation), the added mass provides a $2,822 savings in fuel costs. Analysis source: Passive Solar Design Handbook (USDOE, Jan. 1980).

Installation costs for typical floor and wall finishes

	Finish option	Cost of base	Cost of finish	Total cost	Total cost, adjusted for future savings Moderate climate (Calif.)	Severe climate (Wis.)
Slab on grade 4½-in. slab, 6-in. by 6-in. welded wire mesh over 2 in. sand, over membrane, over 6-in. rock. Edge insulated.	Pre-finished ½-in. hardwood flooring	$3.97	$4.60	$8.57	$8.57	$8.57
	Mortar-set tile	3.97	4.58	8.55	7.02	6.21
	Mastic-set tile	3.97	4.25	8.22	6.69	5.88
	Carpet	3.97	3.33	7.30	7.30	7.30
	1½-in. integral color topping and tooling	3.97	1.55	5.52	3.99	3.18
	Powder-dusted color and tooling	3.97	.81	4.78	3.25	2.44
	Integral color throughout and tooling	3.97	.39	4.36	2.83	2.02
Framed floors ⅝-in. CDX plywood machine-nailed to 2x8 joists (increased to 2x10 for 3-in. concrete topping), includes blocking.	Pre-finished ¾-in. hardwood flooring	2.15	4.60	6.75	6.75	6.75
	Mortar-set tile	2.15	4.58	6.73	5.93	5.51
	Mastic-set tile	2.15	4.25	6.40	5.98	5.76
	Carpet	2.15	3.33	5.48	5.48	6.48
	3-in. integral color topping	2.41	2.13	4.54	3.01	2.20
	3-in. topping with dusted-on color	2.41	2.61	5.02	3.49	2.68
Framed walls 2x6s 16 in. o.c. includes blocking and plates.	½-in. gypboard taped and textured; two coats of paint	1.37	1.55	2.92	2.92	2.92
	⅜-in. premium redwood paneling over ½-in. gypboard	1.37	2.68	4.05	4.05	4.05
	⅛-in. stucco over lath, integral coloring	1.37	3.70	5.07	4.52	4.48

This chart compares the 1982 installation costs per square foot (San Francisco Bay area) for typical grades of floor and wall finish treatments. Unlike wood flooring, carpet or gypboard paneling, thin-mass finishes help to reduce the heating load, lowering total installation costs. Savings are even greater in a more severe climate, such as Madison, Wis.

tially reduces the need for supplementary heat (top chart), but the incremental worth of each additional pound decreases. This is the law of diminishing returns.

The value of increased comfort levels caused by the all-surrounding nature of the mass is more difficult to measure. A remarkable sense of temperature uniformity results, and heat loads can be so low that it becomes practical to use a woodstove for backup heat.

Costs—Although cost is only one criterion in the choice of a building material, colored concrete is invariably an economical option for a finish floor—especially when future savings on heating bills are subtracted from installation expenses (chart above). Over a slab on grade or a wood platform, a concrete finish can cost less than half the price of a carpeted floor.

Concrete is also less subject to chipping than is ceramic tile. It doesn't require refinishing like hardwood or endless maintenance and eventual replacement like carpet.

The economics of interior stucco are not as compelling. Gypsum wallboard remains more economical, even when the present value of the heat savings is deducted from the cost of installing the stucco. However, if alternative wall finishes (such as plywood paneling) are considered, the stucco is only slightly more expensive, and it will never require painting, varnishing or restaining.

Acoustics and solidity—A thin-mass building is remarkably quiet. This comes as a surprise to some people because they assume the hard surfaces will create an echo chamber. Yet, in terms of reflected sound waves, this building is no different from one with wood floors and gypboard walls. Assuming that each type of room contains soft elements, such as furniture, curtains and occasional area rugs, noises will bounce around in a similar manner.

So where does the sense of unusual quiet come from? The key is the ability of the mass to damp out impact noise almost totally. Gypboard will ring like a drum when struck; the interior stucco will give off a gentle thud. Similarly, footsteps upstairs will transmit readily through a frame floor with a wood or carpet finish. A 3-in. slab upstairs will absorb such impact sounds, creating an effective noise barrier between floors.

Along with the improved acoustics, there is a comfortable sense of solidity. The 3-in. slab on the wood frame creates a much stiffer assembly, something you can sense as you walk over it. This is equally true of the stucco walls; they don't give so much when struck. One of the most frequently voiced complaints about current frame construction is that it's cheap and flimsy. Thin-mass buildings make the opposite impressions. The problem seems to be with the finish materials rather than the wood frame.

Retrofits—Can the benefits of thin-mass construction be built into an older home? Probably not. Most existing homes don't have sufficient south-facing glass for thermal storage to make sense, and few existing homes are insulated and weatherstripped well enough to make the added mass worthwhile. Conservation dollars should first be spent on insulation, weatherstripping, caulking and double glazing. There can also be potential structural difficulties. A 3-in. slab is heavy, and shouldn't be added to wood framing without increasing the number of joists or intermediate beams. Problems can also crop up in earthquake-prone areas if the original structure doesn't have adequate shear bracing. Wall outlets may wind up closer to the floor than the code requires, and changes in elevations at stairs become awkward.

A glance ahead—The concepts we're proposing are best suited for new structures and additions to existing buildings. Economically, thin-mass construction makes sense (especially for floors), and the thin masonry sheath improves acoustics and makes for a feeling of solidity. Finally, the thin-mass technique opens up new aesthetic possibilities and challenges us to introduce vigorous new color and texture into our buildings. □

Max Jacobson is a principal architect in the firm Jacobson, Silverstein and Winslow, Berkeley, Calif.

Bricklaying Basics

Tools, materials and techniques for building a brick wall

by Dick Kreh

The art of laying bricks in mortar to form a wall dates back thousands of years, and the tools haven't changed much over the centuries. Whether your project is simple or complex, laying bricks can be one of the most creative and satisfying of all crafts. Bricklaying is not complicated; the key to good work is consistency, accuracy and repetition. This requires strict attention to details, such as the placement of your fingers when holding the trowel. With practice you develop a feel for the work.

You probably already have many of the tools you'll need—a steel measuring tape, a metal square, a ball of nylon line, a brush, a chalkbox and safety goggles or glasses. More specialized bricklaying equipment is described in the sidebar on pp. 106-107. For mixing mortar, you will need a wheelbarrow, a mixing box (for large batches), a water bucket, a 24-in. sq. mortarboard or pan, a mortar hoe (like a garden hoe, but it has two holes in the blade) and a hose with a spray nozzle. If a lot of mortar is to be mixed, utility drum mixers can be rented.

Mortar—A wall can be no stronger than the mortar it is built with. If the mortar is too weak, the wall will fail. If the mortar is too rich, it will be sticky and hard to handle with the trowel, and will be brittle when it hardens.

There are several kinds of mortar, but they are basically all composed of portland cement, lime, sand and certain additives that make them more workable. The simplest to use is prepackaged

Dick Kreh is an author, masonry consultant and building-trades teacher. He makes his home in Frederick, Md.

mortar, which contains all the dry ingredients in the correct proportion; you add the water. Prepackaged mortar is fine for small jobs, but on a project of any size it is expensive. Mortar is mostly sand, and sand is inexpensive.

The most popular mortar for general use is made with masonry cement. It is sold in 70-lb. bags (with lime and certain additives included), and you add the sand and water when mixing. A good proportion to mix is 1 shovelful (part) of masonry cement to 3 shovelfuls (parts) of sand, with enough water to blend it to the desired stiffness. If you mix an entire bag at once (called a batch), use 1 bag of masonry cement to 18 shovelfuls of sand. Masonry-cement mortar is the most economical and has excellent handling properties.

With portland cement/lime mortar, you can adjust the strength of the mix by varying the ingredients. Building-supply stores sell portland cement in 94-lb. bags. The lime comes in 50-lb. bags and must be labeled "Hydrated" or "Mason's Lime," which means that it has been treated with water. This is stamped on the bag. Other types of lime are used for agricultural purposes.

The portland cement/lime mortar mix I use is 1 part portland cement to 1 part hydrated lime to 6 parts sand and water. This is known as Type N mortar and is the standard mix for brickwork. After curing 28 days, this mix will sustain pressures of 750 lb. per sq. in., which is more than ample for most brick masonry.

A word of caution about sand. Buy your sand from a regular building supplier and ask if it is washed. This is important, because with washed sand most of the loam and silt will have been removed. Silt or loam in the sand will prevent

the mortar mix from blending together properly; the resulting mortar will be weak and defective.

Store cementitious materials in a dry place and off the ground so moisture won't get in. If moisture has penetrated into the dry cement and hard lumps have formed, throw it away. If your project is large, it is better to make several trips to the supply house for cement than to keep a lot of it for six months or a year—mortar stored too long tends to lose its strength.

When measuring, proportion the materials carefully, and mix all the dry materials together well before adding clean water. Don't mix any more mortar than you can use in an hour. If the mortar starts to stiffen in the wheelbarrow, you can add enough water to temper (loosen) it, once—this should be only enough water to make it workable, not runny or thin.

Estimating materials—Over the years I have developed some rules of thumb for figuring how many bricks and how much cement, sand and lime to order. These may be of some help to you when you estimate your project.

Figure on using seven standard bricks (a standard brick is considered to be 8 in. long, 3½ in. wide and 2¼ in. high) for every single thickness of 1 sq. ft. of wall area. This allows a small amount for waste, so don't add any more to it for cracked or broken bricks.

One bag of masonry-cement mortar will lay about 125 bricks. One bag of portland cement and one bag of mason's hydrated lime mixed with 42 shovels of sand will lay about 300 bricks. One ton of sand will make enough mortar to lay 1,000 bricks. If you store it on the ground, allow about 10% for waste.

Photos, except where noted: Dick Kreh and Lefty Kreh

Layout—Since bricks vary somewhat in length, you can't just step off the bond (the pattern for layout that will be followed throughout the wall) with a tape measure. You have to lay out the first course brick by brick to a chalkline. This is done most effectively by spacing the bricks ⅜ in. apart (the width of the head joint, or space between the ends of adjacent bricks) without using mortar—this is known as *dry bonding*. A quick way to lay out the head joints is to place the end of your little finger between the bricks. When the first course is aligned, if there is a small gap at the end, either open up all the joints or place a cut brick in the center of the wall or under a door or window. Try to have as few cut bricks as possible. Be sure to lay the face side of the brick out to the front of the wall. (The face side is usually the straightest side, and it always matches the ends of the brick in color.) Leave the bricks in position until the mortar is ready to spread. Then pick them up only two at a time so you don't lose the pattern or spacing. Only the first course is laid out dry. Subsequent courses will be laid following the pattern established by the first course.

Cutting a brick—Bricks can be cut by using either a brick set chisel or a brick hammer. Always check each brick to be cut. If it has a crack in it, discard it because it won't break true. To make an accurate cut, first mark the edge to be cut, and then place the flat side of the blade facing the finished cut. This will ensure a neat, accurate cut. Be careful not to place the fleshy part of your hand against the head of the chisel—the hammer knows no mercy if misdirected. It's a good idea to wear a pair of gloves to prevent accidentally cutting or pinching your fingers.

Most novices have trouble cutting bricks with the brick hammer, but as with so many things, practice makes perfect. Mark the cut with a pencil first, then score across the face of the brick with light pecking blows. Don't try to break the brick yet; the object here is to weaken the brick along the line.

Next, repeat the scoring of the brick across its widest side. It should break cleanly. If not, return to the face side, and repeat the scoring until the brick breaks at that point. Now trim the cut edge with the end of the blade to remove any protruding edges.

Troweling techniques—Cutting and spreading mortar are important skills to master. Of the many different methods used, I feel the easiest is the "cupping" method. It is accomplished in a series of steps or movements.

Start by holding the trowel with the thumb just a little over the end of the handle, near the blade (photo top left). This will provide you with good leverage and balance. It also keeps you from dipping your thumb in the mortar. Your fingers should be wrapped firmly around the handle, but not gripping tightly.

Cut the mortar away from the main pile with a downward slicing motion, pulling it toward the edge of the board (photo top left). Shape the mortar on both sides with the blade of the trowel until it is about the same width and length as

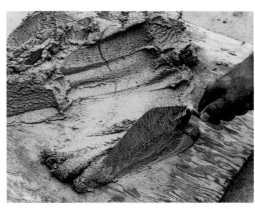

Troweling the mortar. **With a downward slicing motion (top left), cut some mortar away from the pile in the middle of the mortar board, then push it into a rough box shape the size of the trowel blade (top right). Now slide the trowel under the mortar so that it fills the blade evenly (middle photo), snap your wrist down to set the mortar on the blade, and spread the first bed joint (above).**

the trowel blade. One or two motions will do this (photo top right). Then, with a smooth forward motion, slide the trowel under the mortar (middle photo above). Lift the trowel, and at the same time, snap your wrist slightly down to set the mortar on the blade. This keeps the mortar on the blade for the spreading operation.

The mortar is now ready to spread to form the first bed joint (the joint between courses). Move the arm in a sweeping, spreading motion,

keeping the point of the trowel in a fairly straight line and turning the blade sideways at the same time. This is done with a flowing motion for best results. The mortar should roll off of the trowel blade, following the point, in a straight line (photo above). With practice, you will get the same depth of mortar, ⅜ in., for the entire spread. Practice until you can spread about 16 in. at one time.

Mortar spread for the first course of brick on

From *Fine Homebuilding* magazine (December 1987) 43:50-55

Maintaining plumb and level. Check the first brick for vertical alignment with the modular scale rule (top), then level its top (middle photo) and plumb its face (above). Repeating this procedure at the opposite end of the wall gives two points to pull a line from.

A line block hooked over one end of the wall is held in position by a line block at the other end. A dry brick set on top of the line sets it in the plane of the face of the wall.

the base or footing should not be furrowed (indented) with the trowel. It should be solid to prevent water from leaking through.

The first course—Establishing and maintaining plumb and level are critical in any brickwork. Beginning at one end of the wall, lay the first brick, pressing it into the mortar bed with your hand as level and plumb as possible, by eye. Then check with the modular scale rule to see that the top of the brick aligns with 6 on the scale side of the rule (photo top left). Now level the brick from the measured end until the bubble is between the lines on the vial (photo second from top). Then plumb the face of the brick, being careful not to move it out of alignment with the layout line (photo third from top).

Repeat the procedure at the opposite end of the wall so that you have two points to pull a line from. Do this by attaching the line and block over the end of the brick on one end and pulling the line tightly over the end of the brick on the opposite end. Wrap the line around the brick so it won't slip loose. Block (push) the line out to the exterior face of the brick by laying another dry brick on top of it, as shown in the photo bottom left.

With the line in place, pick up two bricks at a time from the layout and spread mortar there. Butter a mortar head joint on one end of each brick as it is laid. This is done by picking up some mortar on the point of the trowel, setting it with a jerk of the wrist on the trowel, and then swiping it on the end of the brick. With practice, one swipe will do the job.

As the bricks are laid in the mortar bed joint, be sure to keep them about ⅛ in. back from the line. This will prevent the wall from being pushed out of alignment by bricks riding against the line. If you lay the bricks too far back from the line, the wall will bow in the other direction. Bricks against the line are called "hard" and bricks back too far are called "slack," and neither is acceptable. After a little practice, you will get the hang of this.

Lay the bricks from each end of the wall to the center. If the spacing doesn't work out perfectly, either open up the head joints a little or you may have to cut a brick. The general rule for cuts in the center of the wall is not to make bricks smaller than 6 in. Sometimes you need to cut two bricks to make the wall look right. Locating the cut brick under a window or door makes it less conspicuous.

If a brick wall is going to leak, it will usually be at the spot where the last brick was laid. Therefore, to prevent this, butter each end of the bricks already in place and each end of the brick to be laid. This may seem time-consuming, but will pay off by preventing any future problems. This last brick is known in the trade as the "closure" brick.

Building the corner or lead—Once the first course has been laid out in mortar, the ends or corners (also called leads) should be built up about nine courses first, instead of trying to lay each course individually. Build the ends of the wall to the desired height, and then the line can be attached with line blocks and filled in be-

Tools and equipment

You won't need much equipment to lay bricks, so it makes sense to buy brand-name, good-quality tools, because they are better balanced and are a pleasure to handle. Bricklaying tools can be obtained at good hardware stores, building-materials suppliers or from mail-order catalogs. One tool company that I have dealt with over the years with complete satisfaction is Masonry Specialty Co. (4430 Gibsonia Rd., Gibsonia, Pa. 15044). Two others are Goldblatt Tool Co. (511 Osage, P.O. Box 2334, Kansas City, Kan. 66110) and Marshalltown Trowel Co. (Box 738, Marshalltown, Iowa 50158). When you need costly equipment or tools, the best solution is to rent them locally. Check the Yellow Pages in your phone book under Rental Service Stores and Yards.

I consider the following tools essential for brick masonry work.

Brick trowels—Trowels for applying mortar to bricks range in length from 10 in. to 12 in. Brick trowels 11 in. long work best. I like a narrow blade, which is called a narrow London pattern, with either a wood or plastic handle. A flexible blade is a must for spreading and applying mortar joints on the edges of the bricks. You can test a blade's flexibility by bending the point against a flat surface. It should flex about 1 in. to be effective. Two very good trowels are made by W. Rose Inc. (P.O. Box 66, Sharon Hill, Pa. 19079) and Marshalltown. However, you may want to pick your trowel out in the hardware store in order to test the flex of the blade.

Brick hammers—Brick hammers are used mainly for cutting bricks. Select a well-balanced brick hammer that weighs about 18 oz. Holding it by the handle, see if it feels comfortable and balanced. The handle may be wood, metal or fiberglass. I like a wood handle. There are many good brick hammers on the market. A few brands that I recommend are Stanley, True Temper and Estwing.

Levels—Of all the tools that you will need, the level is the most delicate and probably the most expensive. It is used to keep your work level and plumb. Levels may be metal, fiberglass or wood. I prefer a wooden level with alcohol bubbles because changes in temperature will not cause the bubble to shrink or expand. Levels come in various lengths. If you are going to get only one, buy a 48-in. level, which fits most situations. If you are going to buy two, also buy a 24-in. level. Any brand name will do, but be sure that the level is true before leaving the store. Do this by holding it in a plumb position alongside another level. The bubbles should read the same, top and bottom. Reverse the edges and check again. Now try the test holding the levels horizontal. Two good levels I like are the American (Macklanburg-Duncan, P.O. Box 25188, Oklahoma City, Okla. 73125) and Exact I Beam (Hyde Manufacturing Co., 54 Eastford Rd., Southbridge, Mass. 01550).

Tool photo, facing page: Michele Russell

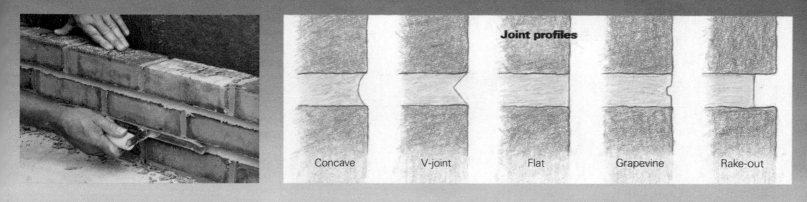

Joint profiles

Concave | V-joint | Flat | Grapevine | Rake-out

Chisels—Chisels are useful for cutting bricks to length, and for removing hardened mortar. For most work, two basic chisels will suffice: a flat broad chisel with about a 3½-in. blade (called a brick set) and a standard mason's cutting chisel with a 1½-in. blade and a long narrow handle.

Jointers—Jointers, or strikers, are used to form, seal or finish the surface of mortar joints on brick, block or stone work. They can add a sense of design, depth or texture to the wall. Jointers should be made of good-quality steel and be reasonably priced. The sled-runner one I use costs about $8.

Jointers for brickwork come in five common shapes (drawing, above right). Convex jointers, the most popular, form a half-rounded, indented profile (photo above left). V-jointers give an angled indented finish that looks great on rough-textured brick or block work. Slickers form a flat, smooth finish and are used for paving steps or repointing old mortar joints. The grapevine profile has a raised bead of steel on its edge that forms a wavy line when passed through the center of the mortar joint. It is very popular for early American brickwork. The rake-out forms a neat raked mortar joint that looks good with sand-finish or rough-textured bricks. You can get one mounted on two skate wheels to keep you from skinning your knuckles. The raking is accomplished by a nail in the center; the depth of the rake is adjusted by a thumbscrew. Oiling around the wheels will keep the tool in good condition.

Mason's rules—Two types of masonry scale rules can be used for brickwork: the modular rule and the course counter rule. The modular rule is based on a 4-in. grid (the basic module for manufactured construction materials). On one side of the rule is the standard 72 in.; the other side has scales for bricks of various heights. The standard mortar joint of ⅜ in. is used for all sizes of bricks. You will probably use scale #6 the most, which means that six courses of standard bricks, including the mortar bed joints, will equal 16 in. in height. Scale #2 is for concrete blocks (8-in. increments). The other scales are used less often. The course counter rule is designed for all standard-height bricks but allows the user to vary the thickness of the mortar bed joints. For most work, the modular rule will suffice.

Line blocks and pins—Line blocks are used to attach a line to the corners of a wall as a guide for laying bricks between them. They are paired wood or plastic blocks with a slot scored on one side and an end for passing the line through. They are held on the wall by the tension of the line pulled tightly between the corners. Building-supply houses often give them away as promotions.

Like line blocks, steel line pins can also hold the string line. They are driven in the mortar head joints, and then the line is wrapped around and pulled tightly from one end of the wall to the other. Line pins are also available on request from most building-supply houses. —*D. K.*

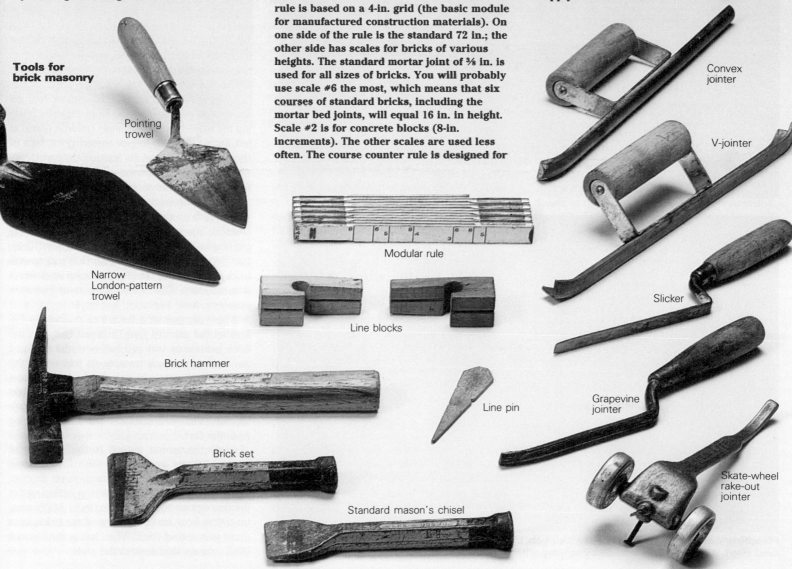

Tools for brick masonry

Pointing trowel

Narrow London-pattern trowel

Brick hammer

Brick set

Standard mason's chisel

Modular rule

Line blocks

Line pin

Convex jointer

V-jointer

Slicker

Grapevine jointer

Skate-wheel rake-out jointer

Putting Down a Brick Floor
The mason's craft is easier when the bricks are laid on a horizontal surface

by Bob Syvanen

Brick floors are experiencing a revival. The material used only for patios or walkways 10 years ago is moving inside the house. There are good reasons for this. The color, texture and pattern that brick brings to a room can't be duplicated with vinyl or wood. And brick is a logical choice for a passive-solar building, because it increases the thermal mass while offering a more finished look than a concrete slab.

Laying a brick floor sounds risky if you're not skilled with a trowel and a brick hammer. But after watching master mason John Hilley lay the floor in a house I'm building in Brewster, Mass., I'm convinced that it's not too difficult. As with any job, if you know the tricks, the battle is half over.

A brick floor laid with mortar is different from a wall or a patio. First, you can ignore plumb and just concentrate on level. Second, you don't have to contend with the shifting, subsiding backyard quagmire that is the substrate for most patios. A brick floor should be laid on a concrete slab, which is flat and solid; or on a wood floor that has been beefed up to take the extra weight of the bricks and stiffened so that its flexibility won't crack the mortar joints.

Brick—If you think that brick is brick, and that your only decision will be how many to buy, you'll change your mind when you hit your first masonry-supply yard. Bricks come in many styles, colors and prices. They run anywhere from 30¢ each to 50¢ each or more. If your floor extends out into the weather, use paver bricks. The surfaces of pavers are sealed, so the bricks won't spall when it freezes. If your floor is inside, anything that strikes your fancy will do.

Used bricks make a very nice floor but they're getting scarce, and as a result, expensive. You never know exactly how much waste you'll get when you scale the old mortar off used brick, but plan on buying at least 3,000 used bricks to get 2,000 usable ones. For new

4x8 bricks, you should figure 4½ bricks per square foot of floor and then add at least 5% for waste.

Since a standard brick weighs about 4½ lb. and you're going to need a lot of them, you'll want them delivered to a convenient place. Most yards bring the bricks on pallets to your job site, and if the delivery truck is equipped with a hydraulic arm, you're even better off. The arm lifts the wood pallets off the truck and sets them down anywhere you say. A skilled driver can just about put the bricks in your back pocket.

Use brick tongs for hauling the bricks from the pallets. Brick tongs are simple tubular-steel contraptions that will carry between 6 and 10 bricks at a time by holding them in compression. At about $16, using tongs beats weaving around the site with a pile of bricks stacked up against your forearm.

Layout—Although there are many patterns or bonds that bricks can be laid in, the *running bond* is still the easiest and one of the most attractive. The joints between the bricks in each course are offset from the joints in adjacent courses by a half brick. This means that each course begins with either a half brick or a whole brick. After sweeping the slab absolutely clean, determine which direction the brick courses will run. You can then begin the rough layout of the floor with a tape measure.

To avoid having to cut a course of narrow bricks at the end of the room, you may need to adjust the width of the joints. Laying out to a full course at the end of the room is time well spent. Do your figuring on paper by adding an ideal joint width, for example, ¼ in., to the width of your brick and dividing the total length of the room by this sum. Then confirm your calculation by *dry coursing*—laying a full row of bricks along the length of the room without mortar to test-fit the layout. Pick carefully for representative bricks, since they can differ considerably in length and width. Remember that thin joints look better than fat

ones when you're making adjustments between courses; and that you'll get another inch at each wall to play with because the wall finish and baseboards that will be installed later will cover that much more of the floor.

Once your dry coursing has been adjusted so that the joints are even, nail 1x or 2x layout boards to the wall studs along each side of the room so that their tops are even with the top of the finished brick floor (drawing, below). Mark the leading edge of each course on the board, and drive a nail into the top of the board at each of these marks for a string line. After laying a course, move the string forward one nail on each side of the room for the next course. This line represents both the finished height of the floor and the leading edge of the course, leaving very little for you to eyeball.

A layout board can also be cantilevered off the top of bricks that have already been laid. This setup works well when an exterior wall takes a jog, making the room narrower, and ending a run of layout boards. Course lines are marked on the end of the board that projects out to the unlaid part of the floor, and the course string is attached so it rides on the bottom edge of the board. This maintains the same finished floor height as layout boards held flush with the top of the bricks.

You should dry-lay a test course along the width of the room, too. You will be starting with a half brick or a whole brick on one end, and adjusting the joint width between the ends of the bricks to determine the length of the brick on the other end. It won't always work out to half and whole bricks, but the less cutting you have to do, the easier the job will be. Don't end a course with a very short brick.

For anything wider than a closet or a hallway, use *control bonds* to make sure that the bricks are being spaced uniformly. These are bricks whose ends are laid to a string as a reference every 6 ft. or so (about 10 bricks) within each course. This in effect breaks a long course up into several small courses.

The control bonds and end bricks are the

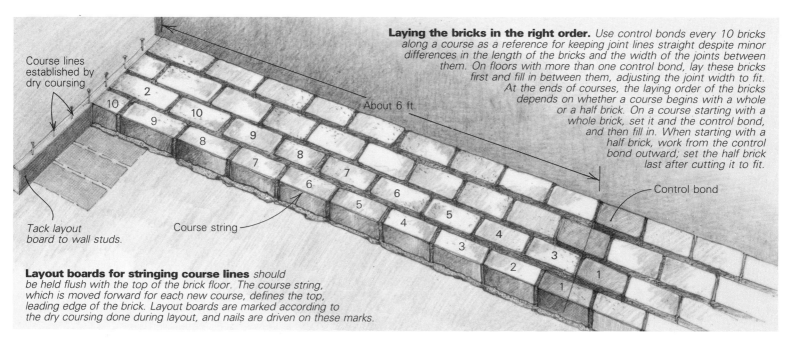

Laying the bricks in the right order. *Use control bonds every 10 bricks along a course as a reference for keeping joint lines straight despite minor differences in the length of the bricks and the width of the joints between them. On floors with more than one control bond, lay these bricks first and fill in between them, adjusting the joint width to fit. At the ends of courses, the laying order of the bricks depends on whether a course begins with a whole or a half brick. On a course starting with a whole brick, set it and the control bond, and then fill in. When starting with a half brick, work from the control bond outward; set the half brick last after cutting it to fit.*

Course lines established by dry coursing

About 6 ft.

Control bond

Tack layout board to wall studs.

Course string

Layout boards for stringing course lines *should be held flush with the top of the brick floor. The course string, which is moved forward for each new course, defines the top, leading edge of the brick. Layout boards are marked according to the dry coursing done during layout, and nails are driven on these marks.*

Photos: Bob Syvanen; Illustration: Frances Ashforth

first bricks to be laid in each course. The rest are filled in to fit. This way, the joints of every other course at the control bond will form a straight line, and the cut bricks at the end of courses, as well as the width of the mortar joints at the ends of the bricks, can be kept fairly consistent from course to course. Try to place control bonds in highly visible spots such as stairways and entries, and string them just as you did the course lines.

Stock the floor once you've completed your layout. This way you can stretch strings and get to know the peculiarities of the room and the slab you'll be working on before you begin littering it with bricks. Using brick tongs, distribute the bricks so that they will be within easy reach when you begin to work. Keep in mind that several layers of bricks on a pallet, or even the whole thing, can be a very different color from the rest of the load. Mix these colors and tones as you stock them on the floor so that your floor doesn't end up with big patches of only dark or light bricks.

Tools—A brick floor is laid with standard mason's tools (photo below). In addition to a 4-ft. level, a tape measure and nylon string, you'll need a brick hammer to break the brick to length at the end of a course. It has a square, flat head and a long, flat chisel peen on the other end, and is made of tempered steel. Brick hammers come in various weights. They have either steel, fiberglass or wooden handles, and cost about $15.

Brick trowels also cost about $15, and are made with wood or plastic grips. There are two basic shapes: the London and the Philadelphia. Most brick masons prefer the London pattern, an elongated diamond shape with its heel farther forward on the trowel than the Philadelphia. London patterns come with either a narrow or a wide heel. The narrow heel is fine for brick since less mortar needs to be carried by the trowel than for stone or concrete blocks.

You'll need a jointer for smoothing and shaping the mortar between the bricks. This tool looks like an elongated steel S, and is gripped in the middle. Each end of the tool has a different profile.

Tools of the trade. Brick tongs, top, make stocking the floor with bricks much easier. The trowel, jointer and brick hammer below it are the basic tools used to lay the floor.

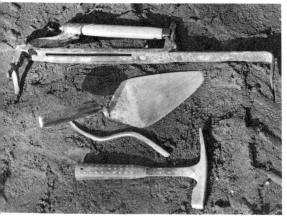

Mortar—The mortar between the bricks in your floor makes it permanent, and provides a visual relief from the brick itself. In this case, it is a mixture of masonry cement, sand and water. Use a shovel and buckets to proportion the ingredients for the mortar. Mix your mortar with a hoe in a mortar box or a wheelbarrow. An easier way to mix is with a mechanical cement mixer. Do your mixing outside where you are free to hose out your mixer at the end of the day, but keep the fresh mortar out of the hot sun.

The amount you'll need depends on two things: how much bedding is required for the slab you are working on, and the thickness of the joints between bricks. If your slab is flat and level, the bed of mortar under the bricks will be fairly uniform. A good bed is ⅜ in. thick, and no less than ¼ in. However, a serious hog, or hump, in the floor can double the amount of mortar, because you will need to bring the bricks for the rest of the floor up to this level with a much thicker bedding.

The width of the joints between bricks is the other factor that affects how much masonry cement and sand to order. Joints look bigger than they really are because bricks are molded and don't have hard crisp edges. A ¼-in. joint after finishing will look ⅜ in. wide, which is a nice size. Figure on one bag of masonry cement for every 100 bricks if your slab is uniform and your joints are ¼ in. wide. You will need 1½ cu. ft. to 2 cu. ft. of sand per bag of masonry cement.

The consistency of your mortar will have a lot to do with your success in laying bricks. Mortar should be firm enough to support a brick as bedding, yet soft enough to compress easily at the joints. A soupy mix will lay down, or self-level, in the wheelbarrow. A mix that's too stiff will support itself even when it's stirred into the shape of a breaking wave. A workable mix is the consistency of whipped cream, and the secret to mixing it that way is adding water to the dry ingredients in small amounts, and lots of practice.

A good mason mixes only the amount of mortar that can be made with one bag of masonry cement at one time. With this amount, the consistency is easy to control. A batch will last about three or four hours before setting up. Temper the mortar every 15 minutes by working it briskly with a hoe or shovel and adding a little water if necessary. The books say not to add water, but all masons do. Whether the mud is mixed with a mechanical mixer or in a tray with a mortar hoe, it's best loaded into a wheelbarrow because it is convenient to move around the work area, and it's easy to scoop out of.

Laying the bricks—You are ready to begin laying bricks when the mortar is mixed and the control-bond strings and the first-course string are stretched. For courses that begin with a full brick, lay the control bonds first, the full end brick second, and then fill in between, as shown in the drawing on the previous page. Start from the control bond and work outward. If the course starts with a half

brick, lay the control bond first, and then work from this brick out toward the ends, cutting the bricks on either end to fit. A little gain or loss that accrues as a result of the inconsistent length of the bricks can be offset by adjusting the size of the last brick.

You cut the last brick in a course with the mason's hammer. Hold the brick in your hand, and hit it sharply once or twice directly over your palm. This usually does it, but some bricks need a shot on both sides. Sometimes they break where you want, and other times you end up with a handful of brick shards. You'll get better with practice, but until you do, order enough extra brick so that each blow isn't critical. If the brick fractures at an angle, set it down on a hard surface and use small, chipping strokes with the hammer to straighten out the line of cut.

Another tool for cutting brick that requires a less practiced stroke is the brick set. This wide chisel is placed on the brick where you want it to fracture, and struck with a hammer (for more on breaking brick, see pp. 105, 117, 133). A brick set will cost you about $6. Plan to waste a few bricks using this tool as well.

You can tell a journeyman from an inexperienced mason just by watching his trowel hand as he picks up a load of mortar and places it. There is a familiarity with the material that is unmistakable. First, using the back of his trowel in the surface of the mortar, he will stroke away from his body. This creates a mound of mortar in the wheelbarrow. With the face of the trowel he scoops a load of mortar in an upward motion, then drops his trowel arm abruptly, ridding the trowel of excess mortar. What remains won't slide off the trowel even if it's turned upside down. The technique is stroke away, scoop up, and settle.

The first trowel of mortar should be placed on the floor for bedding. Thrown is more accurate, though this too takes a bit of practice. Then choose a brick. If your bricks are water struck on one face, this face should be laid up because its slightly glossy surface is less porous and will wear better. With brick styles that are water struck on all sides, or not at all, just choose the best face.

With the brick in one hand and the trowel in the other, pick up a thin line of mortar using the same stroke as before, and wipe it on the edge of the brick. As you get more experienced, it will almost look like you're throwing the mortar on. Unless you overload the brick, the mortar shouldn't slip off while you are handling it. Trial and error will tell you how much mortar to use. Finish buttering the brick by loading up the end that will butt the previously laid brick.

To set the brick in place correctly, all of its joints need to be in compression. This is one of the secrets of good brickwork, and the way to accomplish this is the *shove joint*. The brick should be held out from its ultimate resting place and square to it. As you begin to bed it down, the brick should sit slightly above the string line on the mound of mortar underneath it (photos facing page). Using your thumb, push the side of the brick toward the

From *Fine Homebuilding* magazine (April 1983) 14:32-35

The shove joint. Above, mason John Hilley beds a brick in a running bond. It is buttered on one side and one end with mortar, and will sit well above the course string on the bedding mortar until it is pressed and tapped down. This string indicates the top of the floor, and the leading edge of each course. It is moved ahead one nail on the layout boards at each side of the room after a course is completed. The brick is leveled to the string by a gentle tapping with the butt of the trowel (above right), while pressure is applied with the left hand. This compression is the key to a tight, permanent brick floor. The lines of mortar between bricks are jointed, right, when they reach the consistency of putty, using excess mortar on the surface of the brick. The mortar is fed into the joints and compressed with a jointer. This is done twice on each course to produce smooth and compact mortar joints.

previously laid bricks and use the butt of the trowel to tap the brick down until its face is just below the string. If the brick is too low anywhere, lift it out and load the bed with more mortar. Then press the brick into place again. Also use the trowel handle to tap the end of the brick until the width of the joint between it and the previously laid brick is correct. During all of this you should be pushing the side of the brick with your thumb. This pressure keeps the brick from settling unevenly, and keeps the joints in compression. Don't worry if the mortar doesn't rise all the way to the surface in every spot along the joints. The holes will get filled in later.

Scrape away any excess mortar with the edge of your trowel. This mortar can be allowed to dry a little and be used for jointing. Try not to smear the mortar on the surface of the bricks, but if you do, don't worry. It can be washed off later after it sets up.

Jointing—The brick joints are ready to be filled and jointed when the mortar between the bricks has the consistency of putty. Test it by pressing the mortar with your finger. You can use the scrapings on the surface of the brick to fill the joints if they hold together when you squeeze them in your hand. They shouldn't be soft like fresh mortar, or they won't compress when jointed. Dried or crumbly mortar shouldn't be used either.

The joints should be filled and compressed in two stages. For the first fill, feed the mortar into the joint and press it with the jointer (photo above right). Work your way along the course, filling and pressing. Then go back to the beginning of the course for a second fill and a final jointing. Jointing not only increases the strength and durability of the mortar joints, but it also gives the floor a smoother, more uniform appearance. Do the final jointing with smooth, level strokes. Use your whole arm, not just your wrist.

Cleanup—Let the mortar set for a week before cleaning. You can wait longer, but the brick will get harder to clean. First, scrape off heavy spots of mortar with the chisel peen of a mason's hammer. To remove any mortar smeared on the surface of the bricks, you will need to use a 20% solution of muriatic acid. Mix the acid in a bucket by adding one part acid to four parts water. Always add the acid to the water, and make sure that you are wearing eye protection and heavy-duty, acid-resistant rubber gloves. You will also need running water from a garden hose close by. Open as many doors and windows as you can for ventilation. Muriatic acid is nasty stuff—treat it with respect and caution.

The acid will soften the mortar spots, enabling you to scrub them off the surface of the bricks. For this you will need a short, stiff-bristled brush for hand-scrubbing, and a 9-in. stiff-bristled brush on a long handle. Before you begin, flood the whole floor with water until the bricks are saturated. This keeps the acid action on the surface where it can be controlled, and then washed off as soon as it

has penetrated the mortar smears. The bricks will look shiny when they are saturated, dull when they are beginning to dry out. Keep the bricks saturated during the entire process so the acid won't burn the mortar. Get a helper to operate the water hose.

As soon as you mix the acid solution, test its action in some hidden spot—under the stairs or behind the chimney. The joints will have a soapy appearance as long as muriatic acid is present. It's best to do this job systematically and work small areas. Dip the brush into the acid solution and then begin scrubbing the saturated bricks. As soon as you have covered a few square feet, hose it down until all the foaming stops. Then hose it a bit more, saturate the next area, and move on with the bucket and brush. After the entire floor has been scrubbed and rinsed, hose it down one more time. If any muriatic acid remains, it will continue to break down the mortar.

Brick floors should be sealed, but not until all the moisture has left the bricks and the mortar joints. Wait at least a month. If you get impatient and seal a brick floor with moisture in it, the sealer won't bond. The best product I've used for sealing brick is Hydrozo Water Repellent #7 (Hydrozo Coatings Co, Box 80879, Lincoln, Neb. 68501). Five gallons will do the average floor and will cost a little over $80. Put on two coats, letting each one dry for 24 hours. I follow this with a coating made of equal parts of turpentine and Valspar polyurethane varnish (The Valspar Corp., 1101 South Third, Minneapolis, Minn. 55440). □

Quarter-step running bond

Third-step running bond

Brick Floors

How a New Mexico pro lays a brick floor without mortar or string lines

by Douglas Ring

Brick floors are common in adobe houses here in New Mexico. The deep-red earth tones typical of bricks and their slightly erratic dimensions fit right in with adobe houses. But these aren't the only structures in which I lay brick floors. Some clients want them because they make good heat sinks in passive-solar homes or good radiators in buildings with buried hot-water radiant-heating systems. Still other clients want them because of their looks. After all, one of the most popular sheet-vinyl patterns imitates the look of a brick floor, and the installed cost for the two materials is roughly the same—about $2 a square foot. I do point out to my clients that the two materials wear at different rates. A vinyl floor will probably need replacing in 20 years. But after 20 years the corners which stood a bit high in a new brick floor will be nicely rounded, and if it's been properly finished it will have a patina like a well-used banister. In heavy traffic areas maybe ⅟₆₄ in. will be worn away, leaving another 14 centuries of wear in those places—give or take a century or two.

Choosing the bricks—The phone book will tell you where bricks can be had in your area, and I strongly suggest that you visit each yard to assess the quality and price of the available bricks. Check the bricks for chipped corners and variations in dimensions. Up to 5% chipped corners is typical. If their dimensions fluctuate by more than ³⁄₁₆ in., you'll be spending an inordinate amount of time fitting them together.

We are fortunate to have a local factory (Kinney Brick Co., 100 Prosperity Ave. S.E., Albuquerque, N. Mex. 87102) that makes an inexpensive ($.17 each) low-fire brick suitable for interior use. Most of the brick manufacturers make their pavers (bricks without holes) for exterior use, which means the bricks have to be fired at higher temperatures. This makes them resistant to water so they won't spall when they are exposed to a rain followed by a freeze. Not surprisingly, high-fire bricks are more expensive than their low-fire counterparts because it takes

Douglas Ring is a licensed contractor in Albuquerque, N. Mex.

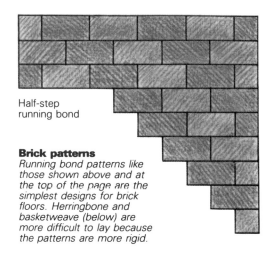

Half-step running bond

Brick patterns
Running bond patterns like those shown above and at the top of the page are the simplest designs for brick floors. Herringbone and basketweave (below) are more difficult to lay because the patterns are more rigid.

Herringbone

Basketweave

more time and gas to cook them. Around here, you can count on spending $.30 to $.50 apiece for high-fire bricks.

The brick you choose should have a smooth surface (roughness is a maintenance problem) and a pleasing color. Dark bricks will absorb more energy from the sun when they are used for solar gain. My favorites are light red or orange, and I keep in mind that the color will darken a little when the bricks are sealed.

Your supplier can tell you how many bricks you will need if you can tell him how many square feet you want to cover. If you want to calculate this number yourself, figure that to cover 1 sq. ft. it takes 4.5 bricks that are 4 in. by 8 in., or 5.2 of the 3⅝-in. by 7⅝-in. bricks. I add 5% to the total to allow for waste, and slightly more for a herringbone pattern. These figures are for bricks that are laid tightly together, without mortar joints between them.

I don't use mortar in my floors for several reasons. Mortar is messy and time-consuming to install, and it will permanently stain porous bricks. It doesn't strengthen the floor, and the mortar joints are slightly lower than the bricks, which makes grooves that collect dirt.

Every brickyard I've ever dealt with delivers their products. The trick is to have them deposit the bricks as close to the job as possible. The bulk of the work involved in laying a brick floor is moving the bricks.

A sand setting bed—Most of the brick floors I lay are atop concrete slabs. On top of the slab I screed a layer of sand, which can be made more level than the slab usually is, as well as accommodate the slight irregularities in brick thickness. The bricks I use are 2¼ in. thick, and I allow about ½ in. of sand between the bricks and the slab as a setting bed. This makes the distance from finished floor level to the slab 2¾ in. Of course, the slab can be more than ½ in. below the bricks, but that means you'll have to move that much more sand around to level your setting bed.

If I'm setting a floor over raw earth, I either remove or compact any soil that has been disturbed, and I fill in any craters with sand. Con-

From *Fine Homebuilding* magazine (June 1986) 33:68-71

versely, if grade is too high I'll take out soil until I'm about 3 in. below finished floor height. Here in New Mexico, moisture coming through the floor isn't a big problem, so we don't have to worry about elaborate drainage systems and moisture barriers under the floor unless the house is cut into a hill. Elsewhere, I expect these precautions would be critical.

Before I start bringing in the sand, I determine the finish-floor height from adjacent floors, doorsills or other fixed landmarks. Where possible, I snap chalklines on walls around the perimeter of the room to indicate the level of the finished floor. Then I add enough sand to bring it to the appropriate level below this line. The line is a big help in making sure that the floor is nice and level where it meets the baseboard. For the middle of the room, I rely on a hand level.

Sand is cheap, but it is also heavy. If I need a lot of it, I have it delivered along with the bricks. If I need more sand I know that I can carry enough to cover 200 sq. ft., 1 in. deep, in my pickup. That equals about ¾ ton.

Screeds and tongs—The best time to lay a brick floor is after the walls of the house have been painted—it isn't easy to remove paint from bricks. Once I've got the sand distributed around the room, I screed it level with the help of metal shelf brackets. These brackets are the kind that are U-shaped in section, with slots to accept the shelf standards. I lay a pair of the brackets in the sand and adjust them to the desired elevation with a level. Then I drag a straight board across two of these level and parallel brackets (photo top). I usually screed about a 4-ft. to 5-ft. wide path across the room in the direction the rows of bricks will run. Screeding done, I remove the brackets from the sand. They occupy so little space that I don't have to add any sand to fill in the grooves. I just smooth them over.

For every 100 sq. ft. of floor you lay, you will have to move about a ton of bricks. The tool you should have for this is a pair of brick tongs. These are like ice tongs only instead of a sharpened point to pierce the ice, they have a metal plate to grip a short row of bricks (photo top). The time they will save you moving only 500 bricks (100 sq. ft) will pay for them. If you have a great distance to move the bricks, a flat wheelbarrow is useful. You can make one of these by replacing the bucket of a regular wheelbarrow with two pieces of plywood. The flat wheelbarrow makes it easier to use the brick tongs to load and unload it.

To shape the odd-size bricks at the end of the runs you will need a 4-in. brick chisel and a 4-lb. sledge with a short handle. To position the bricks you will need a rubber mallet.

Running bond pattern—For beginners, I recommend some variation of the running-bond pattern (drawings, top of facing page). But just because this is an easy pattern to lay down doesn't mean that it isn't attractive. I've been laying brick floors full time for six years now, and this is still the pattern I use most. My favorite is the running bond based on fourths. I call it quarter-step running bond. This means that the ends of the bricks are staggered by a distance

Ring uses metal shelf brackets as screeds to level the sand bed (left). He adjusts each one to the desired height, then uses them to support a straightedge dragged across their tops to level the sand. The brackets are then removed, and the sand is smoothed over. Note the brick tongs being used by a helper. They are well worth the expense if you've got a lot of bricks to carry about.

The herringbone floor shown below intersects the wall and the neighboring course of bricks at a 45° angle, which makes it a floor with a lot of meticulously cut bricks. To make the bevels for the starter course, Ring lops the corners off a batch of bricks and then uses the corners to fill in the triangular voids.

equal to one-quarter of their length. The effect of this arrangement is subtle. It creates a repeating zigzag pattern that I think is more complex and interesting than the standard running-bond pattern based on halves. It's also possible to introduce another geometric pattern into a field of running bond, as shown in the bottom photo.

Most brick floors are oriented so the long dimensions of the bricks are perpendicular to the longest walls in the room. This is to avoid the frustration of trying to make long rows of bricks parallel to the walls. Variations in the bricks make perfectly parallel rows hard to achieve.

Assuming you use the quarter-step pattern, you will need equal numbers of bricks cut into quarters and halves and three-quarters for starter bricks. One brick can provide quarter-length and three-quarter-length starters for a row. The cuts don't have to be perfect because a baseboard will cover up any cuts right next to the wall. If you don't have a baseboard, make sure that the top edge of the brick is a crisp cut (see the sidebar on the facing page). It doesn't matter what the cut looks like below the surface.

Begin laying a brick floor by placing the starter courses in the corner of a room on a screeded bed of sand, as shown below. These starter pieces are the beginning of a quarter-step running-bond floor. The bottom photo shows how you can highlight a geometric pattern in a field of bricks by using bricks of different colors.

Setting the bricks—Cut a variety of starter pieces and begin to set them on a screeded setting bed in the corner of a room (photo below left). Don't worry too much about getting the size of the pieces exact. This is a handmade floor, and I think it's a waste of time to try and make it look like printed linoleum. First put a whole brick in the corner and tap it gently with the rubber mallet straight down—about ⅛ in. to ¼ in. Then put a ¾-length brick next to it, followed by a ½-length brick and then a ¼-length brick. Then the sequence starts again. Do not tap the small pieces hard or they will sink too far. As you add more bricks to these rows, slide them straight down against their neighbors so that sand is not trapped between the bricks.

Keep adding rows of bricks, several rows at a time, screeding more sand when you run out of a flat place to lay them. When you finish laying all the bricks you can fit into the room, you will have two adjacent wall edges finished, and two walls with cuts still to be made. Do the cuts now. Waiting to do these cuts at the end of the entire job is like waiting until the end of dinner to eat your lima beans.

Aesthetics and adjustments—Before you tap each brick into place, make sure it doesn't have any unsightly nicks in it. If it does, turn it over or put it aside for cutting. You will have to use your judgment about what sort of defect is acceptable, as almost all bricks are a little flawed. I often save the worst ones for the parts of the kitchen that I know will be covered by cabinets.

As you lay the bricks, try to have a sense of what is level. Think of something like a calm ocean. If you try to level each brick with an instrument, the job will be tedious. If the sand is level, approximately the same amount of tapping will be sufficient for each brick. If you have to pound to get the brick even with its neighbor, chances are the floor is heading downhill. If this is happening, tuck some sand under the last brick with your fingers, fill any hole with a small handful of sand and continue. Have a level on hand but don't be a slave to it. Adobe building tradition in the Southwest encourages a handmade look, and a brick floor fits right into this sensibility. A more formal house wants a flatter floor, especially in the dining room. But how a brick fits with its neighbor is more important than whether or not one end of the room is ¼ in. higher than the other.

Not only can you tip a brick accidently, you can also tip them on purpose. You can use this to advantage to change levels from one room to another to compensate for misplaced door sills, or even to make ramps instead of steps.

The straightness of the rows of bricks is another matter for individual interpretation. Most bricklayers want to get out the string and follow the straight line to ensure parallel lines. This is okay, and you will certainly achieve straighter rows of bricks by using a string line. But this is not required, once you realize that straight does not necessarily mean better. Most brick floors curve a little because of variations in the bricks and the walls. It adds to the charm.

Often you can curve the lines of bricks to good advantage. I've done floors in rectangular

rooms where I started the bricks in one corner and went to the opposite corner with a gentle S-curve. Then I filled in the rest of the floor, maintaining the gentle curves. On a floor like this, keep the curves smooth and large. Tight curves make bigger spaces between the bricks.

There is no rule that you must use the same pattern throughout a single house. In the bottom photo on the previous page, you can see where the pattern changes from quarter-step running bond to herringbone. Basketweave and herringbone are traditional patterns for brick paving that are more difficult than the running bond because the bricks are locked together in two directions. While both basketweave and herringbone patterns arrange bricks perpendicular to one another, herringbone is particularly challenging because of all the 45° cuts involved at walls or other courses of bricks.

Finishing and maintenance—When the last bricks are in place, straighten any rows that you find offensively crooked by twisting a trowel in the cracks, or by replacing oversized bricks. This is also the time to lift any bricks that are too low. Using two trowels to pinch a brick from the sides, lift it out, and add a bit of sand to the bed to make it flush with its neighbors.

Once you've made the necessary adjustments, sweep fine sand into the cracks between the bricks—an average-sized room will take about three or four shovelfuls. This is an exciting part of the job. The sand filters into the cracks as though the floor were a giant hourglass, locking the bricks tightly in place.

The floor is now basically complete. Interior floors should be sealed to resist staining. Standard practice has been to coat brick floors with liquid plastics, but my experience with refinishing floors has brought me to the conclusion that it is best to seal the bricks with something that penetrates deeper than plastic. This way you walk on the brick itself, which is almost indestructible, instead of on a thin layer of plastic. I mix my own sealer for this purpose (Ring Brick Floor Sealer, available from Ring Brick Floors, 2631 Los Padillas S.W., Albuquerque, N. Mex. 87105). One gallon costs $15, plus $5 handling, plus UPS shipping, and complete instructions for use are on the can. The mixture consists of about 80% linseed oil, along with thinners to help it penetrate and some additives to help it dry. This concoction penetrates the bricks to a depth of about ⅛ in.

For normal maintenance, wash the floor with a mixture of water and vinegar—about ¼ cup vinegar per gallon of water. Don't use soap because it can leave a residue. Give it regular sweepings with a large dust mop sprayed with a conditioner like Velva-Sheen or Conquer-Dust (available at janitorial-supply stores). The dust-mop conditioner will keep the floor from looking dull but will not build up like wax. Liquid wax is a curse upon brick floors. It builds up and gets milky and dark. For a slightly higher shine, use Indian Sand Treewax (available at hardware stores). It is a brick-colored paste wax that will not turn milky or yellow. It won't build up because it is too hard to apply. Dust-mop maintenance is the same regardless of finish. □

The bevel on the tip of the brick chisel should be on the waste side of the cut, and the shank of the chisel should be tilted slightly away from the workpiece. In this position, it will create a clean edge and an undercut in the finished piece. Here Ring uses the removed portion, inverted and trimmed, to complete the angled face for a starter course in a herringbone floor.

Cutting bricks with a chisel

The fastest and most practical way to cut a large number of bricks is with a 4-lb. hammer and brick chisel. For what it would cost you to rent a diamond saw for one day you can waste about 100 bricks, and by the time you have cut 100 bricks you will be pretty good at making pieces. I make the larger pieces first so if I break the piece that I am making, I can make smaller pieces from the fragments.

If you are new to cutting bricks, start by making half-bricks. Put a brick on a small pile of sand so that it has a solid support. This will keep the brick from breaking from the bottom up because of a lumpy seat.

You don't have to measure for the cutline. Just place the chisel in the middle of the brick, perpendicular to its length, and whack it hard with the hammer. Watch what you are doing—mistakes can be painful. When you have mastered this (halves are easy), try one-third/two-thirds and one-quarter/three-quarters. The brick will break slightly away from the bevel of the chisel, making the piece on the flat side of the chisel more suitable for use in a floor. This undercut piece is less likely to have outcroppings protruding beyond the plane of the top edge. Since only the top edge will show, it doesn't matter what the rest of

the brick looks like as long as it doesn't stick out.

Bricks like to break across smaller cross sections and nearer to their middles, so rather than make a 45° cut all the way across a brick to start a herringbone course, I just take off a corner, as shown in the photos above. I use the triangular piece that I have

removed from the brick to complete my 45° bevel.

If the triangle comes out with excess brick on the underside, I trim it by chiseling it from the bottom, and I direct my blows so the pieces break off short of the top.

When you get to the end of a run of herringbone, you will often

need a long beveled piece to fill in the remaining gap (photo below). After much trial and error I have arrived at a sequence of cuts that will usually get me this piece.

First I cut the brick to near its finished length, which reduces its mass at the crucial area of the cut—the tip. I leave it about ⅛ in. long, because it often chips a bit at the corner. The second cut should go right to the tip and can be slanted back to start paring down the remainder. Too much angle here will lose you your point. The third cut trims away more brick. The fourth cut is on the finished line, with the flat of the chisel always toward the piece you want. The final cut is one sharp blow to cleave away the remaining fragment near the point.

For cuts the full length of a brick, don't hit the chisel hard enough to break the brick through the first time. Whack it on both ends with increasing firmness until it breaks. You can even hit it on the end to start it cracking the way you want. If ten blows have not opened it up, hit it harder. If the piece you want is the length of the brick and less than 1½ in. wide, it is better to use two short pieces. An extra joint in the floor or even an extra row of joints is "authentic."
—D. R.

To cut a brick with a long beveled edge, begin by shortening the brick to about ⅛ in. longer than the finished length. The second and third cuts remove the bulk of the remaining waste. The fourth cut makes a finished edge along the desired line, but because the cut is longer than the blade of the chisel, a fragment usually protrudes near the point. The fifth cut removes this protrusion, resulting in the beveled brick shown above.

Building a Fireplace
One mason's approach to framing, layout and bricklaying technique

by Bob Syvanen

I have been involved in building as a designer and carpenter for over 30 years, but building a fireplace has always been a mystery to me. I recently had the chance to clear up the mystery by observing, photographing and talking to my mason friend, John Hilley, as he built three fireplaces. I now understand more clearly than before what I should do as a carpenter and designer to prepare a job for the mason. I also know I can build a fireplace.

The job actually begins at ground level, with a footing (drawing, facing page). A block chimney base carries the hearth slab, upon which the firebox and its smoke chamber are built. The chimney goes up from there.

The importance of framing—As a carpenter, I've had to reframe for the mason too many times. This is usually because the architect or designer didn't realize how much space a fireplace and its chimney can take up, and how this can affect the framing around and above it. We'll be talking about a fireplace built against a wall, which is a pretty simple arrangement, but planning is still important.

Most parts of the country have building codes that specify certain framing details. In Massachusetts, where I live, code requires that all framing members around the fireplace and chimney be doubled, with 2 in. of airspace between the framing and the outside face of the masonry enclosing the flue.

The modified Rumford fireplaces that Hilley usually builds are my favorites because they don't smoke, they heat the room about as well as a fireplace can, and they look good. The firebox is 36 in. wide by 36 in. high, and the two front walls, or pilasters (returns) are 12 in. wide, for a total masonry width of 60 in. From the fourth course above the hearth, the rear wall of the firebox curves gently toward the throat. It's harder to lay up than a straight wall, but I think it looks a lot better. The back hearth is 20 in. deep and about 18 in. wide at the back—not in line with Count Rumford's proportions (see pp. 130-133), but the minimum allowed by the Massachusetts code.

To figure the full masonry depth, you have to add to the 20-in. back hearth 4 in. for the back-wall thickness, 4 in. for the concrete-block smoke-chamber bearing wall, and 4 in. for the concrete-block substructure wall, for a total of 32 in. Thirty-six inches is better, because it gives extra space for rubble fill between the back wall and the block. Using

these dimensions, the chimney base is 36 in. by 60 in. Add a front hearth depth of 24 in. (16 in. is minimum), and clearance of 2 in. on each side and rear, and you get a total floor opening that's 64 in. wide by 62 in. deep. In situations like this one, where the fireplace is on a flat wall and the chimney runs straight up, with no angles, the framing is simple—double the framing around the openings and leave 2 in. of clearance around the masonry.

To locate the flue opening in the floor above the fireplace, find the center of your layout and drop a plumb line. This determines the side-to-side placement of the flue. Its depth is determined by the depth of the firebox. The flue will sit directly over the smoke shelf, and is supported in part by the block and brick laid up behind the firebox's rear wall. The framing for the chimney depends on the flue size. An 8x12 flue requires a minimum 18x22 chimney (a 1-in. airspace all around, inside 4 in. of masonry). Once the ceiling opening is framed, you can establish the roof opening by dropping a plumb bob from the roof to the corners of the ceiling-joist opening.

Wood shrinkage is something you should take into account when you're framing around the hearth. I think the hearth looks and works best if it's flush with the finished floor. Since it is cantilevered out from the masonry core (see below), and isn't supported by the floor framing, shrinking joists and beams can leave it standing high and dry. I've seen fireplaces built in new houses where the 2x10 floor joists rested on 6x10 beams. The total shrinkage here could leave the hearth an inch above the finished floor. A better framing system is to hang the joists on the beams and thereby reduce the shrinkage 50%.

From footing to hearth—The fireplace really begins at the footing, which is usually a 12-in. thick concrete slab 12 in. larger all around than the chimney base, and resting on undisturbed soil. The footing for this fireplace, therefore, is 48 in. by 72 in. Between it and the concrete hearth slab is a base, usually of 8-in. concrete block if it is in the basement or crawl space. To make sure the hearth comes out at the level you want it, the height of this base has to be calculated to allow for the 4-in. thick reinforced-concrete hearth slab, the bed of mortar on top of it, and the finished hearth material—in this case, brick.

Before pouring the hearth slab, the opening

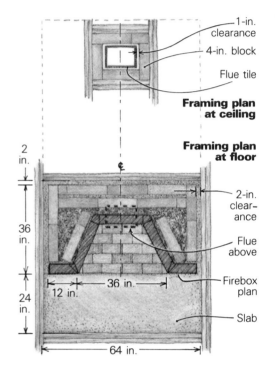

Framing plan at ceiling

Framing plan at floor

in the top of the concrete-block base is covered with a piece of ½-in. plywood that is supported by the inside edges of the blocks, leaving most of the course exposed for the slab to bear on. Cover the holes in the block with building paper or plastic, and build the formwork, secured to the floor joists, to support the cantilever at the front of the hearth. Then pour your 4-in. slab over a 12-in. grid of ⅜-in. rebar located 1 in. from the top.

Once the hearth slab has cured, it's time to lay up the structural masonry core that will support the chimney. Only the firebox, pilaster and lintel bricks will be visible on the finished chimney, so Hilley used 4-in. concrete block for the core. The blocks should be laid at least 4 in. from the face of the firebox brick and far enough in from the line of the front wall to allow for the pilaster bricks. Hilley sets a brick tie in each course to tie the pilasters in with the block.

Before beginning the brickwork, Hilley nails vertical guide boards (drawing, p. 120) to the face of the studs that frame the walls on each side of the fireplace opening, from floor to 12 in. above the lintel height. These boards are the thickness of the finished wall, and they locate the face of the fireplace. He marks off the brick courses up to three courses above the lintel on each guide board, starting from the

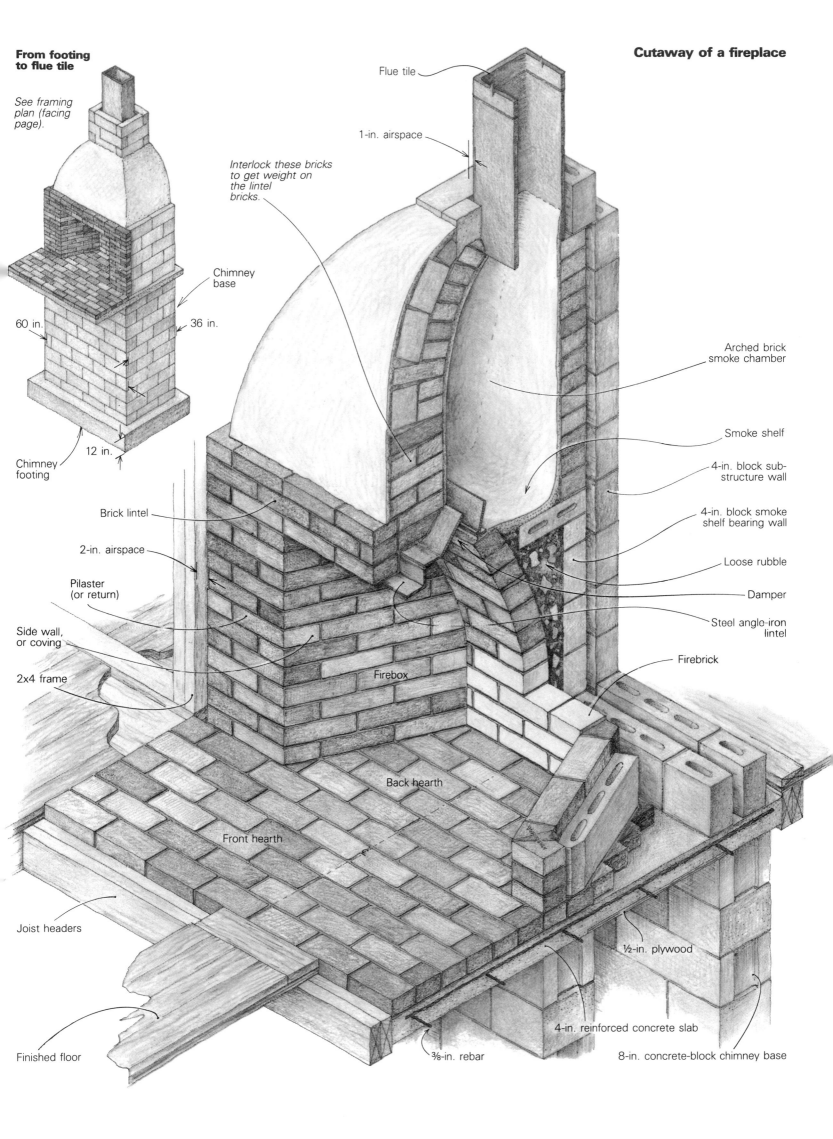

From footing to flue tile

See framing plan (facing page).

Chimney base

60 in.

36 in.

Chimney footing

12 in.

Cutaway of a fireplace

Flue tile

1-in. airspace

Interlock these bricks to get weight on the lintel bricks.

Arched brick smoke chamber

Smoke shelf

4-in. block sub-structure wall

4-in. block smoke shelf bearing wall

Loose rubble

Damper

Steel angle-iron lintel

Firebrick

Brick lintel

2-in. airspace

Pilaster (or return)

Side wall, or coving

2x4 frame

Firebox

Back hearth

Front hearth

Joist headers

Finished floor

½-in. plywood

4-in. reinforced concrete slab

⅜-in. rebar

8-in. concrete-block chimney base

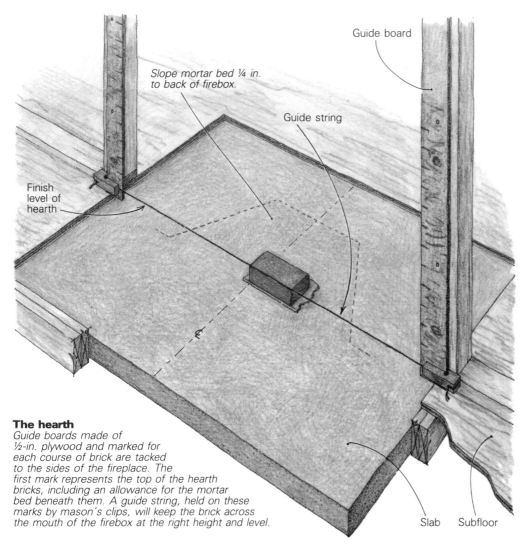

Slope mortar bed ¼ in. to back of firebox.

Guide board

Guide string

Finish level of hearth

The hearth
Guide boards made of ½-in. plywood and marked for each course of brick are tacked to the sides of the fireplace. The first mark represents the top of the hearth bricks, including an allowance for the mortar bed beneath them. A guide string, held on these marks by mason's clips, will keep the brick across the mouth of the firebox at the right height and level.

Slab   Subfloor

Firebox layout

Mitered corner
|← 9 in. →|← 9 in. →|
20 in.
|← 18 in. →|← 18 in. →|

Square corner
¢
½ in.

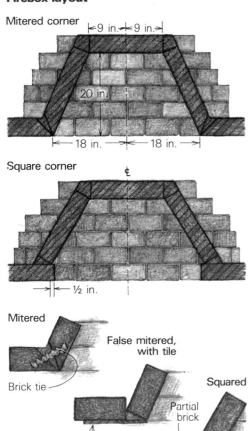

Mitered

False mitered, with tile

Brick tie

Squared

Partial brick

Tile face

Corner details

The firebox **is being laid to the penciled layout, starting with five courses of the back wall. Notice the curve starting at the fifth course of the back wall. The cut-brick piece for the front mitered corner will be alternated from front wall to sidewall on each course to maintain a strong bond. The V-shaped gap at the rear will be filled with rubble. Brick ties every couple of courses hold the joints together. The brick ties in the concrete block will secure the brick front wall (or return). The small torpedo level will be used to level the back wall.**

hearth, which on this job is 1 in. above the subfloor. Once the guide boards are marked, Hilley uses a guide string on mason's blocks to control the height and alignment of the brick courses as he lays them up.

Hilley picks sound, hard used brick for the firebox and hearth. The hearth is laid to the guide string in a good bed of mortar (drawing, top left). The firebox walls will be laid on this brickwork, so it extends beyond their eventual positions. Hilley likes to slope the hearth toward the back wall about ¼ in. to keep water from running into the room if any rain finds its way down the chimney. As with all brickwork, small joints look best, so pick your bricks for uniform thickness (see articles on pp. 110-113 and 114-117).

Laying out and building the firebox—With the hearth laid, Hilley finds the centerline of the opening, and marks off 2¼ bricks on each side for a 36-in. opening. Standard bricks are 8 in. long by 3¾ in. wide by 2¾ in. deep, but these measurements can vary, especially with used brick. Hilley uses bricks instead of a tape or ruler for an accurate layout, because 4½ used bricks (two times 2¼), laid end to end, don't always total exactly 36 in. The line of the back wall is 20 in. from the front line, and its length is figured by counting a little more than a brick on each side of the center line. Hilley pencils these lines on the brick hearth.

The lines for the diagonal sides of the firebox are drawn between the ends of the front and back lines. Where the side line meets the front line at the juncture of pilaster and firebox wall, you can draw either a mitered corner, or a square corner (drawing, bottom left). I like the look of the mitered corner, and I think the time it takes to cut the bricks is worth it. Cutting brick with a masonry blade in a skillsaw is easy when the brick is held securely between two cleats nailed to a plank. Both pieces of the cut brick are used, so cutting halfway through from each side is a better way to go.

One way to achieve a mitered look without cutting is to start a full brick at the front corner and butt the front return brick to the back corner of the starting brick. The triangular gap in front can be filled with mortar and covered with a tile facing, finish parging, stone, or the like, as shown in the drawing at left.

When Hilley is doing a square-cornered fireplace, he brings the side walls to a point ½ in. back of the edge of the return. This gives a neat line, which is very important with used brick because its width can vary from 3½ in. to 4 in.

Firebrick isn't required when you're building a firebox like this one, but Hilley uses it because heat-stressed common brick sometimes fractures violently. Most people don't like the look of firebrick in a Colonial fireplace, so he uses it only for the first six or eight courses—just high enough to cover the hot spot of a fire. You can see this blackened hot spot on the back wall of any fireplace. After a few fires, the firebricks soot up and blend in with the used brick in the rest of the

From *Fine Homebuilding* magazine (April 1984) 20:54-58

fireplace. Hilley doesn't use refractory cement with the firebrick, but he does keep his mortar joints under ¼ in. thick.

Hilley begins by sprinkling sand or spreading a piece of building paper on the brick hearth. This simplifies cleanup later. Then he lays up four courses of the back wall plumb, level, and parallel to the front—a small brick wall about 20 in. wide by about 11 in. high. The fifth course is a tad longer. It's also tilted or rolled in slightly by troweling on more mortar at the rear of the joint than at the front. This is the beginning of the curved back wall (photo facing page).

Next, five courses of the mitered side and front wall are laid up using the angle-cut brick at the front corners and by cutting and butting the rear brick to the back wall. The way to do this at the back wall is to score each end brick in the back wall with the tip of the trowel as you hold the brick in the rolled position. The coving is plumb, so the trowel should come off the bricks of the coving below and follow through in a plumb line, as shown in the drawing below. The scratch is very visible, and cutting is done with a brick chisel or the sharp end of a mason's hammer.

The two pieces of angle-cut brick at each front corner should fit together tightly where they show, and the V-shaped gap behind should be filled with mortar and a piece of brick. Hilley also likes to use a brick tie across this corner every couple of courses. This corner can get out of plumb easily, so a constant check with a level is a must. If a running bond

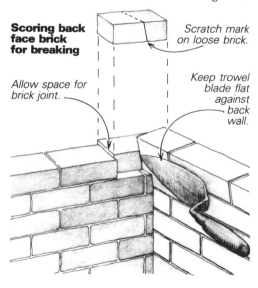

Scoring back face brick for breaking

Allow space for brick joint.

Scratch mark on loose brick.

Keep trowel blade flat against back wall.

is to show on the lintel course over the opening, you will have to watch the bond on your pilasters so that it will flow right into the bond on the lintel course.

Continue by rolling a few courses of the back wall, then building up the side walls. The roll will produce a gentle curve up to the damper, and it will make the back wall wider at lintel height than it is at the base. Each back-wall course is a little longer than the one below it, which is why the end bricks have to be marked in place for cutting. When a back-wall course needs to be a tad longer than two bricks, Hilley stretches it by setting a half-brick, or less, over the middle of the back

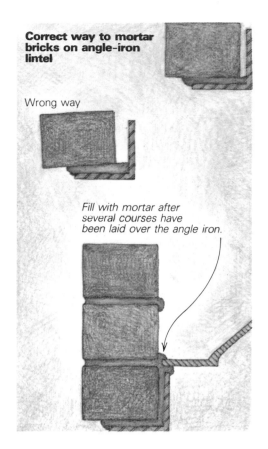

Correct way to mortar bricks on angle-iron lintel

Wrong way

Fill with mortar after several courses have been laid over the angle iron.

course below. The stretch, in other words, is accomplished in the middle of the course, not at its ends.

It is important while you're laying up the firebox to keep the side walls plumb. (In a square-cornered fireplace, the front and back walls are laid up first, a few courses at a time. The side walls are filled in.) You also must keep the back wall parallel with the hearth bricks. To do this, eyeball down the face of the back wall as it is laid, or measure from front to back on each side.

At the top of the firebox, the width of the opening from the outside face of the lintel brick to the rear face of the back-wall brick should be around 16 in. Hilley's formula for the amount of roll to give each back-wall course is simply experience. This is how most masons work. I'm always amazed at the way they seem to come out exactly where they want to be with exactly the right-sized opening, with no measuring at all. A novice might want to make a cardboard template to use as a guide, or spring a thin strip of wood against the first few courses to see how the curve projects up to lintel height.

Standard firebrick is thicker than used brick, so the back-wall courses will be higher than the side-wall courses. But the height should even out by the time you reach the lintel because the upper back-wall courses are tipped or rolled forward. As the back wall approaches lintel height, you can see how its courses relate to those of the front and side walls. By varying the joints, the wall heights can be adjusted to match.

When the firebox is at lintel height, Hilley fills in the space between the concrete-block wall and the back face of the firebox almost to the top with loose rubble. The rubble acts as a

The lintel. Side walls, back wall, and angle-iron lintel are at the same height to support the damper. The first course of bricks over the lintel overhangs the flange of the angle iron, and these bricks have to be laid up carefully so they won't roll forward. Pieces of building paper tucked at the ends of the angle iron serve as expansion joints.

heat sink, and more important, keeps the firebox positioned while allowing for expansion. A little mortar thrown in now and then will keep some of the rubble in place if a burned-out brick ever has to be replaced.

The lintel—A very important step in fireplace building is the proper installation of the angle-iron lintel. In this 36-in. fireplace, Hilley used 3-in. by 3-in. angle iron, which he installed with its ends bearing 1 in. or so on the pilaster bricks with a minimum of mortar—just enough underneath to stabilize it. The lintel's ends must be free to expand, and to ensure this Hilley tucks rolled-up scraps of building paper at each end. They act as spacers, keeping mortar and brick away from the angle-iron ends, and allow it to move.

The bricks in the first course above the lintel overhang the steel, and they have to be laid carefully (photo above) so that they won't roll forward. To help keep them from rolling, Hilley doesn't trowel any mortar behind them until a few courses have been laid, as shown in the drawing above. This eventual filling in, though, is important. Hilley feels that it prevents distortion of the angle iron from excess heat.

The damper—The damper should be sized to cover the firebox opening. The opening should be about as wide in front as the damper's flange, and from 2 in. to 5 in. narrower at the rear, depending on the damper's shape. The front flange rests on the top edge of the angle iron, and the side and back flanges rest on the firebox brick. The damper should be set in a thick bed of mortar on the brick and angle-iron edge, after three lintel courses are laid up, as shown in the photos at

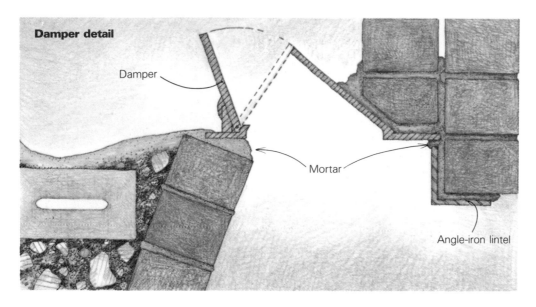

Damper detail

Damper

Mortar

Angle-iron lintel

The damper is mortared in place after three lintel courses are laid up. The space between the back wall of the firebox and the concrete-block core is ready for loose rubble fill, as shown in the drawing above.

The smoke shelf behind the damper is a 1-in. mortar cap over 4-in. concrete blocks on top of the loose rubble fill behind the firebox. The damper side is higher than the rear so any rainwater will drain away from the opening.

Laying up the smoke chamber is not fussy work. Hilley uses soft brick and concrete block, and then he parges the smoke shelf and chamber walls with mortar.

center left. As with the angle-iron lintel, it is important to keep masonry away from the ends of the metal to allow for expansion.

Smoke chamber—The smoke chamber is the open area behind the damper, where cold air coming down the chimney bounces off the smoke shelf at the bottom and is deflected upward, along with smoke rising from the firebox. As a base for the smoke shelf, Hilley lays a flat course of 4-in. concrete block on top of the rubble and concrete-block back wall. He sometimes lays a few concrete blocks, dry, directly on top of the loose rubble behind the rear wall. Then about 1 in. of mortar is smoothed out to make the smoke shelf's surface. Rainwater will puddle up here, so pitch the shelf away from the firebox and trowel it well. (Accumulated water will eventually evaporate or be absorbed into the masonry.)

The smoke chamber (drawing, p. 119) is formed by rolling the bricks of each course inward until the opening at the top is the size of the chimney flue tile. Hilley rolls the bricks a few courses at a time, alternating the corner bricks to maintain a bond.

Where the rolled brick courses meet at a corner, Hilley breaks off a piece of the lead corner for a better fit. He uses soft, spalling used bricks for this work. They are easy to shape, and it's not fussy work. In fact, Hilley had me hold up a sagging wall while he finished an adjacent supporting corner. A wall will collapse if laid up too much at one time.

Hilley says rolling the bricks to meet an 8x10 flue should give you a smoke chamber 24 in. to 36 in. high. Don't reduce from damper size to flue size too fast, and keep the smoke chamber symmetrical. Hilley once built a fireplace with the flue on the right side of the smoke chamber. This created unbalanced air pressures in the chamber and caused little puffs of smoke on the right side of the firebox.

The inside face of the smoke chamber is parged with mortar. (Be sure you leave enough clearance for the damper to open.) A piece of building paper or an empty cement bag laid on the damper before parging will keep things clean. You don't want your damper lid locked in solid with mortar droppings. The smoke-chamber walls must be 8 in. thick, so Hilley builds out their lower part with interlocking brickwork, and the upper part with flat-laid 4-in. concrete block. Then he parges the whole business with a layer of mortar (photo bottom left).

The rolled brick and outer block shell of the smoke chamber transfer the flue and chimney weight to the lintel, keeping the lintel bricks in compression. The first flue tile sits on top of the smoke chamber, fully supported by the brick, and the chimney is built around it. Brickwork against a flue will crack as the hot flue expands, so there must be at least a 1-in. airspace between the tile and the chimney shell. If the chimney is concealed, the masonry can be concrete block. □

Consulting editor Bob Syvanen is a carpenter in Brewster, Mass. Photos by the author.

Laying Brick Arches

A masonry inglenook
becomes the warm center of an architect's new house

by Elizabeth Holland

When Ed Allen and his wife Mary started thinking about designing their own house seven years ago, they knew they wanted something big and barnlike, yet New England simple and cozy. Ed, an architect, had been fascinated by domes, vaults and arches during his studies, but it wasn't until the couple spent six months living in Liverpool during the winter of 1975-1976 that an inglenook became part of their house plans.

"We spent a lot of cold evenings huddled around a fire, trying to keep warm in an unheatable English house," remembers Allen. "We kept sketching our ideas and dreaming about how we wanted to have a house where we wouldn't be so cold."

The sketches showed the influences of the English and Welsh houses they had seen, particularly of their visit to the Welsh Folk Museum at St. Fagan's. This is where the Allens encountered the inglenook.

It is thought that the word *ingle* comes from the Gaelic *aingeal*, meaning fire or light, and was originally applied to open fires burning on primitive hearths. In medieval times, it came to mean a fireplace. The inglenook was a corner or a small room near the chimney where the family would gather before the heat of the flames. This became the central idea in Ed Allen's plans, and the inglenook ultimately became the dominant design element of the house he was to build.

Allen spent a long time working out the exact dimensions of the inglenook, and was working on a 4-in. module. Then the house was designed around that, from the inside out. "I was always trying to keep it as small as I could," Ed says, "but it kept growing as I made sure all the spaces around it were the proper size."

The Allens' inglenook is all brick, part of a 50-ton masonry mass that encompasses flues and fireplaces, and divides the house into two equal parts on each floor. The inglenook was designed to accommodate four or five people comfortably. The interior dimensions are 7 ft. 4 in. wide and 7 ft. deep from front to back, not including the fireplace.

According to Allen, it was all worked out logically—the dimensions of what he wanted things to be, plus the dimensions of the necessary brick. For example, the archway is 16 in. thick because the walls contain an 8-in.-square flue plus 4 in. of brick on each side.

To make the layout easier, Allen designed the brickwork to be modular, with 7⅝-in. long bricks and ⅜-in. mortar joints, for an overall length of 8 in. Later he discovered that only about two-thirds of the assorted Full Range Belgian bricks he ordered were the right size. The rest were up to ¼ in. too long. This required some cutting when he got to his closers, the last bricks in each course (for tips on cutting brick, see pp. 105, 117, and 133).

Mortar color and tooling can significantly affect the look of brickwork. Allen chose a standard dark masonry cement. On the horizontal mortar joints, he used a flat-joint finishing tool to make a weathered joint, flush with the brick at the bottom and cut back at the top. This joint casts a shadow on the mortar joint, accentuating the pattern of the brick. The vertical joints, however, are gently concaved.

Choosing a bond—Bonding, the overlapping patterning of the bricks, knits the various wythes (thicknesses of brick) together. For a single wythe, a simple running bond can be used. But structural brickwork is usually at least 8 in., or two wythes, thick. "An 8-in. thick wall can use any of a variety of bonds, and some of them are quite beautiful," says Allen. "I was planning to use an English Garden Wall bond because if you're doing an 8-in. wall where one side is going to be concealed, you can save a lot of time by laying up the concealed wythe in 4-in. concrete blocks."

An English Garden Wall bond consists of three courses of stretchers (bricks with a long edge showing) and a fourth course of headers (bricks with their short ends facing out and

---

Bricklaying tips for amateurs

After years of studying and teaching brickwork, and laying bricks, Ed Allen is convinced that interior brickwork—including arches—is well within the grasp of a reasonably careful amateur.

"Once you get into masonry, it's just like putting up a wood frame for a house—absolutely routine, very secure, very simple," he explains. "There's just no end to what you can do with it."

Bricklaying is a relatively straightforward concept that gets fairly complex in practice. Here are some tips for beginners:

- You can't learn brickwork on your own. Learn from someone who knows how to do it. You can pick up the rudiments in a day or two. "Probably 90% of the success in laying brick is getting the mortar the right consistency and using the trowel properly. Learning to mix mortar to the proper consistency and to use a trowel are things that simply can't be gotten from a book."
- Plan in advance and know your dimensions. Determine the heights at which the arches will spring, and the heights of the arches. Consider the placement of flues when determining thickness.
- An arch supports a vertical load by transforming it into a diagonal load. Make sure your design has enough mass on either side of the

arch to absorb the thrust and keep the arch from spreading.
- The labor involved in laying brick is almost directly proportional to the number of corners. Called leads, the corners are laid up first, four to six courses at a time. Care in laying up the leads pays off in level courses of the right height. The bricks that fill in the flat stretches between corners, aligned with strings pulled taut between the leads, are laid relatively quickly. It's best to eliminate as many corners as possible. You've always got to decide, of course, whether it's worth the extra labor to get it the way you want it.
- Practice by building the foundations for whatever you are going to build. This gives you a chance to develop techniques before laying up courses that show. If you're still having trouble, get some help.
- It's easiest to work at waist level, or roughly between your knees and shoulders. Arrange the scaffolding accordingly. As you go higher, your work slows down because it becomes more cumbersome to transport heavy and bulky materials.
- Brickwork is not as precise as carpentry. The irregularity of it is part of the charm, and a real plus for amateurs. Nevertheless, you should strive for precision, so things don't get too far out of whack. —*E.H.*

---

their lengths extending in across the two wythes). On the concealed wall, a course of concrete block takes the place of three courses of brick plus their mortar joints. The fourth course is composed of the headers, lying across both the brick and block wythes.

A header course is laid so the bricks straddle the vertical mortar joints in the stretcher course below. Since Full Range Belgian bricks are narrow, spacing the bricks properly would have required extra-wide joints. This led Allen to change from a full header course to what's called a Flemish header course (drawing, right), where headers and stretchers alternate.

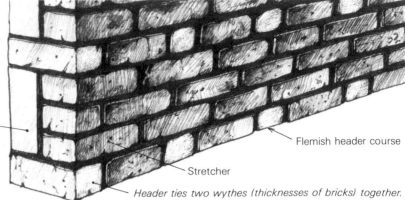

Concrete block is 4 in. thick and as high as three courses of brick with mortar joints. Using block where it won't show saves time and money.

Brick bonding

Flemish header course

Stretcher

Header ties two wythes (thicknesses of bricks) together.

Building the arch.

Allen and a friend did all the masonry work themselves, in two stages. First, in May, they laid up the block foundation for the masonry core. The brickwork began in July, and from then until mid-autumn they laid 7,000 face bricks, an undetermined number of concrete blocks and concrete bricks, and several hundred sections of flue tile. The graceful curve of the inglenook's archway is inviting, a welcome contrast to the sturdy straight lines of the brick walls. Within the inglenook and overhead, the arch repeats, each time in a slightly different form.

"Arches are really fun—they're a wonderful structural form," Allen reflects. "I think everyone has an immediate, positive emotional response to them. By the time we finished the arches we decided they were quite easy to build and not terribly time-consuming, a conclusion quite contrary to our initial expectations."

Step one: the centering form—The brick walls are laid up to the level called the springing of the arch, the point where its curvature begins. Now the centering is built—the wooden form over which the bricks in the arch will be laid.

There are many ways to build a centering. Mine consists of two identical curved trusses with a single rectangular piece of ¼-in. Masonite nailed securely to the top of them (drawing **A**, below). You should check the centering's fit by holding it between the brick sidewalls.

Step two: patterning the arch—Lay the centering on its side on the floor. Stand the bricks you're going to use on end, all around the curve (**B**). You should have an odd number of them. Using an even number results in a mortar joint positioned at the crown of the arch. This looks bad and makes the structure weak. Because of the curvature of the arch, the mortar joints will be wedge-shaped. They should be about ³⁄₁₆ in. wide at their narrowest point, next to the centering. Shuffle the bricks around until the spacing looks good. You could have wedge-shaped bricks specially made, but these are usually expensive, and must be ordered well in advance of when you'll need them.

With a pencil, mark the thicknesses of the mortar joints on the centering itself. Take away the bricks and use a square to run the joint lines across the curved surface of the form. When you're finished, the centering will be marked to show you where all the bricks belong, and how thick the mortar joints should be. This step is crucial—without it you will end up with uneven joints, and you'll have to trim bricks to fit odd spaces.

Step three: placing the centering—Lift the centering between the existing brick walls. The bottom of the centering should be just a couple of inches below the spring of the arch (**C**). Support it by four lengths of 2x4, cut just a little longer than the distance from the floor to the spring of the arch, so they can be angled under the four corners of the centering and wedged in place. It's best to align the centering so its front edge is even with the brick walls; then bricks in the arch can be laid to the front edge of the form. If your centering is wider than the arch, pencil a line on it to indicate where the front ends of the arch bricks should be laid. Level up by tapping the bottoms of the wedged 2x4s. Once the centering is level and in the right position, drive shims into the small gaps between the centering and the walls at the four corners, to hold the form firmly in place.

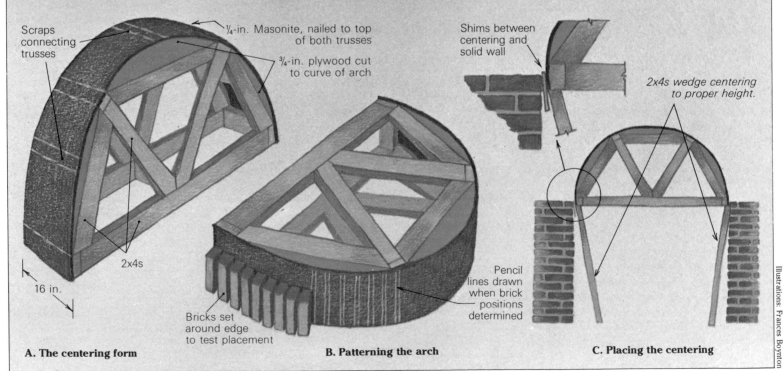

Scraps connecting trusses

¼-in. Masonite, nailed to top of both trusses

¾-in. plywood cut to curve of arch

2x4s

16 in.

Bricks set around edge to test placement

A. The centering form

B. Patterning the arch

Shims between centering and solid wall

2x4s wedge centering to proper height.

Pencil lines drawn when brick positions determined

C. Placing the centering

Illustrations: Frances Boynton

D. The ends of the bricks in the arch must be in the same plane as the wall. Check for alignment with a trammel board tacked to the center of the horizontal truss member.

E. Bricks are set between the pencil lines that were drawn on the centering during a test-fitting on the ground. This guarantees a proper fit and mortar joints of uniform size.

F. Bricks have been trimmed with a mason's hammer where wall meets arch. It looks best if the width of the curving joint between wall and arch remains constant.

Step four: laying up the arch—Lay bricks from both lower edges until they meet at the top. As they go up, check the bricks with a level to make sure their ends are in the same plane as the wall, or with a board tacked to the center of the truss (**D**). If you follow your pencil marks, you'll end up with the right number of bricks, and uniformly wide joint spaces (**E**).

Step five: finishing the wall—With the centering still in place, lay up the flat walls around the sprung portion of the arch. If the centering is removed too soon, the arch might collapse, because the mortar is still fresh and because a semicircle is not the strongest arch form. It could well bulge out at the sides and drop in at the center. Once the walls are laid up around the arch, however, it can support a lot of weight.

In the Allens' house the vaulted ceiling over the inglenook carries a concrete slab floor above it.

Always work in from the leads at the outer end of the wall, keep the vertical joints lined up properly and use taut string lines to keep the courses perfectly level.

On each course you will have to cut the last brick to fit, where it intersects with the arch. A diamond saw is the most efficient tool for this, very precise and sure. But a diamond saw is expensive and to Allen's eye the cut is too sharp-edged and cold. An abrasive masonry blade ($5 to $6 in a hardware store) in a circular saw is a good alternative. "It's a messy operation and it doesn't cut the brick as well as a diamond saw," he says, "but you can score the brick deeply and then crack it with a hammer to get a pretty clean break."

For his own arches, Allen used a mason's hammer to cut the bricks. This takes some skill, he cautions, and results in a somewhat ragged break and a lot of wasted bricks because you can't always get the bricks to break exactly as you want. The flatter angles that are required near the top of the arch are particularly difficult to make with a hammer.

A ragged cut on a few bricks, though, is less important than making sure that the curved mortar joint between the arch and the entire brick wall around it is a constant thickness (**F**). A curved joint that varies in thickness is unattractive, and once it's there, you can't do much about it.

Once the courses are laid up around the arch, tap out the 2x4s carefully, drop the centering gradually, then remove it and admire your arch.

Tooling the joints—In typical brickwork, after the brick is laid in the wet mortar, the mason cuts off the excess mortar at the face of the brick. In one to three hours the mortar will be thumb-print hard, the proper consistency for tooling with a V-shaped or rounded metal rod called a jointer or striking tool. This produces a clean and attractive joint. (For exterior brickwork, tooling is doubly important because it helps compact the mortar at the face, making it much more weather-resistant.)

But with an arch, the bottom mortar joints can't be tooled when the mortar is thumb-print hard, because the centering is in the way. By the time it is removed, the mortar is too hard for tooling.

Yet old arches and vaults have well-tooled joints. Allen was puzzled. Books on the subject were no help. Several masons suggested rubbing the joints with full-strength muriatic acid. The acid quickly dissolved the excess mortar that had been stuck between the brick and the centering, but the joint remained undefined and fuzzy.

Allen eventually learned that before the era of portland cement, arches and vaults had been laid up with lime mortar, which sets very slowly. He says that even if the centering were kept in place for several weeks, when it was removed the lime mortar would still be soft enough to be tooled. Although lime mortar is not as strong as portland cement, the arches themselves are structurally stable enough to be able to compensate for the weaker mortar. Allen recommends either using lime mortar alone for the arches, or using it for only the bot-

tom part of the joint that meets the centering, and then using mortar made with portland cement for the work above this.

Living with the inglenook—The Allens' inglenook can hold up to six people within its candlelit confines. The snug spot sits off a spacious country kitchen, and is used more often on social occasions than on weekday evenings. "It can get a little crowded," Allen says, " but in that kind of space it doesn't feel crowded. I think people are accustomed to drawing close around a fire." ☐

Elizabeth Holland lives in West Shokan, N.Y. She writes about the design and construction of energy-efficient buildings, and is contributing writer for this magazine.

Renovating a Chimney
New flue liners convert fireplaces for woodstove use

by Joseph Kitchel

Installing new flue lining in an old chimney is somewhat akin to digging a basement under a finished house—it's not impossible, but it's a lot easier to do beforehand, during the original construction. Unfortunately flue liners were invented a long time after many chimneys had been built, and the older the chimney, the greater the need for new liners. Cast from fireclay or fireproof terra cotta, they provide a safe, effective exit route for smoke and combustion-related gases. Unlined chimneys are hazardous, ineffective and troublesome by comparison.

The chimney we had to work on was part of a common wall in a three-story Brooklyn rowhouse, built between 1860 and 1870. Although it served three fireplaces, we planned to eliminate the one on the third floor and convert the other two for woodstove use. This meant installing two separate flues in the chimney space, one for each stove. We decided to begin the job on the second floor, adding the first-floor flue sections at a later date.

Getting set—Before starting a job like this, it's important to know the type of stove to be used, because the location of the stove's exit pipe determines where you install the thimble fitting in the new flue. I had decided on a Lange 6303A/B, a Danish woodburning model with an exit pipe that can be adapted to run from either the back or the top of the stove. I chose to use the horizontal exit, since this would allow me to install the thimble in the first flue section rather than farther up on the chimney wall. A tile or slate hearth would be added after the new linings were installed, so I took this into account when determining the height of the stove pipe and the location of the thimble fitting.

I also wanted to install the cleanout door in the same bottom flue section. Most chimney cleanouts are in the basement, but since this was a second-floor installation, I thought it best to locate the cleanout on the same level. The cleanout door is now hidden behind the old cast iron cover plate I removed from the original fireplace. It was one of three such covers provided for fireplace-to-stove adaptations, and I incorporated it in the rebuilt chimney to retain a bit of the appearance of the original fireplace.

Another important decision in planning is whether to use round or square liners. Smoke rises in a swirling motion, so the corners of square flues become cold spots where creosote and soot can accumulate. Consequently round flues are generally considered to be superior, even though they are more expensive and a bit more difficult to install. Round linings are manufactured in 24-in. sections, with inside diameters ranging from 6 in. to 18 in. (even sizes only). Since our stovepipe was only 4½ in. wide, we chose 6-in. linings. Transporting and handling the flue sections requires a gentle touch, since the castings are extremely brittle. The only other materials we needed for the job were portland cement, sand and white cement (added in small quantities to the face mix to approximate the color of the old mortar).

Before beginning the actual work, I covered the floor around the chimney with some foam rubber carpet padding followed by overlapping sections of vinyl-covered canvas wallpaper. Unconventional as it might sound, I've found these two protective layers an ideal combination: The vinyl-covered canvas sheds water (and mortar), while the foam cushions the floor against the inevitable falling bricks. Chimney reconstruction is messy, especially when a substantial number of bricks must be removed, as was the case here. Protecting the surrounding work area from the ravages of the job is well worth the effort.

Digging in—Our plan was to install the flue liners in the old chimney passageway and then brick them in, restoring the wall to its original appearance. First we had to remove the facing brick layer to expose the inside of the chimney. A hammer and a sturdy, narrow cold chisel are the best tools for this job. To cope with the occasional stubborn or awkwardly placed brick, I also kept a pry bar close at hand. Generally, the old mortar was brittle and could be chipped away easily. Working from floor to ceiling, we removed the chimney wall in a staggered pattern so that the re-laid bricks wouldn't look so conspicuous. We saved all of the face bricks to use again. The entire brick wall had been

The first step in installing new flue liners is removing bricks from the second-floor fireplace to expose the chimney cavity. Hammer, railroad spike and pry bar are essential tools. A vinyl-covered canvas dropcloth over rubber carpet padding protects the floor. At right, Kitchel aligns two pieces of angle iron, which will be cemented in place across the face of the cavity to reinforce the chimney before the new liner is installed.

From *Fine Homebuilding* magazine (August 1981) 4:38-41

painstakingly cleaned of plaster at an earlier stage in the renovation, and its rosy pink color would have been very hard to match with new brick. Besides, there was always the expense of new brick to consider, and the trouble involved in hauling it up to the work area.

Inside the chimney we found thousands of brick fragments. Some had been built into the chimney to achieve the turns and separations between old flue channels. Others were simply the result of our chiseling work or age-related crumbling. All this rubble had to be removed so the cavity would be clean when we installed the new flue sections. An old railroad spike turned out to be a good tool for cleaning off the bricks. Its wide head was easy to hit and its broad, flat point broke the soft mortar away quickly.

The original chimney was constructed so that the flue of each fireplace started in the center of the chimney. As it progressed upward, the flue had to angle to the right or left to pass the fireplace above and thus make room for its flue. We wanted to keep the new flue sections as straight and plumb as possible, to make the chimney safe and efficient, and also to make the masonry work easier. Since we planned to eliminate the third-floor fireplace and its flue, both new flues could run straight up, with no sidesteps or bends. We located the first-floor flue on the extreme left side of the chimney; the flue for the second floor would run straight up the middle, and the resulting space on the right could be used as a tunnel for some electrical cables. In general, old chimneys provide a very good vertical tunnel through which wiring or plumbing can be run; keep in mind, however, that adequate separation between the conduits (in the form of airspace and solid masonry) must be provided to prevent excessive transfer of heat.

Even though we didn't plan to install a second stove on the first floor for some time, we needed to build both linings into the chimney from the second floor up. Then we could later complete the job from the floor below without disturbing the second-floor brickwork. With this in mind we dug out the chimney cavity to a point 14 in. below the second floor.

Installation and reconstruction—Although conventional mortar mix is available for brick-laying jobs like this, we are used to concocting our own mortar from one part portland cement to three parts sand. Our mortar board, a piece of exterior-grade plywood bordered with 1x2s, was set directly on the canvas dropcloth.

We cemented two 4-ft. lengths of 3-in. by 3-in. by ¼-in. angle iron across the cavity opening below the second floor. Positioned parallel to one another (the L pointing up) and joined to the cavity wall, these iron beams form a reinforcing baseplate for both flues. We set the first-floor liner between the Ls and surrounded it with a course of cement blocks, building the cavity up level with the second floor.

Because both the cleanout door and the stovepipe thimble are designed to fit into a square flue liner, we had to use an 8-in. square liner as the base section for the second floor flue. Cutting openings in the ceramic pipe to receive these fittings (photo top right) is a delicate job. You can

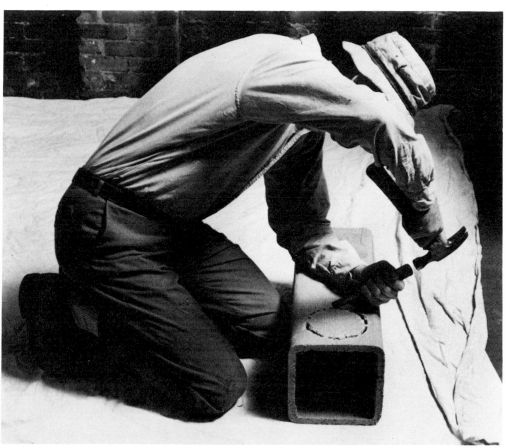

The first section of liner is cut to receive the thimble fitting, a tricky business since the ceramic material is very brittle. Closely spaced holes are first drilled along the cut-line; then the waste piece is knocked out with a hammer and chisel, as shown.

After the cavity has been built up to floor level, the first flue section can go in, left. The square opening at the bottom of the flue will receive the cleanout fitting. When both flue sections are in place, the chimney is bricked in around them, starting with a layer of cement blocks, right. The round flue section will eventually be connected to a new lining from the floor below; the square section in the foreground is for the second-floor lining. The back walls of old chimneys like this one usually require parging, or stuccoing over, before the new lining and interior brickwork can be laid in.

A round flue section is cemented to the square base, left. To ensure a good fit, a square piece of wire lath with a hole cut in the middle is set in the mortar between the two liners. Right, joints between round liner sections are sealed completely with mortar. The remaining interior brickwork extends to within an inch or so of the liner, allowing room for heat-induced expansion. Face ties in the last interior course will be bent into the face course as the wall is rebuilt.

A plywood sheet, top, temporarily replaces the third-floor hearthstone. The face course below has been relaid using the original bricks and repointed to match the style of the original chimney. Above, bolts sunk in the reconstructed face course just below the third floor secure a new chimney girt, to the wall. This will support a new section of floor that replaces the third-floor hearthstone.

use a rotary saw equipped with a masonry blade for the straight cuts. We made the round opening by drilling closely spaced holes with a masonry bit and then carefully chiseling out the tile between the holes.

The next step was to brick in the chimney around both liner sections. Again we used cement blocks, since they wouldn't be visible and were easy to position and level. The solid masonry around the hearth would also hold the heat well.

Laying the first round flue section on top of the square base section was a bit like fitting a round peg in a square hole. To make the joint secure, we cut a 6-in. hole (the round pipe diameter) in a square piece of wire lath and placed it between the sections. Then we covered the joint with mortar and bricked in the cavity to within an inch or so of the liner. We filled this space with loose, dry rubble on the theory that the liners could expand and contract more freely, thereby minimizing the chance of cracking. Bricks were soaked in a bucket of water before we laid them into the wall. The extra water held by each brick lets the mortar cure more slowly, providing a stronger bond that won't powder with age.

In laying up successive flue sections, we established a working plan: First, we'd check the back wall of the cavity for loose joints. More often than not this course of brickwork (the face brick of the room next-door) would need parging (stuccoing over). Then we'd set the new liner section in a bed of mortar on top of the lower section, lay up the bricks around it until we were one course above the top of the liner, and repeat the operation. Since the first liner of the left-hand flue had been set 14 in. below the first liner of the right-hand flue, the joints of the liners were automatically staggered. We laid up all the liners and their supporting brickwork first, inserting face ties in the last layer of back brick to ensure a good bond with the face course.

In reconstructing the chimney face, we had to work into the irregular pattern deliberately created when we removed the brick. Whenever we could, we matched color and texture in fitting the new brickwork into the old. We even tried to imitate the rather haphazard pointing style of the original wall. The mortar had to be allowed to slump a little in the joint before being struck off flush with the brick face. If voids appeared, we left them. This took a little self-discipline, but the result was successful.

When we reached the ceiling of the second floor we had to remove a brick arch, which supported the marble hearthstone of the third-floor fireplace. This arch had originally been installed so that the hearthstone did not touch any of the floor joists. To make it safer and easier to work on the third floor, we temporarily covered the gap left when the hearthstone was removed with a sheet of ¾-in. plywood. Later on, when the masonry work was complete, we could frame in a new section of the floor, since no stove or flue was planned for this level. With all this in mind, I sunk three bolts in the reconstructed chimney face just below the floorline to hold the new chimney girt that would support the flooring, as shown at left.

Work on the third floor went smoothly. We

moved our cushioned dropcloth upstairs before tearing out the bricks and continuing to lay in the flue sections. We stopped just short of the ceiling. This was also the roof of the building, and we didn't want to disturb the brickwork at this juncture because the flashing joint between chimney and roof was fine—the bricks were all sound and the flashing didn't leak. We figured we'd just be asking for trouble by disturbing the seal. Instead, we decided to leave the face course of bricks intact and remove only the interior cross-bricking that divided the original flues. These interior partitions had to come out in order to keep the new flue sections running plumb, since the chimney narrowed slightly near the roofline and the two original right-hand flues jogged to the left. Working from the inside, we cleared out as much of the chimney cavity as we could reach before we went topside to complete the operation.

On the roof—The chimney wall continued above the roofline, separating our flat-roofed building from the higher gable-roofed building next door. Above the flashing, we discovered that the brickwork was not in very good condition; even the wall on either side of the chimney needed to be rebuilt. We carefully dismantled the damaged areas and cleaned out the chimney cavity above the roofline. Now all we had to do to finish the job was to install the remaining flue sections and to rebuild the chimney and wall around them.

To protect against downdrafts caused by wind deflecting off the gables, a chimney should extend well above the roofline. We built ours up about 3 ft., topping off the two flues with the bell end of a sewer-tile section and a wall-capping tile. Decorative chimney tile manufactured specifically for this purpose is four to six times more expensive than the substitutes we used. (The wall-capping tile has perforations cast into it, since each cylindrical section is meant to be divided in two. We used the paired pieces together to make the final flue piece.) High temperature resistance would not be a consideration here, since the stoves we were planning to install have internal baffles that prevent most of the heat from going up the chimney along with the smoke.

We extended the tiles about 16 in. above the brick line and topped the brickwork with a sloped cap, mixing the mortar just as we had throughout the job. The exposed chimney had been quite weathered before we rebuilt it, so we decided to give the entire chimney face above the roof line a stucco finish, as additional protection from the elements.

We completed the lining and rebuilding in November of last year and quickly installed the woodstove on the second floor. Since then, the heating bill for our gas-fired steam system has been reduced by about two-thirds. We still have the option of putting another woodburning stove in on the first floor—and the satisfaction of having improved an old chimney without altering its original character. □

Joseph Kitchel, 42, of Brooklyn, N.Y., is a prop builder, cabinetmaker and renovator.

The chimney cavity, top, hollowed out at the roofline in order to accommodate the two flues. Only the top part of the chimney has been removed; the flashing joint where chimney meets roof remains intact. Center, flue-pipe installation proceeds on the roof just as it did inside the house, the only difference being that some of the adjacent wall has to be rebuilt as well. Once the final flue sections have been bricked in, the chimney gets a sloped cap, above, so that water will run off. The stucco finish on the exposed brick provides additional protection from the elements.

Rumfordizing Brick by Brick

How to convert an energy-wasting fireplace to an efficient heater

by Kent Burdett

Most modern fireplaces don't do a very good job. Many smoke so badly that they can't be used, and almost none are efficient heaters. In fact, many of them draw more heat out of the house than they return, sucking in warm room air and sending it up the chimney. But it's possible to convert one of these mere ornaments into a functioning and efficient fireplace. An American Tory named Benjamin Thompson, later called Count Rumford, demonstrated the relevant principles two centuries ago.

Rumford proved that the key to an efficient fireplace is a properly proportioned firebox, with important dimensions based on the width of the opening. (The parts and proportions of a Rumford fireplace are shown on the facing page.) Both the firebox's depth (distance from opening to fireback) and the width of the fireback should each equal one-third the opening's width. This makes for a shallow firebox with covings angled at a sharp 45° to reflect the fire's radiant heat into the room. The fireback,

which must be vertical to a height equal to one-third the opening's width, begins to slope forward from that point to a small throat above the lintel. The sloping back reflects more heat, and the small throat results in a more forceful movement of air up the chimney. It also leaves room for the smokeshelf, a necessary feature where descending cool air and ascending hot air circulate to set up a strong draft.

Rumford's workmen renovated so many smoking fuelwasters in England that a new word entered the language. His wealthy customers didn't just have their fireplaces improved, they had them "rumfordized." Once you know the principles, you can rumfordize your own fireplace. Fireboxes are not structurally connected to the masonry of the chimney, so you can tear out an unsatisfactory one and replace it easily.

Materials—For this job you'll need clean sand and water, masonry cement for the rubble fill behind the new firebox, firebricks and fireclay

to join them. Experience provides the best way to judge just how much of each you will need for a specific project. For a fireplace 24 in. wide and 30 in. tall, I used 84 firebricks (2¼ in. by 4½ in. by 9 in. each), along with ¼ yard of sand and 2 sacks of cement. You can buy clean sand by the fraction of a yard at most lumberyards.

There are two kinds of fireclay, premixed and dry. The premixed costs twenty times as much as the dry variety, and can't be scraped or chipped from the faces of firebricks after it dries. Dry fireclay (available at ceramic supply houses) is sold in 50-lb. sacks, but one sack costs less than a single gallon of premixed.

As for the firebricks, try to buy what you need from the same lot. They will be fired to the same hardness, and there will be fewer small variations in their dimensions. You can find them at masonry supply houses and many lumberyards.

You probably already own or have access to most of the tools you'll need: a lightweight hammer, a tape measure, a try square, a level, a soft

---

Benjamin Thompson, Count Rumford
by Simon Watts

An efficient fireplace was hardly Benjamin Thompson's only contribution to civilized living. He was an extraordinary American whose adventurous life and considerable scientific achievements remain largely unknown. He was an ingenious inventor, always trying to improve clothing, coffee pots, eating habits, lamps and whatever else crossed his path. Perhaps he was overshadowed by his great contemporary, Benjamin Franklin, but I suspect his obscurity has more to do with his having been on the wrong side of the American Revolution.

Born in 1753, Thompson showed his scientific bent early. While still in his teens he experimented with gunpowder and electricity, and was already keeping a detailed journal of his observations, which became a lifelong habit. On at least one occasion his scientific curiosity nearly cost him his life. Attempting to repeat Franklin's famous experiment, he constructed a 4-ft. kite and flew it in a thunderstorm. Going the more prudent Franklin one better, he soaked the kite string in water to make it a better conductor. The results were suitably dramatic: Watching from the house, his family was amazed to see the youthful experimenter outlined in fire. He later remarked in his diary, "It had no other effect on me than a general weakness in my joints and limbs and a kind of listless feeling. However, it was sufficient to discourage me from any further attempts."

In 1772 Thompson was invited to teach school in Concord, N.H. Within a few months he married a wealthy young widow, and for the first time was financially independent. He settled down to manage his wife's estates and pursue his scientific studies, but those were restless times. Rebellion was in the air, and the colonists were taking sides. Thompson remained loyal to the king, barely escaped being tarred and feathered, and fled to England, where he was put in charge of recruiting, equipping and transporting British forces in North America.

After the war he went to Bavaria, where he was given the job of reorganizing the Elector's woe-begone army. With characteristic

thoroughness, Thompson spent several years making a detailed study of the army, and finally came up with a plan so comprehensive that it stunned his critics speechless. Thompson's report focused on the army's two major expenses—food and clothing. Questioning the existing cloth, he set about experimenting with different materials, such as fur, feathers, cotton, wool and jute, to find out which were the cheapest and most effective for soldiers' uniforms. He devised ingenious experiments to compare thermal conductivity, and was the first to suggest that it was not the material, but the air trapped in the fibers, that provided insulation. He then designed a new cloth and looked around for a firm to weave it. None of the existing companies was willing to cooperate, and they all thwarted his every attempt to set up a new factory.

This setback prompted Thompson to make his most spectacular experiment in social reform. At that time Munich was plagued by professional beggars so numerous and well organized that they practically ran the city, even intimidating the police. Thompson saw an opportunity both to staff his factory and to rid the city of its beggars. On New Year's Day, 1790, he made his move, and before nightfall every beggar had been arrested and locked up in what was euphemistically called "The Poor People's Institute," but which was actually a workhouse. Thompson ran the Institute with a firm hand, and within a few months the former beggars had been trained to produce cloth of an acceptable standard.

By 1791 Thompson had become a general in the Bavarian Army, as well as Minister of War and Minister of Police. In 1792 he was made a Count of the Holy Roman Empire, and adopted the name of Rumford—the original name for Concord, N.H.

If you're interested in reading more about this singular man, try Sanborn G. Brown's *Count Rumford: Physicist Extraordinary* (§18.25 from the Greenwood Press, P.O. Box 5007, Westport, Conn. 06881).

Simon Watts is a writer and cabinetmaker in Putney, Vt.

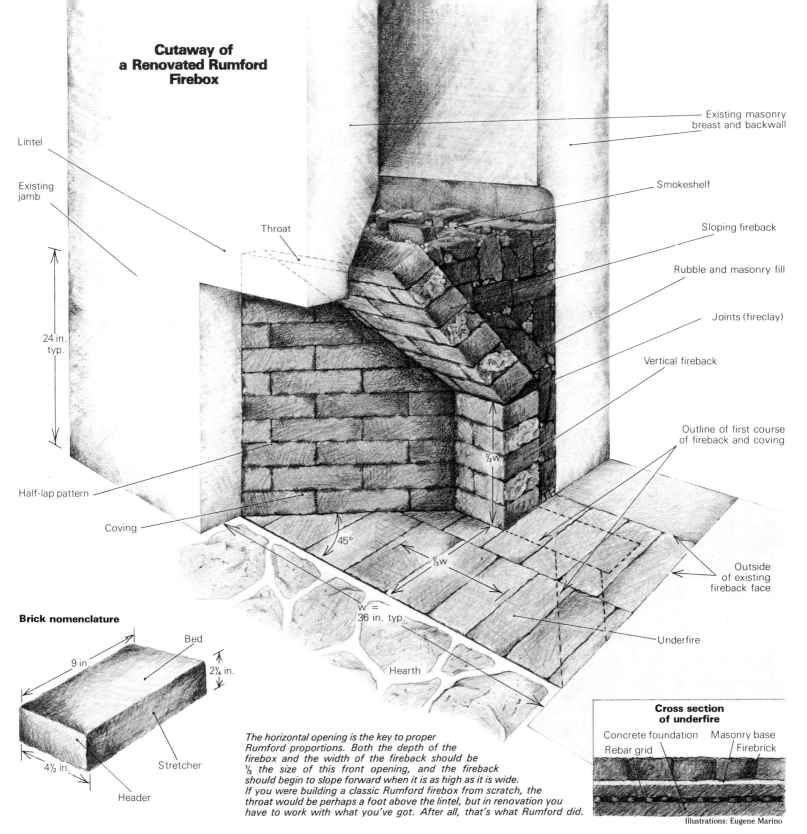

Cutaway of a Renovated Rumford Firebox

Lintel

Existing jamb

24 in. typ.

Half-lap pattern

Coving

Brick nomenclature

Bed

9 in.

2¼ in.

4½ in.

Stretcher

Header

Throat

45°

⅓W

W = 36 in. typ.

Hearth

Existing masonry breast and backwall

Smokeshelf

Sloping fireback

Rubble and masonry fill

Joints (fireclay)

Vertical fireback

⅓W

Outline of first course of fireback and coving

Outside of existing fireback face

Underfire

The horizontal opening is the key to proper Rumford proportions. Both the depth of the firebox and the width of the fireback should be ⅓ the size of this front opening, and the fireback should begin to slope forward when it is as high as it is wide. If you were building a classic Rumford firebox from scratch, the throat would be perhaps a foot above the lintel, but in renovation you have to work with what you've got. After all, that's what Rumford did.

Cross section of underfire

Concrete foundation   Masonry base
Rebar grid   Firebrick

Illustrations: Eugene Marino

brush, a string, a shovel, a hoe and a wheelbarrow for mixing cement. You will also need a 4-in. brick set (a chisel for breaking bricks), a 12-in. mason's trowel and a cold chisel or all-purpose masonry chisel. All these tools can be bought at most hardware stores.

Underfire—Often, all you have to do to remove an original firebox is to reach up, grab the top back brick, and pull (photo next page, top left). If the bricks don't tumble right down, use a hammer and an all-purpose chisel. This is a dirty job that stirs up a lot of dust, so be sure either to cover the furniture or to move it out of the room, and seal the doors to the rest of the house with tape. Save both the old firebricks and the rubble

behind them. You may be able to use them later. If there is already a damper mechanism at the throat, leave it. Cement from the chimney bricks or tile will hold it in place.

After pulling down the old fireback, covings and rubble fill, remove the bricks of the existing underfire. Using a chisel and hammer, firmly tap the masonry foundation beneath. If it seems solid, you can go ahead and lay the new underfire. If the foundation is cracked, loose or crumbling, however, you will have to replace it. Chisel it out to a depth of about 4 in., then lay a grid of ½-in. rebar 4 in. on center before pouring a new foundation of concrete. You will want the finished underfire to be level with the outer hearth, so take careful depth measurements and

leave enough room above the foundation pour for a base and the underfire brick. If a sound foundation is too high for the bricks you are using, you will have to chisel it out. When the foundation is ready, pour a base ¼ in. to ½ in. deep for the new underfire. Mix two parts of clean, dry sand with one part of masonry cement, then add enough water to create a pourable mixture. Spread it over the foundation.

Lay the firebricks for the underfire with no fireclay between them. The hearth takes a lot of abuse, and cracked brick can be broken away from the base cement and replaced easily if it has been installed this way.

Lay the first row of bricks beginning with the ones on the extreme right and left of the open-

Photos: Gary Hensarling

Fireboxes aren't structurally connected to the masonry of the chimney. You can often tear one out, as at left, by reaching up and tugging. Above, the underfire should be installed dry, without fireclay between the bricks. Be sure each brick is set level. (The bottle is covering an old gas line that will become a fresh-air intake.)

When the fireback is only one brick long, the most efficient way to butt coving and fireback is to cut the coving brick to fit along the fireback's header, left. When the fireback is longer, cut the coving brick to fit against fireback stretchers. Above, firebox is filled in behind with rubble and masonry. A fairly wet concrete mix will flow to the bottom.

ing. Once these are in place, you can rest your level across them to be sure they and subsequent bricks are perfectly horizontal. Next, use a plumb bob to find the opening midpoint, and center a carefully leveled brick on it. Place and level bricks alternately on either side of the central one until you reach the two at the edges of the opening. The fit will probably be less than perfect, and these outside bricks will have to be trimmed. They will eventually be covered by the masonry of the new firebox, so all the bricks that show will be the same size. This looks good, and it also makes them easier to replace, if that ever becomes necessary.

Once the first row is set in place, complete the rest of the underfire. You don't need to cover the entire masonry base you've poured, just enough of it to provide a solid surface on which to build the fireback and covings. The rest will be cov-

ered with rubble and masonry as you build up the firebox and fill in behind.

When the underfire is laid, draw in the lines of the fireback and covings. This is when the critical psychological problem arises. If you're not used to Rumford dimensions, the outline of your new firebox will look too shallow. You'll wonder if wood will fit, and whether such a fireplace could possibly draw well. Don't worry. A Rumford can be up to two-thirds more efficient than a squat, deep, modern fireplace. But be prepared for kibitzers telling you it won't work. This is such a predictable nuisance that I prefer to do this part of the job without an audience.

Fireback and covings—The night before you plan to lay the firebricks along the lines you've drawn, mix 2½ to 3 gal. of dry fireclay with enough warm water so that when a brick is

dipped in, ¹⁄₁₆ in. to ⅛ in. of the mixture will adhere to it. The next day, your technique will be to dip each brick's bed and headers into the fireclay and lay up the covings and fireback with joints ¹⁄₁₆ to ¹⁄₃₂ in. wide.

Begin by laying up the first course of the firebrick. You will build up the firebox using a half-lap pattern. The fireplace in the photos above has a front opening just 27 in. wide, so I used a single 9-in. firebrick to form the vertical fireback. Firebacks are usually more than one brick long. For a half-lap pattern, start by placing a brick on each side of the midpoint of the line you've drawn on the underfire. Then lay bricks along the line to a point at least half a brick length beyond its end. The next course will start with a brick centered above the midpoint line, laid up so that each brick overlaps half of each of the two bricks beneath it. It's a good idea to

Marking and Breaking Brick

For a simple straight cut, mark the brick as shown in figure A, and lay it on a cloth sack or cement bag filled with clean, screened sand.

Hold the chisel as shown, and give it a tap just heavy enough to score the brick along one of the lines you've drawn — don't attempt to break it with one blow. Score

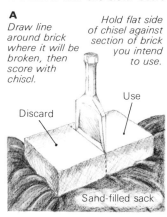

A
Draw line around brick where it will be broken, then score with chisel.

Hold flat side of chisel against section of brick you intend to use.

Discard

Use

Sand-filled sack

the other three sides, being careful to hold the chisel firmly against the brick, so it doesn't pop up after you strike it. The brick will break along the proper plane.

For the angled cuts required when the lower courses of firebrick and coving meet, the breaking technique is the same, but scribing

the lines to score along is somewhat more complicated (see figures B and C).

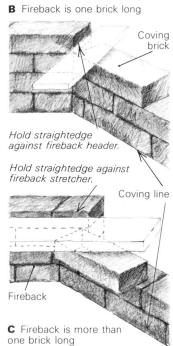

B Fireback is one brick long

Coving brick

Hold straightedge against fireback header.

Hold straightedge against fireback stretcher.

Coving line

Fireback

C Fireback is more than one brick long

Once the fireback begins sloping, scribing lines becomes a two-step process with the straightedge

because the angle must be cut through two planes (see figure D).

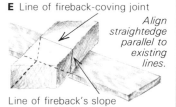

D
Coving brick
Sloping fireback brick
Straightedge
Draw line here as in figure C.

Transfer angle of straightedge to this point on coving brick.

To draw lines on the third and fourth sides, set your straightedge beneath the brick, align it parallel

E Line of fireback-coving joint

Align straightedge parallel to existing lines.

Line of fireback's slope

to the existing lines as shown in figure E, mark the edges, and connect the dots. Break the brick the same way as for other angles, striking the brick perpendicularly. —K.B.

Establishing the sloping fireback

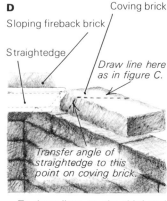

Lintel
Rubble and masonry fill
Firebrick
Throat
2-in. wide stick to form throat
Board indicates fireback's angle of slope
Wedge of fireclay
Fireback begins to slope when height equals depth

The throat of an efficient fireplace should be only 2 in. deep. Its depth and location determine the slope of the fireback. Use a stick to form the throat, and lean a board against it from the top of the vertical fireback. Then lean a firebrick against the board and fill in with a wedge of fireclay.

---

pre-mark the center of each brick, so that they can be set quickly in place. At either end, use a half brick, so that the courses come out the same length. Alternate these methods as you build the fireback up course by course.

With a single-brick fireback, the simplest place to join coving to fireback is along the fireback bricks' headers (photo facing page, center left). On longer firebacks, fit the deepest coving brick against the stretchers of the extended fireback bricks. Both of these techniques require cutting coving bricks at angles. Building up the firebox will require a number of such angled cuts, and even, as the fireback begins to slope forward, double-angle cuts. The drawings above show how to deal with this, probably the most technically difficult part of building a firebox.

After this brick is cut to the proper angle, set it aside, and begin laying bricks from the front. Set more bricks along the coving line until you're close enough to the fireback to bridge the gap with the brick you've set aside. Transfer the measurement of that distance to the short stretcher of the brick, cut it to length, and lay it in place.

The second course of the coving should begin with a half brick at the front to achieve the half-lap pattern, but the rest of the procedure is the same. Right and left covings should be mirror images of each other.

As you build up the fireback and covings, fill in behind with rubble and concrete (photo facing page, center right). You may be able to use much of the rubble you pulled out when you tore down the original firebox. You can also use broken brick you know you won't need to build up the new firebox. I usually use a concrete mixture of three parts sand to one part cement, though I often use a two to one mixture if it's already

around and handy. Add enough water to achieve a fairly wet consistency, so the concrete will flow to the bottom of the rubble.

Throat—When the fireback is as high as the firebox is deep, it's time to start sloping it forward. It is at this point that you have to decide how wide a throat your fireplace should have, because this will determine the angle of the slope. Almost all modern fireplaces have much too large an opening at the throat. Vrest Orton, in his book *The Forgotten Art of Building a Good Fireplace* ($3.95 from Yankee, Inc., Depot Square, Peterborough, N.H. 03458), says that the opening should be just 4 in. deep. I believe that even this is too much for anything smaller than a fireplace of a baronial hall. For most installations, 2 in. is more suitable.

Exactly where behind the breast will the throat fall? This is where the difference between building a fireplace from scratch and renovating one in the space allowed becomes most evident. In a classic Rumford, the throat will be perhaps a foot above the lintel. However, you probably won't have room to work much higher than a few inches above the lintel, so you'll have to form the throat there.

Begin by finding a stick of wood with a 2-in. dimension, and stand it vertically in the front of the firebox to mark the width of the throat. Remember that a milled 2x4 isn't the right size in any plane, but you can rip it to a true 2 in. along its nominal 4-in. face. Next, cut a flat board, perhaps a piece of plywood, just long enough to extend at an angle from the front of the fireback's highest course to the vertical stick at the point you want to form the throat.

This board describes the fireback's slope. Take a firebrick and, keeping in mind the half-lap pat-

tern you want to continue, reach behind the board and set the brick's front stretcher against the board's back. Fill beneath the brick with a wedge of fireclay, let it set up for 10 to 15 minutes, then take down the board and remove the vertical stick. Lay up the rest of the course, using your level as a straightedge to make sure that the rest of the bricks slope at the same angle, and allowing each wedge to set up. Once the slope is formed, subsequent joints on the fireback need no time to cure. Build the covings up course by course with the fireback. The fireback widens as it slopes forward, so you will have to extend each course a half brick or so beyond each end of the previous one.

As the fireback widens the covings get shorter, but the fireback's slope means that the final coving brick on each course must be cut at angles through two planes rather than just one.

If there was no original damper, you'll have to insert a new one before you build the firebox too high. Damper mechanisms can be bought at many lumberyards and brickyards. Get one to fit the width (not the 2-in. depth) of your fireplace's throat. You can wedge it up out of your way as you work to finish the covings and fireback. The rubble and masonry fill behind the fireback will create a flat smokeshelf on which the damper will eventually rest. You needn't fasten the mechanism down; just fill with concrete any gaps between metal and masonry.

You may light a fire in the firebox at any time during construction with no ill effects. Remember, though, that you wouldn't race your car until it's warmed up. Build a small fire at first, and only gradually stoke it up to a roar. □

Kent Burdett is an Oklahoman who has been installing and renovating fireplaces for 15 years.

A Russian Fireplace

Laying up a masonry woodstove with baffles and tons of thermal mass

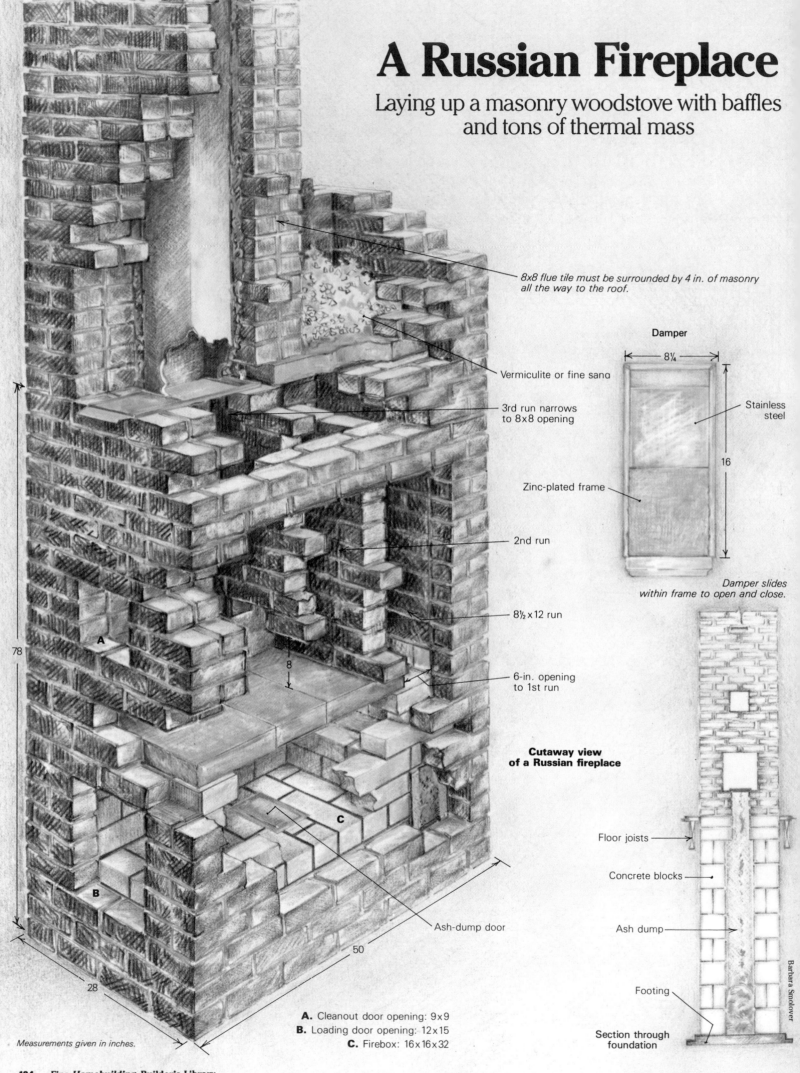

8x8 flue tile must be surrounded by 4 in. of masonry all the way to the roof.

Vermiculite or fine sand

3rd run narrows to 8x8 opening

2nd run

8½ x 12 run

6-in. opening to 1st run

A

B

C

78

8

28

50

Ash-dump door

Cutaway view of a Russian fireplace

Damper

8¼

16

Stainless steel

Zinc-plated frame

Damper slides within frame to open and close.

Floor joists

Concrete blocks

Ash dump

Footing

Section through foundation

Barbara Smollover

Measurements given in inches.

A. Cleanout door opening: 9x9
B. Loading door opening: 12x15
C. Firebox: 16x16x32

The revival of wood burning has kindled new interest in getting as much energy as possible from every log. Though there are lots of cast-iron and sheet-metal stoves on the market, the masonry stove may well be the most efficient means of heating with wood.

I first became interested in masonry stoves after reading an article in the February 1978 issue of *Yankee* magazine. It was through this article that I came to meet and work with Basilio Lepuschenko of Richmond, Maine, who had designed a masonry heater that is popularly called a Russian fireplace. Basilio and Albie Barden of the Maine Wood Heat Company have been responsible for generating most of the recent interest in masonry stoves. For a history of masonry stoves, see box on p. 137. The masonry heater in the drawing on the facing page is a modified version of the stove designed and copyrighted by Lepuschenko.

Masonry stoves have little in common with the traditional open fireplace. With their enclosed fireboxes and baffle systems, they are more like modern airtight metal woodstoves. But there are functional differences. Most obvious is that the masonry heater can store heat. Even the smallest unit contains about 64 cu. ft. of mass, and once the external walls heat up to their maximum 150° to 200°F, the brick will slowly radiate warmth for nearly 48 hours.

A stove of this design has to have a minimum of three runs and two baffles. The smoke must go up, pass down over the hot firebox, and go up again. Larger units have four baffles, and smoke makes an additional pass down and up before exiting up the stack.

Another difference is that a masonry stove is not designed to burn continuously. It is fired up only once or twice a day. A quick, hot fire with relatively small, split wood is ideal. After the wood has been reduced to ashes, a sliding damper located at the end of the last baffle is closed, trapping all of the heat inside, allowing it to warm the room throughout the day or night. Temperatures of up to 1,200°F ensure the secondary combustion of gases in the baffle system and eliminate the danger of creosote buildup. (The only time I've ever seen creosote build up in a masonry heater was in one with baffles much longer than usual. Baffles longer than about 48 in. let the temperature drop enough for creosote to begin to form.)

Firing up this way, wood consumption is drastically reduced. Even the most ardent wood-burners should be willing to forsake their dream of a 24-hour burn for a 60% reduction in the amount of wood required to heat their homes.

Masonry stoves are relatively simple to build, but designing a heater into the floor plan can present some difficulties. Proper clearance from combustible material like framing and siding is a must. The minimum is 36 in. on all sides.

In order to take full advantage of the stove's ability to radiate heat, it should be centrally located. The stove shown here is 50 in. long and 28 in. wide. It is complete at about 6½ ft. above the floor, and from there, a single 8x8 flue tile extends to join up with those from other fireplaces.

I like to see floor plans designed around the masonry stove rather than awkward attempts to squeeze one in somewhere like an additional closet. Masonry heaters work so well that I dislike compromising their efficiency for mere architectural considerations, but my feelings in this regard border on masonry madness and should be viewed as such. Suffice it to say that space and clearance can be a problem. Ten tons of hot brick can't go just anywhere.

There is no single correct way to build a masonry stove. Flue configurations are limited only by the builder's imagination, and many stoves differ in this regard. The basic design of Lepuschenko's Russian fireplace has proven itself efficient. Since my introduction to Basilio I have built four stoves to his general design, though no two were exactly alike.

Every masonry stove has its own design problems and requires some modification to fit within a given house or masonry system. Brickwork isn't nearly as easy as it looks. If you haven't got the patience to plan precisely and proceed carefully, or if you have a low frustration threshold, you may want to leave the construction of your masonry heater to a mason.

One of the main differences between this stove and those of Lepuschenko's design is the thickness of the walls. Because of strict building codes in Connecticut, where I work, the walls of this stove are 8-in. double-brick thickness throughout. The ones I learned to build with Lepuschenko in Maine were a single 4-in. brick thick in the baffles above the firebox. The double thickness increases the thermal mass substantially but also lengthens the time the unit takes to heat up. I personally think a single 4-in. brick wall is the way to go, and that additional thickness is unnecessary for either safety or thermal storage.

Another variation in this particular unit is the use of high-temperature refractory cement in the laying of the firebox. I lay all the firebrick used in the construction with a premixed, air-setting, high-temperature cement, which provides a strong bond with little deterioration after extended firing. I also use either precut arched firebrick or 12-in. by 24-in. by 3-in. refractory slabs to form the ceiling of the firebox. The arched firebox ceiling provides greater strength than the refractory slabs, and I like to use it in five-run masonry stoves.

The dimensions of a masonry stove are based upon the length of the individual bricks, the number of baffles and the location of the stove. There are some key factors to consider, however, when designing a masonry heater.

Firebox dimensions are important. The firebox of this stove is 12 in. wide by 16 in. high by 32 in. deep. Height and width should remain the same even on larger stoves, although the depth can be increased to 40 in. Substantially larger fireboxes and the resulting larger fires could cause increased thermal stress and minor hairline cracks in the mortar joints. In masonry heaters, bigger is not always better.

A three-run Russian fireplace built on Lepuschenko's model will heat an area of 600 sq. ft. or more. Significantly larger areas would require a five-run heater. Both work well.

Simplicity and availability of materials are important features of Lepuschenko's design. The firebrick slabs and precut arched firebricks are available only from a refractory-materials supply firm, as is the high-temperature cement. All other materials are sold by masonry suppliers. No special tools are required, although I often use a diamond-blade brick saw to make the necessary cutting faster and more precise. In most cases, a mason's hammer or a brick set will do (for brick-breaking tips, see pp. 105, 117, and 133). The standard cast-iron cleanout and loading doors used on the stove can be bought at most masonry supply yards. The sliding damper is stainless steel in a zinc-plated frame. This is one item that is not readily available. I have a friend with access to sheet-metal fabricating equipment, who can make one up for me in about two hours.

Construction—First make sure the floors have been framed so there's enough room for the masonry to pass through. A plumb line, a ruler and a little thought are all you need. Once the layout has been established from basement to roof and all the necessary carpentry has been done, it's time to start.

Foundation work for a masonry stove is conventional—using the same materials (concrete blocks) and procedures you'd follow for a traditional fireplace. For this job, we poured a footing 12 in. deep in the basement, 5 in. larger all around than the external dimensions of the heater. Then we built up to the first floor level with 8-in. block laid to the same dimensions as the heater. The whole center of the foundation becomes an ash dump, so we left an opening for a cleanout door. If the stove is being built on a slab, there's a lot less work, but you still need that 12-in. thick footing.

Once the block foundation is at floor level, you need to cap it. Often you can just span the gap with 4-in. solid block, leaving an opening for the ash dump, and building up courses to floor level.

The courses of brick begin above the floor joists. I use plumb lines to keep the brickwork true. I think this makes the construction both quicker and better looking. This is also why I lay corners first and use a horizontal line to lay the brick between them.

The firebox sits about 12 in. above the floor. Its firebrick base is laid and leveled in standard mortar (I mix mine with one part portland cement to one part lime to four parts clean sand, adding water as needed). The opening for the ash dump should be close enough to the loading door to allow access, and it must fit flush with the top of the firebox floor. To accomplish this, I notch the firebrick ¼ in. deep, using a contractor's saw with a carborundum blade.

Next, I lay six to eight courses of the external brick walls, leaving an opening for a 12-in. by 14-in. loading door at the front. It's important that this brick be laid with full joints, and that any excess mortar be removed, and the remainder jointed smooth on the inside as well as out. This will ensure a tight fit between the firebox and the external brick.

The firebricks of the firebox walls should be

Paul Lang of Newtown, Conn., has been a mason for eight years.

laid out dry to determine its exact depth. They are then laid in refractory cement, which can be thinned and the bricks dipped, or left as is and applied with a trowel. I trowel it on because it is easier for me.

The firebrick must be set tight against the exterior walls, because any airspace will act as an insulator. There should be no cement or mortar between interior and exterior bricks. In this stove, I used a layer of ⅛-in. mineral wool as an expansion joint between the firebrick and the external brick. This was an experiment. I've read that master Finnish stove-builders employ joints of this type to allow for some movement of the firebrick without hindering the conduction of heat.

At the rear of the firebox, I build a shelf two or three bricks high and about one brick's length deep (photo left). On this job, I used cement block to form the shelf, then covered it with firebrick. This is the beginning of the flue passage. This first vertical run is the hottest spot in the stove. It must always be built to double brick thickness, even if the rest of the stove is only one brick thick. I raise the external brick walls another few courses—enough so I can install the 3x3x¼ angle iron 20 in. long that spans the top of the loading opening. Then I continue the firebox walls to a height of about 16 in. before laying its ceiling.

The ceiling of the firebox can be formed with cantilevered firebrick, refractory arches or firebrick slabs. I chose slabs for this stove because they are quick to install, they fit nicely within its dimensions and they are strong enough for a heater this small. I use refractory cement for this, too. The first slab is set tightly in place against the front wall just above the loading opening, and all three slabs butt together to form a continuous firebox ceiling that stops 6 in. to

Above: Lang sets firebrick at what will be the hottest part of the masonry heater: the beginning of the first run. The mineral wool between the exterior common brick and the firebrick along the side of the stove is experimental—an attempt to eliminate stress cracking. The ash-dump door is set in the floor of the firebox, while the loading opening is spanned with angle iron. A cast iron door will be hung after the brickwork on the stove is completed.

Left: The entire heater is of double-wall construction. Lines dropped from above ensure plumb at each corner. The two cantilevered bricks at the center are temporarily supported by scrap. A third brick will span the gap to form the base of the second baffle, allowing the smoke to pass beneath. The opening just begun in the front of the stove, top, is the cleanout.

Facing page: The first two vertical runs are sealed off with refractory slabs, surrounded by mortar and common brick. Several courses of brick will cover the slabs, then the outer walls will be laid up to the ceiling. Sand or vermiculite will be poured into the cavity as a firestop.

8 in. from the back wall to leave room for the first vertical run.

Next I lay six to eight more courses of external brick, incorporating a cleanout opening in the front wall just above the firebox ceiling. As with the loading opening, the cast-iron door can be installed later. This opening is helpful in cleaning up fallen mortar during construction, though its primary function is to allow access for inspection and cleaning of the first two runs during use. Properly built and used, masonry heaters should require no cleaning at all. Only the repeated burning of green wood could cause creosote to build up inside.

I lay out the base of the baffles dry to position them so each of the three vertical runs will be the same size. Runs should be 8 in. by 12 in., though small variations are all right. The first baffle is built right on top of the firebox ceiling. The second requires an 8-in. opening at its bottom. The easiest way to form this opening is with cantilevered bricks (photo facing page, bottom). Both baffles must be tied into the side walls of the heater. I make this bond by laying a baffle brick into the interior side wall every four or five courses. In a heater with walls only a single brick thick, the 4-in. butt end of these bricks will show on the outside. External and internal walls go up several courses at a time, with the baffles continuing upward within them.

This masonry stove will draw well with an ordinary 8-in. by 8-in. flue tile. The final run is narrowed to this dimension with cantilevered brick, and the damper is set in place. The damper and frame are only ⅜ in. thick, and fit into the mortar joint of the external brickwork.

A flue tile is set atop the damper, and the brick is laid around the flue to ensure a tight fit. In this stove, the clay flue tile had to angle sharply to clear the framing of the second floor. Ordinarily,

the flue would continue straight up, and be surrounded by 4 in. of masonry.

The unit is ready to be capped off about 4 ft. from the top of the firebox. The first baffle ends 8 in. from the top, and the second continues all the way up. At this point, the first two runs of the heater are sealed off. I use firebrick slabs, as shown in the photo below, but cantilevered brick would do just as well. The slabs are quicker to install but more expensive.

The firebrick slabs should be covered with a couple of courses of brick so that there is at least 8 in. of masonry over the vertical runs. At this point the brick can stop, but I usually like to continue the external walls up to the ceiling because the stove looks more complete that way. The resulting interior space can then be filled with fine sand or vermiculite to act as an additional firestop.

After the stove has been completed, stoutly resist the temptation to fire it up immediately. A great deal of moisture is trapped inside the masonry and must be allowed to evaporate completely. The mortar must cure slowly. The ideal curing period is three or four months, though few owner-builders can be so patient. Firing early can cause the moisture to be driven out quickly, causing possible stress cracking. Fires should be small at first, allowing a break-in period of about three days.

Whether you choose to build a masonry heater or oversee the construction of one, try to keep one thing in mind. The internal brick of the heater should be laid with the same care as the external walls. Though the flue passages will never be seen after the heater is completed, they are really the heart of the stove. The reward for careful craftsmanship can be an object of functional beauty that will heat your home for many years to come. □

A History of Masonry Stoves
by Albie Barden

In Europe the shift from a hunting-gathering existence to an agricultural society produced both permanent shelters and bake ovens. The primitive European oven was also a wood-fired masonry heater. A fire was built directly in the clay and stone oven. Once the fire burned out, coals and ashes were removed, and bread was put in to bake.

In Central Europe and the Alps, masonry heaters and bake ovens have been built for several hundred years. These have a firebrick core and a ceramic exterior, often surrounded by warm benches reserved for the elders of the family. Sometimes the heaters are fired from a kitchen and project into a dining-living room. Sometimes they are fired from a hall or some other service area.

In Scandinavia, the tall rectangular and cylindrical styles of the heaters that evolved serve no baking function. These massive, tile-covered radiation heaters were placed in rooms that required heating. In Russia, Lithuania, Poland, Czechoslovakia and Finland, heaters more commonly used brick or whitewashed stone than the fancier tile used for heaters in Germany, Austria and Switzerland.

The brick heaters currently becoming popular in the United States and Canada have their origins in the simple brick heater, or *grubka*, found in nearly every home of the great Russian land mass. Their efficiency and rustic charm attracted American attention at least once before, nearly 200 years ago when Ben Franklin and John Adams saw them in Europe.

In the late 1700s, models of the European brick stove were brought to Salem, Mass., by John Dodge, an enterprising sea captain who quickly patented and began marketing the idea, until his business was brought to an abrupt end by a heart attack. Apprentices carried on his work for a brief time, and elsewhere in New England other heaters of brick, stone or soapstone were built.

In the 20th century, a modest tradition of Russian-style masonry heaters was sustained by a small White Russian community in Richmond, Maine. At the same time Sam Jaakkola, a Finn in western Maine, built several Finnish-style heaters, the first masonry heaters I ever saw. In contrast to the Russian-style heaters, which are end-loading and work well as room dividers, the Finnish-style heater is front-loading and doubles as an open fireplace, with large doors you can open to see the flames.

The best research on brick-and-mortar heaters is being done in Finland, where a leading architect-designer, a major foundry, brick and firebrick manufacturers and a cement firm are all working together to design and test projects using vocational schools as construction sites and students as their labor.

In the United States, the New Mexico Energy Institute in Albuquerque and the Southeast Community College at Beatrice, Nebraska, both have home-grown research programs on masonry-heater construction.

Albie Barden's Maine Wood Heat Company sponsors workshops on masonry heaters, sells plans and materials,and publishes the Masonry Stove Guild Newsletter. *(For information, send a self-addressed, stamped envelope to RFD 1, Box 640, Norridgewock, Maine 04957.)*

The Fireplace Chimney

Flashing and capping are the tricky parts of the job

by Bob Syvanen

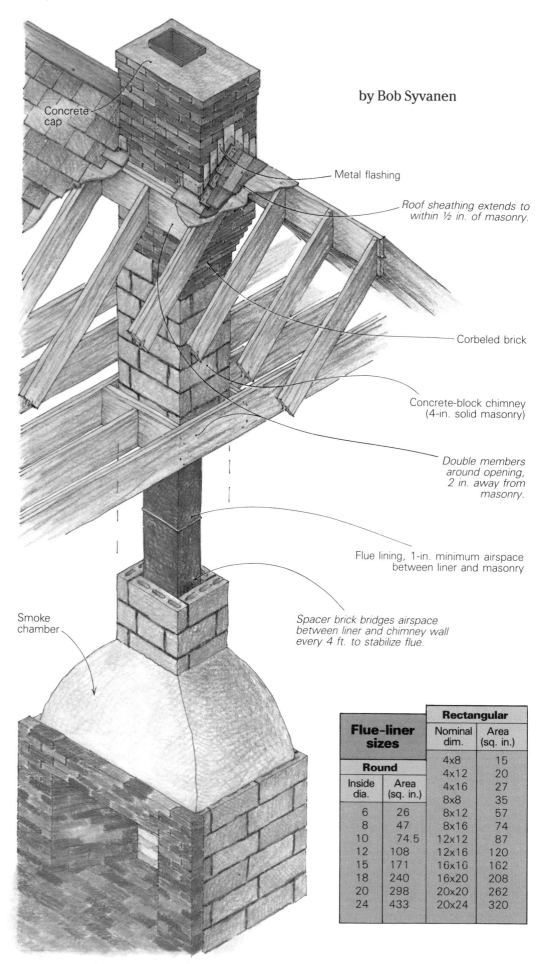

Concrete cap

Metal flashing

Roof sheathing extends to within ½ in. of masonry.

Corbeled brick

Concrete-block chimney (4-in. solid masonry)

Double members around opening, 2 in. away from masonry.

Flue lining, 1-in. minimum airspace between liner and masonry

Smoke chamber

Spacer brick bridges airspace between liner and chimney wall every 4 ft. to stabilize flue.

Flue-liner sizes		Rectangular	
Round		Nominal dim.	Area (sq. in.)
Inside dia.	Area (sq. in.)	4x8	15
		4x12	20
		4x16	27
		8x8	35
6	26	8x12	57
8	47	8x16	74
10	74.5	12x12	87
12	108	12x16	120
15	171	16x16	162
18	240	16x20	208
20	298	20x20	262
24	433	20x24	320

Throughout most of history, masonry chimneys were just hollow, vertical conduits of brick or stone. These single-wall chimneys have many problems: they conduct heat to the building's structure; they aren't insulated from the cold outside air, and so allow severe creosote buildup as the cooling gases condense; and they suffer from expansion and contraction, which lead to leakage of water (at the juncture of chimney and roof) and smoke (through cracks in the masonry).

In present-day chimneys, ceramic flue tile carries the smoke. It can withstand very high temperatures without breaking down, and also presents a much smoother and more uniform passageway, which means easier cleaning and thus less chance of chimney fires. Brick or concrete block, laid up around the tile but not in contact with it, serve as a protective and insulative layer.

Ceramic flue tiles can be either circular or rectangular in section, and are available in a number of sizes (see the chart below left). Brick, block and mortar are the only other materials you need to build a chimney (to find out how to build a fireplace, see the article on pp. 118-122). John Hilley, a mason I've worked with for the last eight or nine years, uses type S mortar throughout the entire chimney, but some codes require refractory cement to be used for all flue-tile joints.

Requirements and guidelines—Before you start building a chimney, you've got to consider sizing, location and building-code requirements. I'll be talking about masonry chimneys that are built above fireplaces, but most of the construction guidelines also apply to masonry chimneys that are meant to serve a woodstove, a furnace or a boiler.

The Uniform Building Code ($40.75 from I.C.B.O., 5360 S. Workman Mill Rd., Whittier, Calif. 90601) contains basic requirements for chimney construction, and most states and municipalities have similar rules tailored to meet particular regional needs. I have a 1964 U.B.C. that reads the same as the 1984 Massachusetts State Building Code. It calls for a fire-clay flue lining with carefully bedded, close-fitting joints that are left smooth on the inside. There should be at least a 1-in. airspace between the liners and the minimum

Consulting editor Bob Syvanen lives in Brewster, Mass. Photos by the author.

Illustrations: Christopher Clapp

Just below the roofline, the chimney wall shown above changes from 4-in. thick concrete block to brick. Block is much faster to lay up than brick, but it's usually used where appearance isn't important. The brick wall has been corbeled out against a temporary plywood form, enlarging the chimney above the roof and also centering it on the ridge. From this stage on, a good working platform on the roof is essential, right.

4-in. thick, solid-masonry chimney wall that surrounds them. Combustion gases can heat the liner to above 1,200°F, and the space between liner and chimney wall allows the liner to expand freely. Only at the top of the chimney cap is the gap between liner and wall bridged completely, and here expansion can be a problem. We'll talk about this later.

Chimney height is pretty much dictated by code. According to my Massachusetts Building Code, "All chimneys shall extend at least 3 ft. above the highest point where they pass through the roof of a building, and at least 2 ft. higher than any portion of a building within 10 ft." This rule applies at elevations below 2,000 ft. If your house is higher above sea level than this, see your local building inspector. The rule of thumb is to increase both height and flue size about 5% for each additional 1,000 ft. of elevation.

Exterior chimneys must be tied into the joists of every floor that's more than 6 ft. above grade. This is usually done with metal strapping cast into the mortar between masonry courses. No matter where your chimney is located, it should be built to be freestanding. The chimney can help support part of the wood structure, but the structure shouldn't help support the chimney. And there should be at least 4 in. of masonry between any combustible material and the flue liner.

The U.B.C. also has standards for flue sizing. For a chimney over a fireplace, the interior section of a rectangular flue tile should be no less than $\frac{1}{10}$ the area of the fireplace opening. If you're using round flue tile, $\frac{1}{12}$ the area of the fireplace opening will do because round flues perform slightly better.

Rectangular flue tile is dimensioned on a 4-in. module—8 in. by 8 in., 8 in. by 12 in.,

8 in. by 16 in., 12 in. by 12 in., and so on. Standard tile length is 2 ft., though you can get different lengths from some masonry suppliers. When a mason talks about a 10-tile chimney, he usually means that it's at least 20 ft. high.

Using the flue-sizing formula isn't a guarantee that your chimney will draw properly. The chimney height and location, the local wind conditions, the firebox type, and how tight the house is are all factors that influence performance. John Hilley has found that on Cape Cod (at sea level), a 10-tile chimney atop a shallow firebox will work fine with flue sizing as low as 7% of the fireplace opening. Generally, short chimneys won't draw as well as taller ones. If you've got any doubts about what size flue tile to use, it's a good idea to ask an experienced mason or consult with your local building inspector.

Chimney construction—The fireplace chimney starts with the first flue tile on top of the smoke chamber of the fireplace (see the drawing on the facing page). As described in the article on pp. 118-122, the smoke chamber is formed by rolling the bricks of each course above the damper to form a strong, even, arched vault with an opening that matches the cross section of the chimney's flue tile. Since the smoke chamber will carry the weight of the chimney above it, it's got to be soundly constructed. If the rolled bricks of the chamber form an even, gradual arch, there should be no problems.

To begin the chimney, seat the first flue tile on top of the smoke-chamber opening in a good bed of mortar. Building paper or an empty cement bag on the smoke shelf will catch mortar droppings. You can remove the

paper by reaching through the damper when you've finished the chimney.

Lay the chimney wall up around the flue tile, maintaining a 1-in. minimum airspace between tile and chimney wall. It's best to lay one flue tile at a time, building the chimney wall up to a level just below the top of each liner before mortaring the next one in place. Use a level to keep the liners and the chimney walls plumb. The mortar joint between tiles should bulge on the outside, but use your trowel to smooth it flush on the inside.

At every other liner, Hilley bridges the 1-in. airspace with two bricks or brick fragments that butt against the flue tile. These should extend from opposite sides of the chimney to opposite ends of the liner. This stabilizes the flue stack without limiting its ability to expand and contract with temperature changes.

If the chimney wall will be hidden behind a stud wall or in an attic, you can use 4-in. thick concrete block, which can be laid up much faster than brick. You can then switch to brick just below the roofline. Below the roof, chimney size is usually kept to a minimum to save space. If a larger chimney is desired above the roof, the chimney walls can be corbeled in the attic space (photo above left). The corbel angle shouldn't be more than 30° from the vertical (an overhang of 1 in. per course). A temporary plywood form set in place at the desired angle works well as a corbeling guide.

Coming through the roof—Roof sheathing should extend to within $\frac{1}{2}$ in. of the chimney, with structural members around it doubled and no closer than 2 in.

The next step in this part of chimney construction is to set up a good working platform on the roof (photo above right). Staging for

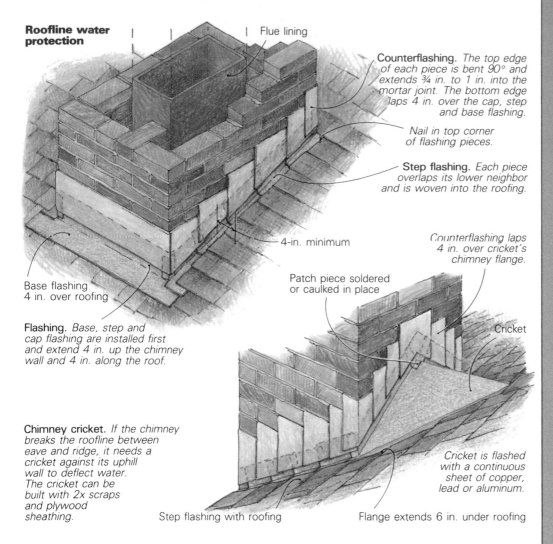

Roofline water protection

Flue lining

Counterflashing. *The top edge of each piece is bent 90° and extends ¾ in. to 1 in. into the mortar joint. The bottom edge laps 4 in. over the cap, step and base flashing.*

Nail in top corner of flashing pieces.

Step flashing. *Each piece overlaps its lower neighbor and is woven into the roofing.*

4-in. minimum

Counterflashing laps 4 in. over cricket's chimney flange.

Patch piece soldered or caulked in place

Base flashing 4 in. over roofing

Flashing. *Base, step and cap flashing are installed first and extend 4 in. up the chimney wall and 4 in. along the roof.*

Cricket

Chimney cricket. *If the chimney breaks the roofline between eave and ridge, it needs a cricket against its uphill wall to deflect water. The cricket can be built with 2x scraps and plywood sheathing.*

Cricket is flashed with a continuous sheet of copper, lead or aluminum.

Step flashing with roofing

Flange extends 6 in. under roofing

chimney work must be steady, strong and roomy. You can rent steel staging or build your own from 2x material. Either way, check the soundness of your staging well before you load it up with bricks and mortar. And before you bring any mud up on the roof, lay down a dropcloth to keep the roof clean.

To get to the staging, I put a ladder on the roof with its upper end supported by a ridge hook. It's a good idea to put some padding between the eave and the ladder rails so that the edge of the roof isn't damaged as you trudge up and down. You'll need a second ladder to get from ground to roof, unless the roof is steeply pitched. In that case, you can simply extend one ladder (if it's long enough) from ridge to ground.

Flashing—You're bound to find generously caulked flashing if you look closely at the chimneys in your neighborhood. I've lost track of the patch jobs I have done trying to stop chimney-flashing leaks. What sometimes happens is that rain gets blown in behind this flashing during a storm. Infrequent leakage doesn't mean the flashing job was poorly done, and caulking is a good stopgap in cases like this. But if your flashing leaks regularly in rainy weather, it probably wasn't installed correctly in the first place.

The Brick Institute of America (1750 Old Meadow Rd., McLean, Va. 22102) recommends flashing and counterflashing at the roofline. This creates two layers of protection,

with the counterflashing covering the base and step flashing on all sides of the chimney.

The Brick Institute recommends the use of copper flashing, but today you'll see more widespread use of aluminum, since it's much less expensive. Through-pan flashing (explained at right) is usually done with lead.

If your chimney straddles the ridge, first install the base flashing against the two chimney walls that run parallel with the ridge. Then step-flash the sides. Each flashing piece should extend at least 4 in. onto the roof, and is held with one nail through its upper corner.

Install counterflashing over the step and base flashing. The bottom edges of the counterflashing overlap the base and step flashing by 4 in. (drawing above left). The top edges of the counterflashing are turned into the masonry about ¾ in. They can be cast into the mortar joints as the chimney is built, or tucked into a slot cut with a masonry blade after the chimney is finished.

If the chimney is located against the side of the house or in the middle of a sloping roof, then you need to build a cricket against the uppermost chimney wall (drawing, above right). Otherwise, water will get trapped here, and you'll eventually have a leak. I use scrap 2xs and plywood to construct the slope, then cover the cricket with building paper and flash it with a large piece of aluminum or copper. The cricket flashing should extend 6 in. under the shingles and be bent up 4 in. onto the masonry, where it's covered by counter-

Through-pan flashing

Driving rainstorms can cause a lot of water to penetrate a chimney through cap and brick. Even when perfectly installed, conventional flashing can do little to stop this kind of penetration. The best way to drain out water that gets between the outer brick wall and the flue lining is to install through-pan flashing. Through-pan flashing is just what it sounds like—a continuous metal pan sloped from the flue lining to the roof. Weep holes between bricks just above the pan provide drainage.

There's some controversy about the effect that through-pan flashing has on the strength and stability of the chimney, since it breaks the mortar connection between bricks in adjacent courses. According to the Brick Institute, "if there is insufficient height of masonry above the pan flashing, wind loads may cause a structural failure of the chimney." This might be a valid warning, but I've never seen this kind of structural problem here on Cape Cod, which has its share of windy weather. And there's no arguing that through-pan flashing creates a more complete water barrier at the roofline than the conventional flashing scheme. Unless you live in earthquake or hurricane country, I can't see any reason not to use this system.

You can use copper or lead for through-pan flashing. John Hilley prefers lead because it's less expensive than copper, more malleable and generally easier to work. A utility knife will cut the stuff. Lead isn't supposed to last quite as long as copper, but I've seen 50 year-old lead pans that are still in good condition.

Building the curb—Before you can install the pan, you've got to build a curb where the chimney walls come through the roof. This is done with a combination of angled bricks or blocks and mortar, set in a form made from 2x lumber (top photo, facing page). The side walls of the curb should match the slope of the roof, and be around 1½ in. above the roofline. The lower and upper walls of the curb should be 4 in. to 6 in. above the roofline, parallel with it, and level. Like chimney walls, curb walls should be at least 4 in. deep.

Top the curb with a smooth layer of mortar, and the next day, rub the surface with a brick and round the corners to soften any sharp edges that could pierce the lead. Then install base, step, and cap flashing against the curb. With through-pan flashing, the pan takes the place of the counterflashing that is usually attached to the chimney walls.

The lead pan—If the joint between flue tiles falls just below or just above the roofline, then one piece of lead works well since you can roll it out on the base, find the outline of the flue, and cut holes for the next flue section to fit through. Alternatively, two or more sheets of lead can be used to make the pan. Just overlap the joints 6 in.

To determine the size of the pan, add 20 in. to the length of the chimney's lower wall and 24 in. to 32 in. to its side-wall measurement. Measure the side wall by following the angled curb with your tape measure. The chimney shown here is 64 in.

wide by 32 in. deep, and it required a sheet of lead 84 in. by 60 in. (7 ft. by 5 ft.). Lead sheets usually come in even foot widths and different lengths. For this job, Hilley trimmed a 6-ft. by 8-ft. sheet to size. These dimensions work for a chimney straddling the ridge. When the front and back walls of a chimney are on the same side of the ridge, the lead on the back or upper side of the chimney should be long enough to extend under two shingle courses plus one inch (more on a steep roof).

With the curb finished and flashed, and the lead cut to size, the next step is to install the pan. Roll the lead out over the curb and position it symmetrically (photo center left), then press down gently to find the outline of the flue tile.

Hilley cuts the hole for the tile about 2 in. smaller than the outside dimensions of the flue. Then he makes relief cuts in each corner and folds up the lead so that the flue can slide through it.

You have to be careful when handling the lead sheet; it's surprisingly easy to tear and puncture. The easiest way to carry a sheet is to roll it up. Never form lead with a hammer. Use your hand, a block of wood or a rubber-handled hammer handle. If you do pierce or tear the pan while installing it, pull the hole up above the surrounding pan so that water will drain away from it. You can also mend a hole by parging it over. As you position the pan, keep in mind that the object is to direct water away from the flue.

Once the flue tile that extends above the pan is in place, pull the lead up the sides of the flue to achieve the necessary outward slope. You can stiffen the top edge of the pan around the flue by folding it over. This helps to prevent sagging. After the lead is formed up tightly around the flues, parge the lead-to-flue joint with mortar.

Laying up the brick—Lay the first brick course over the pan in a thin bed of mortar (photo center right). Be careful to follow the curb under the lead. Make a few weep holes at the lowest points of the pan. You can do this by temporarily inserting a twig or rope in the mortar between bricks, or by leaving a vertical joint between bricks open. You'll have to cut the bricks that fit just above the sloping sides of the pan. Use either a mason's hammer, a chisel or a masonry blade. Once the courses on both sides of the ridge join, finish the chimney just as you would if there were no through-pan flashing.

Finish the through pan by creasing its exposed corners, trimming the pan edges and bending them down against the roofing. Before you start this part of the job, carefully lift the lead and sweep out any loose rubble. Creasing the corners so that they look nice is hard to do. The easiest way to start the crease is by slipping a short length of wood or angle iron under the pan at the corner. Hold the wood or metal edge so that it bisects the 90° corner, and gently start to form the lead over it with your hand. Then, if necessary, use a scrap 2x4 or a similar tool to form the lead into a more defined crease, as shown at right. Sharpening the crease should force the pan down against the roofing. A second crease, close to the first one, will force the pan down even farther. —*B.S.*

The curb for the through-pan, above, is a combination of bricks and formed concrete, built directly on the chimney walls where they intersect the roofline. Curb walls should be at least 4 in. thick. The side walls of the curb are sloped to match the roof pitch, and 1½ in. above the roofline. End walls are level, parallel with the ridge and 4 in. to 6 in. above the roofline. At left, a lead sheet is cut to fit over the flue tiles and carefully rolled and bent over the curb. This forms the through-pan that will drain water away from the center of the chimney and onto the roof. Below, the pan is parged to the flue, and the first course of the chimney cap is mortared to it, bearing squarely on the curb.

The last step is to crease the corners of the pan, trim its edges and bend them down against the roof. Wood blocks and gentle hand pressure are used in this final forming.

flashing. Once the cricket is finished and flashed, flash and counterflash the three remaining sides as mentioned earlier.

The cap—The chimney cap covers the airspace between flue tile and chimney wall, stabilizing the flue and directing water away from the rest of the masonry (drawing, below left). It's good to corbel out the chimney's top brick courses or to install a cap that overhangs the chimney walls. This will direct runoff onto the roof rather than onto the brickwork.

You can buy precast caps in a few sizes or cast your own in place. Installing or forming a cap is more complicated than it sounds because of the way the ceramic flue behaves. Cross-sectional expansion of heated flue tile is accommodated within the 1-in. airspace between tile and chimney wall. The concrete chimney cap bridges this gap, and if it's mortared directly to the tile, you're bound to have cracking and breaking problems. Even if the upper tiles stay cool and don't expand widthwise, the tiles near the fire will expand along their length, forcing the entire flue upward.

One way to accommodate vertical movement of the flue is to create an expansion joint between brick courses just below the cap. The topmost flue tile, cast to the cap, forces it upward as the flue stack expands. The cap's weight closes the expansion joint as the stack cools. Hilley creates his expansion joint by sprinkling a thin layer of sand on the brick course before laying down the mortar (photo top left). The sand prevents the mortar from adhering to the brick directly beneath it, but doesn't affect the bond to the bricks above.

The chimney top's first corbeled course creates an inside ledge that will support a form for the concrete cap (photo center left). You can use corrugated sheet metal or scrap wood for the form. If you use wood, as Hilley does, make sure it sits loosely on the brick ledge, and soak the wood in water before pouring the cap. This way, the wood won't swell and force bricks out of their bond.

The last flue tile should project above the level of the last brick course at least 2 in. so the concrete cap can slope down toward the edges of the chimney (photo bottom left).

Forming the concrete cap is the last step. Hilley casts the cap directly to the topmost flue tile, relying on the control joint several courses below to accommodate flue-stack expansion. An alternate method is to cast an expansion joint around the topmost flue tile. To do this, pack some ⅜-in. dia. backing rope around the top flue tile where the cap will fit, and then caulk the space with a flexible, non-oil-base sealant after the cap is cast. □

The chimney cap

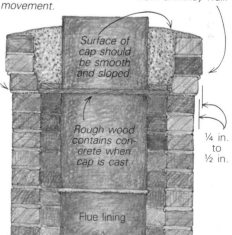

Expansion joint opens and closes with flue movement.

Top three to four brick courses are corbeled out to bring drip edge away from chimney wall.

Surface of cap should be smooth and sloped.

Rough wood contains concrete when cap is cast.

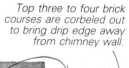
¼ in. to ½ in.

Flue lining

Capping the chimney. **Top, John Hilley creates an expansion joint with a loosely spread layer of sand beneath the mortar. Located several courses below the chimney top, this joint will widen and shrink with the normal expansion and contraction of the flue tile. The wood will be part of the concrete cap's bottom form. At center, the cap is cast between the flue and the top three brick courses. At left, the completed cap slopes away from the flue.**

Maine Stonework

A personal approach to laying up and pointing rock

by Jeff Gammelin

My stonemasonry is inspired by the ledges and outcroppings of coastal Maine, and my approach is the result of studying these natural formations, in which the stones lie upon each other with at least one point of contact. The joints are tight, and I try to create, through a minimal use of mortar at a given point, the feeling that there is at least one spot where each stone rests directly upon another. This point of contact is one of the basic distinctions of my stonework.

When my wife and I began building with stone in our home, we realized our focus on making a collection of individually beautiful stones was shortsighted. The stones by themselves were beautiful, but more important, they needed to complement each other with regard to shape, color, texture and size. Using a wide variety of stones, from string-bean size to 500-lb. mantel stones, provides a contrast that suggests the interdependence so evident in natural rock formations.

My crew and I use freeform arches and serpentine lines within the mass because they imply strength, and we accomplish a feeling of

Stone is an elemental part of nature. It is the stonemason's task to take stone from its natural context and to use it in forms that participate harmoniously with their surroundings. This translation from the natural to the controlled reveals the individuality of the stonemason's art. Every mason I've seen lays stone differently. My methods spring from the style of my work. For example, I want to keep the individual stones as natural-looking as possible once they are laid. To do this we begin by laying up dry a small group of stones using small rock chips as shims. My crew and I work mostly with granite, and a bit with basalt and sedimentary stones. If stones need to be shaped, we do it with carbide-tipped points (9-in. long steel bars with a point on one end and a flat for striking on the other), hand

sets or chisels. Large granite stones that need to be worked for steps, mantels or lintels are cut. This is done by snapping a line, tracing it several times, drilling 2-in. deep holes every 4 in. to 8 in., and inserting half-rounds and wedges in the holes. The wedges are driven until the stone breaks. A hand set and point are used for trimming. We soften the resulting sharp edges by rubbing or crushing them with the stone hammer or by filing them with a rasp. Hammer marks are hand-rubbed to darken them.

Before we lay a group of stones, we often mark each one with chalk to show its position relative to the others, and to show how far the mortar line should extend to the face of the stone. This helps us control shadow and texture, and is useful on corners and in niches.

Mortar and pointing—We usually use dark masonry cement which, along with the varying degree of horizontal stone surface exposed, produces an interplay of shadow and texture. We work with about five different viscosities of mortar, depending on temperature and humidity, the type and size of the stones, and on whether it's interior or exterior work. Basically we like a rich, plastic mortar—wet enough for a thin vertical application but dry enough to support the stone. A good understanding between mason and tender is important to a good mix. Making a batch of mortar ideally suited to the conditions takes a lot of practice, but the influence it has on the smoothness of the operation cannot be overestimated.

The mortar is spread just thick enough to sup-

From *Fine Homebuilding* magazine (May 1985) 26:69-71

restful substantiality by using strong, solid-looking stones at the corners.

Our stonework falls roughly into two categories: composition within a rectangular framework, and the creation of more organic or sculpted forms. Working within the rectangular form of a fireplace is like painting on blank canvas. The composition proceeds within definite limits.

We try to develop a mosaic quality, blending size, shape, color and texture into a composition in which the stones have a life and character that derives from their participation in the whole. We concentrate on two things in establishing the design. First, because the observer's eye starts by moving along the joint lines, we manipulate joint depth, and the continuity, size and articulation of the joint lines. Second, because the eye soon moves along the stone masses themselves, we place individual stones to create larger forms. These forms should have an organic quality that reflects the nature of the material. The details of the stonework should be mutually dependent and intrinsically related.   —J. G.

port the stone evenly at all intended points. A tighter joint is generally a stronger joint. The less the stone is moved around after it is laid in mortar, the neater the joint and the less likelihood the mortar will run water over the face of the stone. It's important to prevent mortar from running onto any surface that will be exposed in the finished wall or chimney. While cleaning with muriatic acid can rectify some mistakes, the overall impression is sharper and more controlled if messes are avoided. After the group of stones is nestled into place, it is left undisturbed for a few hours until the mortar in the joints becomes somewhat leatherlike.

We then tool the joints with a thin piece of flexible steel. We usually don't add mortar, but we sometimes scrape it away in small areas,

making sure it is well compressed to seal the stonework. Exterior stonework requires more attention, because we want to be sure we eliminate all of the recesses or cavities that could trap moisture.

Our procedure is in contrast to the common practice of pointing at a later date by adding a differently formulated pointing mix. I'm critical of that method for two reasons. First, the pointing mix is usually brought out to the face of the stone, and cups over a bit of the stone's exterior surface. There's a great danger that this mortar, exposed to the elements, will lose its seal with the stone and direct water behind the outside surface. The freeze-thaw cycle can result in spalls or cracks that may need attention.

Another, more serious problem is that if the

wall is going to be pointed later, the mason usually doesn't pay much attention initially to forming a good, thorough seal between the stones. In fact, the mortar is usually left coarse so that the pointing mix will adhere. So once water penetrates the outside surface of the structure and is directed to the space between the stones, there is little to prevent it from traveling deeper into the work.

With our method, the pointing is neater, quicker and ultimately stronger, because it is integral with the mass. And because we point up each day's work before we pick up and go home, problem areas are dealt with while they are still fresh on our minds.   □

Jeff Gummelin lives in Ellsworth, Maine.

Masonry Heater Hybrid

Cross a traditional fireplace with a masonry stove for something home owners can warm up to

by G. Karl Marcus

In the spring of 1985, I was asked to design and build a hybrid masonry heater—part stove, part open-hearth fireplace. This went beyond my experience with traditional masonry stoves, and left me scratching my head for a way to satisfy the design criteria. That's when I met Frank Piwarski. He had recently opened his own design and consulting firm, called Ultra-Fire, and was designing everything from fireplace inserts to commercial waste-wood incinerators. He'd never been involved with masonry stoves before, but Piwarski understood combustion mechanics well enough to design a firebox with a gasifying combustor and a heat exchanger that gave my client just what he wanted.

During the construction of this hybrid prototype, Brad and Cheri Miller approached me about helping them design and build a similar stove in their owner-built home. They had read about masonry stoves, were impressed with the concept of heat storage and had planned for a fireplace of some sort on their main floor.

The Millers and I discussed the size and shape the structure should take, and we settled on a single glass loading door, a chimney behind the firebox and a raised stone hearth across the front. There would also be a warming bench and a woodbox on the kitchen side. The stove would be finished with grey river rock. I could see it. Standing in the dim light of evening, amidst bare walls and the smells of new construction, we shook hands and planned for a mid-October commencement.

Foundation and firebox—Working together, Miller and I started in the basement and laid up a foundation with 8-in. block and 12-in. by 12-in. pumice flue liner. We installed a cast-iron clean-out door and a 6-in. thimble, so that a stove

The stonework and glass door lend this hybrid masonry stove some of the charm of a traditional fireplace (facing page). The stove's thermal mass and system of flue baffles enable it to store and radiate heat over long periods of time. A heat exchanger installed inside the firebox also gives it the capacity for quick heating response. A circulating fan blows outside air through the heat exchanger and into the living room through openings over the fireplace door. Three cast-iron cleanout doors on the left side of the stove provide access to the flue baffles (left). Above the cleanouts is a handle for closing the damper. The back of the stove, facing into the kitchen, features a stone bench, a nook for wood storage, and a warming shelf (above).

could be added later, if the Millers decided to use their basement apartment.

Forming and pouring the slab for the firebox and hearth was no small task. We had to cut a main supporting beam for the floor, and cantilever part of the slab over foundation walls.

Also, we ran two air ducts between the floor joists from the garage to precise locations in the slab. One was a 1½-in. black iron pipe to carry combustion air to the firebox, and it needed to be on the centerline to connect to the combustor housing. Similarly, the 6-in. galvanized duct carrying cool air to the heat exchanger had to

fall just inside the front wall of firebrick and immediately behind the left wall of the firebox.

Miller is a structural engineer with the U. S. Forest Service in Missoula, Mont., and his specialty is bridge design. He designed the slab to carry 2,000 lb. per sq. ft. That sounded good to me, since the front wall of the stove would not stand directly above a foundation wall. The raised hearth would sit over the northwest foundation wall and the front wall of the stove would rest about 20 in. back, over an area of short span between block walls.

We began construction of the firebrick core

by measuring back from the flue 4 in. for the partition-block surround and another 1 in. for an expansion joint between chimney and stove. The back wall of the stove would rise along this line. We snapped a second line 27 in. into the living room from this line and parallel to it to establish the outside edge of the front wall.

The precast components I make are sized to be compatible with firebricks. Standard fireplace firebricks measure 2½ in. by 4½ in. by 9 in. My slabs, which create horizontal flue baffles, measure 2½ in. by 9 in. by 27 in. Their length determines the depth of the stove. With brick walls

From *Fine Homebuilding* magazine (October 1987) 42:68-73

12-in. by 12-in. pumice flue liner

4-in. partition block with wall ties

Masonry heater hybrid
This stove combines precast horizontal flue baffles and high-tech accessories to provide both heat-storage capacity and quick heating response. A 1-in. pipe running through the floor delivers outside combustion air from the garage to the gasifying combustor in the floor of the firebox. Hot air rising through the flue runs then heats the masonry mass. Outside air also cycles through the heat exchanger via a 6-in. duct. Heated by passage over the fins inside the heat exchanger, this air moves into the living room through openings over the fireplace door.

Damper

Precast-slab flue baffles, 2½ in. by 9 in. by 27 in.

Heat exchanger

Damper handle

Cast-iron cleanout-door opening

Inner lining with fins

Fire-brick

Expansion chamber

Lintel

Gasifying combustor

Stainless-steel punch plate

Fused-alumina combustor plate

Smelter brick

Firebox

Air inlet

Combustor housing

6-in. support slab

Galvanized-steel bottom plate

Stone veneer over ¼-in. expansion joint

Overall dimensions (before stone veneer): 6 ft. wide by 8 ft. 7½ in. high by 27 in. deep with a 15-in. high hearth (One brick: 2½ in. by 4½ in. by 9 in.)

4½ in. thick, this makes the interior of the flue 18 in. from front to back.

We had agreed that for purposes of design and traffic flow through the house, the brickwork should be 72 in. wide. I like to design in multiples of 9 in. because this simplifies the construction. Since the width of the brick is exactly half its length, proper lapping, or "breaking," of joints can be accomplished easily with either full or half-length bricks.

Miller and I made the firebox larger than is typical for most masonry stoves so that more and bigger logs could be loaded at one time, and so that loading would be required less often. The firebox is 27 in. wide and will easily accommodate 24-in. logs. We extended the front wall of the firebox 4½ in. into the hearth area to enlarge the burning pit and to make the firebox easier to load and clean. This deviation gave us a firebox depth, front to back, of 22½ in.

The gasifying combustor on the floor of the firebox would sit under a pit of coals and drive off volatiles from the fresh load until the wood turned to charcoal. The stainless-steel heat exchanger and "big screen" glass door would provide immediate heat, while the masonry baffles absorbed and stored much of the rest.

The sidewalls of the firebox are a single brick thick. Since flue gases exit near the upper right corner of the combustion chamber, I shifted these walls 4½ in. left of center. This maximized an 18-in. sq. expansion chamber on the right side of the firebox while leaving a minimal 9-in. chamber on the left to house the heat-exchanger service duct.

With these features in mind, Miller and I snapped chalklines representing all the walls to make certain everything was where it should be. Then we thinned a 100-lb. can of Sairset refractory mortar (A.P. Green Refractories Co., Green Blvd., Mexico, Mo. 65265) to dipping consistency (creamy yet with enough body to form ¹⁄₁₆-in. mortar joints) and began setting bricks.

The dark bricks used in the lower portion of the stove came from the Anaconda Copper Mining Company's smelter in Anaconda, Mont. They are very dense, weighing 10 lb. each. By the end of the day, they seemed a lot heavier.

Miller and I stood on opposite sides of the fireplace with a tub of mortar between us, taking turns dipping first the bottom, then an end of each brick. Dipping the bricks, rather than troweling mortar, saves time and material. Mortar consistency is the key. Sairset ready for dipping is like thick pancake batter. We would gently mush the buttered end against the last brick, tap a few times until mud oozed out the joint and then grab another. A plumb line dropped at one corner of the stove gave us a starting point.

At the sixth course, along the left wall of the firebox, three bricks were hung out 2 in. to form a support for the heat exchanger. The L-shaped exchanger would cover the left wall and ceiling of the firebox. Its job is to heat cool, fresh air from the garage and blow it into the house through the vent over the glass door. An elbow was fixed to the 6-in. duct, and an opening made through the left wall of the firebox. High-temperature stove wires coiled in the duct connect ceramic switches in the heat exchanger to

fans mounted in the garage. One switch controls the combustion-air fan connected to the gasifying combustor. The other operates a 265-cfm cooling fan connected to the heat exchanger.

We stopped the front wall of the firebox at the sixth course, 15 in. above the concrete. The solid stone hearth would finish at this height.

In the rear wall of the firebox, I set a rectangular plug (about 9 in. by 18 in.) made of refractory cement. For times when little heat is needed (summer fires, for instance), the plug can be removed to allow heat and gases to bypass the upper flue baffles and pass directly up the chimney. For practical reasons, we should have fabricated a metal damper instead, operable from outside the firebox, which would have allowed the Millers to use it as a bypass for easy startup.

At the ninth course in the right sidewall of the firebox, we located the exhaust port through which hot gases and flames would pass into the expansion chamber and flue runs. The opening measures 7½ in. high, the height of three brick courses, by 18 in. long. Using refractory cement and crushed firebrick, I cast lintels and set them over the openings in both sidewalls of the firebox. When properly cast and cured, these components are rated safe by the cement manufacturer at temperatures up to 2,700°F. My stoves never see sustained temperatures this high, but there's something to be said for the peace of mind gained by overbuilding.

Flue baffles—We continued up with bricks to the fifteenth course, nine courses above the hearth level. I used an angle grinder with a carborundum blade to cut notches ⅜ in. deep in the top of the bricks at the upper front corners of the firebox. An angle-iron lintel 3½-in. by 3½-in. by ⅜-in. thick sits in the notches.

Next, a heavy bed of mortar was troweled on the bricks and the first row of slabs was placed. This course of slabs forms the roof of the firebox as well as the floor of the first baffle. Six slabs, side by side, create each of the five baffle levels in the stove, leaving a 9-in. by 18-in. opening at alternating ends of each run for the unrestricted flow of flue gases into successive baffles (drawing, facing page).

For cleaning access to the flue runs, I left 9-in. wide by 7½-in. tall openings in the brickwork, and covered them later with cast-iron cleanout doors. In this stove, three cleanout ports are located in the left end wall of the structure, centered at U-turns in the baffle system (photo left, p. 147). This allows easy access to both an upper and lower flue run through each port.

After the first row of slabs was in place we continued up with our walls. We ran out of smelter brick while building the first baffle and switched to common fireplace firebrick. We made all the flue runs in this stove 7½ in. tall (three courses). Each fourth course is a row of slabs that bind the long walls together and increases structural stability.

The fireplace cuts off a corner of the living room and divides it from the kitchen. Because it sits at a 45° angle to the room, the right rear corner of the stove actually stands in the kitchen. This meant that the fifth baffle, which finished off 4 in. below the kitchen ceiling, would

be the last full-length flue run we could build. A half-length sixth run at the left side of the stove would get us back to the chimney.

Where this corner of the stove extends under the sheetrocked kitchen ceiling, we laid a course of common brick at the stove's perimeter. Behind this brick, we sealed the tops of the baffle slabs with an inch or so of regular mortar, and covered this with 2 in. of loose vermiculite masonry insulation. We used partition block to fill in over the sixth baffle, leveling the right side of the stove with the short final flue run.

We installed a sliding damper, made locally of ¼-in. plate steel, in the last baffle. It rides in angle-iron tracks held firmly against the brick by stainless-steel wire. As a safety feature, we made the Millers' damper so that it doesn't close all the way; it leaves a ¼-in. space to allow carbon monoxide to escape should the damper be closed prematurely. The connections between stove and chimney were made with modified flue liners and refractory cement.

Although the physical labor was intense, building the firebrick core for this stove was surprisingly fast and simple. Including time spent packing materials, working mortar to the proper consistency and checking the drawings for construction errors, Miller and I built this structure in three working days. We used about 700 firebricks, 39 full slabs, 4 half-slabs and 200 lb. of refractory mortar. I figure the core of the Millers' stove, excluding the chimney and veneer, weighs close to 7,500 lb.

Through the roof—Once the core was complete, we turned our attention to the chimney and blockwork on the kitchen side of the stove. The Millers wanted a stone bench to the left of the chimney and a tall, narrow woodbox to the right (photo at right, p. 147). A second compartment over the woodbox was planned as a warming shelf for bread dough.

We placed a ⅝-in. sheet of rigid fiberglass between the chimney and stove as an expansion joint. This allows the stove brick to get hot and expand without affecting the structural integrity of the stack. Before starting the stonework, we covered all other vertical faces of the fireplace with ¼-in. thicknesses of this same rigid fiberglass insulation, except for the corners, where we went a little heavier with the material.

Portions of the chimney to be veneered were blocked up with hollow-core 4-in. partition blocks. Wall ties set in the block at 16-in. intervals connect the stone to the block. Partition block for the woodbox was worked into the right side of the chimney, laid against the back wall of the stove and returned into the kitchen 20 in. to remain flush with the chimney.

Two concrete slabs 3½ in. thick by 20 in. square were precast and set into the chimney/woodbox structure to form the warming shelf and its roof. One course of block filled with debris and sand became the base for the stone bench.

We were able to miss the valley in the roof with the chimney, but could not avoid cutting a doubled-up 2x12 beam at the north edge of the kitchen ceiling. The cut ends of this beam were temporarily supported by wood posts on hy-

After laying up the horizontal flue baffles and firebrick core, the author mocked up keystone arrangements over the fireplace to determine which stones would work best.

A stainless-steel heat exchanger covers the left wall and ceiling of the firebox. Outside air, driven by a circulating fan, is heated as it passes through the heat exchanger and moves into the living room through the vent over the firebox door.

Two ceramic sensors are mounted inside the heat exchanger. Wired with high-temperature stove wire, they serve as thermostats to control the circulating fans for the combustor and the heat exchanger.

Positioned to miss the valley, the stone-faced chimney rises 5 ft. above the roof.

An airtight glass door permits a view of the fire without compromising the heating efficiency of the stove. Openings in the bottom corners of the door frame allow secondary combustion air, preheated by the frame itself, to enter the firebox. Hot air from the heat exchanger blows into the room through the vent above the door.

draulic jacks. Eventually, both ends of the beam were supported by the stonework.

In the loft, a short section of stud wall had to be removed to let the chimney pass through. We used 4-in. solid block in this section because the back side of the chimney, located in an attic wing over the kitchen, was not to be veneered. Recent changes in the Uniform Building Code require a masonry flue to be wrapped either with an 8-in. hollow-core block or 4-in. solid block where no additional veneer is planned.

We ran block through the opening in the roof, flashed it and went up with 5-in. stone veneer placed directly against the flue liner. The chimney stands about 5 ft. above the roof and extends just over 2 ft. beyond the peak (photo facing page, bottom left). We finished the chimney on a beautiful, sunny Friday afternoon. By Monday, a Canadian cold front had blown in, signaling the end of Indian summer.

Stone veneer—Somewhere along the line, Brad, Cheri and I had taken a day and collected six tons of stone for the veneer, so when cold weather hit, we were ready. Of the many local stones I use, none looks more common on the ground or more subtly attractive laid up than the grey, pre-Cambrian quartzite the Millers chose.

We collected the stone along a dirt road in a narrow canyon where it had eroded off the parent rock and tumbled in the stream a short distance. This brief encounter with the stream rounded off the sharp edges but left the stone's angular nature intact. With this type of rock, it is relatively easy to obtain the look of washed river stone while maintaining tight, uniform joints.

Corners were no problem, either. Quartzites that aren't too massively bedded often break straight through the beds, providing an endless supply of 90° angles. Throughout the construction, we were able to use natural corners.

The colored mortar, deeply struck joints and fairly rough placement of stones are design features I've copied and embraced in my work as invaluable aids. A mortar joint darker than the stone it encases will often go unnoticed. A stark joint offends the eye.

We spent a lot of time striking joints. I buy new, flat, wooden-handled jointers and saw them off so the steel is 3 in. or 4 in. long. Twice a day we went over the rock we'd just put up, compacting more than cutting away, careful to leave a smooth, square face on the mortar.

Miller and I laid the six tons of stone that cover his fireplace in twelve working days. The creek stones weren't large enough to make decent seats for the raised hearth and stone bench. So we used massive rocks from the talus slopes that collect at the base of cliffs.

During the early stages of the stonework, we recognized a number of rocks spread across the floor as obvious candidates for keystones. We played with several arrangements on the floor, but what finally materialized over the fireplace door seemed almost to invent itself—a wonderful accident (photo facing page, top left).

On the other hand, the catenary arches over the woodbox and warming shelf were planned while picking the stone, and they seem somewhat forced, I think. Two long narrow stones were set level with each other in the front wall to support a mantle to be seasoned and mounted at a later date. We capped the flat area over the warming oven and the top of the stove with 2-in. rock, overhanging the veneer a couple of inches, like the nosing on a countertop.

Heat exchanger and gasifying combustor—With the stonework complete, installing the high-tech accessories was the final step in what had been a long and exhausting construction process. The baffled, double-wall heat exchanger had been fabricated to our specifications in a local welding shop. And even though we'd been extremely careful with our measurements, we all breathed a sigh of relief when the unit was slid in and connected to its air-supply duct with no major hangups (photo facing page, top right).

The high-temperature stove wires coming from the fans were routed over the back side of the heat exchanger and run into the hot-air delivery opening. Ceramic sensors were wired in to automate the fans (photo facing page, middle right). The sensor servicing the combustion air fan turns it on at just over room temperature, and keeps it on until temperatures in the heat exchanger rise to 180°F.

The circulating-fan sensor kicks on its 265-cfm fan when the exchanger reaches 120°F and keeps it running until the air passing over the switch falls to 110°F. The inside face of the heat exchanger is 14-ga. stainless steel. The outside is a single piece of 14-ga. black steel. Three baffle fins 1⅞ in. by 20 in. were welded to the stainless-steel liner (drawing, p. 148).

Primary combustion air comes from the garage through the 1-in. pipe to the gasifying combustor that is set in the floor of the firebox. The combustor is simply a dedicated air source that is located underneath the fuel load. It works together with the deep-pit firebox to generate high combustion temperatures and minimize thermal drag (which results in slow startup). The intense heat of the fire triggers a chemical change that converts molecules of the solid wood into a 2,200°F superheated gas. These gases move through the fuel load and "fire off" upon contact with secondary combustion air, introduced through the door frame.

Rectangular steel tubing (2 in. by 3 in.) mitered and welded into a 15-in. by 19-in. rectangle makes up the combustor housing. At one end of the housing, six ¼-in. dia. holes were drilled in the inner wall of the tubing. Next, the fused-alumina combustor plate (9 in. by 13 in. by 1 in. thick), which looks like a metallic sponge and diffuses the combustion air, was placed inside the combustor housing. A light-gauge piece of stainless-steel punch plate fitted over the combustor plate keeps fine ash from clogging the pores of the combustor. The bottom of the preheater is closed off with light-gauge sheet metal and self-drilling metal screws.

A low-cfm blower with the capacity to develop a relatively high static pressure is used in this system to deliver the optimum amount of preheated, low-turbulence combustion air for anticipated fuel loads. Seven to ten cubic feet per minute of primary air are supplied for an expected fuel consumption rate of approximately 7 lb. to 10 lb. per hour. Theoretically, a 70-lb. load should burn seven hours or more. In practice, we're seeing slightly elevated consumption rates due, we think, to air infiltration around the door and heat exchanger.

Tubular steel frames create the simple but effective fireplace door (photo facing page, bottom right). The jamb is made of 2-in. by 3-in. tubing, mitered at the corners and ground smooth before finishing. The door frame proper is made of 1-in. by 2-in. tubing and is attached to the jamb with a piano hinge. Japanese glass ceramic is sandwiched between rope gaskets and held against the inside of the door with stainless-steel brackets. A flat gasket fixed to the inside of the door frame compresses to form an airtight seal when the door is locked shut.

The turned wooden handle has a double catch for safety. One twist opens the door a crack and retains it, allowing enough excess air in to sustain intense blazes of short duration, the traditional method of firing masonry stoves. A second twist and the door swings open 90°, providing convenient access to the firebox.

The large frame against which the door closes doubles as a secondary combustion-air preheater, pulling room air off the hearth through openings at the bottom corners of the frame. Air ascends the vertical sides of the frame, fills the hollow tubing above and behind the door, and is forced down the glass through slots nibbled in the tubing. This feature encourages unburned hydrocarbons starved for oxygen to "fire off," producing beautiful, extended flames in a low-turbulence, high-heat environment.

The floor of the firebox was made with 2 in. of refractory cement over 3 in. of tamped sand. It finished out about 10 in. below the hearth, creating a firepit surrounded by four brick walls. This feature helps establish a deep coal bed while minimizing the ash removal.

Driving out the water—The fireplace was finished around the middle of January, 1986. Initially, it seemed to burn a lot of wood without storing much heat in the mass. However, after the first month or so, the amount of heat that was radiated back to the living space by the stone seemed more in keeping with the amount of fuel being consumed. This may have been due, in part, to the Millers' method of firing. Any high-mass, heat-storing fireplace takes some time getting used to.

But the real reason for poor heat efficiency during a stove's start-up phase probably lies with the water trapped in the masonry. Until the water in the structure is driven out by heat, which takes time with a stove this big, the mass can't possibly attain its true heat-storage capacity.

A four-hour fire in the uncharged mass will raise the surface temperature of the stone well above body temperature in twelve hours. Long after the fire burns out, the heat exchanger continues to cycle, producing hot air for two or three minutes at a time. The damper is never adjusted during a fire. Miller leaves it open, even between fires, closing it only when he leaves for extended periods. □

G. Karl Marcus is a mason in Missoula, Mont.

A Stone Cookstove and Heater

This dual-purpose unit combines masonry, steel and a long smoke path

by Jonathan von Ranson

Since turning homesteader in the woods of western Massachusetts, I have had a few definite surprises. Having a boss from my former newspapering days show up at the door was one. Deciding one morning to go ahead and move into the stone house that took us three-and-a-half years to build was another. The greatest—and in some ways the most pleasant—surprise is the recurring one of operating a stone cookstove. It cooks our meals all year round, and in winter it's our 800-sq. ft. house's main source of heat.

We found out about masonry cookstoves as we were researching what kind of cast-iron one to buy. ("We" consists of Susan and me and our three teenagers, Erik, Kristin and Joel). Our research led us to Albie Barden of the Maine Wood Heat Co. of Norridgewock, Maine (see p. 137). He invited us up to see his masonry stoves, and we made the journey to central Maine one February day.

Barden fired up the imposing brick range in his kitchen. The oven climbed to 350°F in about 15 minutes, with only four or five sticks of wood. In his parlor is the heater version from which the cookstove takes much of its inspiration. He lit a fire in that, too. People who've built masonry stoves like to show how well the draft pulls: down and around and through . . . "as long as it goes up at the end," says Barden, watching your face.

Not ones to make hasty decisions, Susan and I were more than halfway home again when we turned around and drove back to buy the Finnish cast-iron parts Barden imports for the stove.

The Finnish connection—In this country, masonry heaters have come to be known as Russian fireplaces, because the first ones built here were put together by Russian émigrés in the style of the brick stoves of their homeland. But in fact, the Finns are in the forefront of masonry-stove technology, and it is they who have developed the special mortars and fittings that allow cast iron to be wedded to masonry so that a wood-fired heater can also be a cookstove.

In all masonry stoves, the smoke path is long. The hot gases travel past virtually every stone, leaving some of their heat in each before going up the chimney. In the stove we built, the smoke makes a 5-ft. loop through the base of the chimney to warm that mass after it's already traveled about 14 ft. in a serpentine 8-in. by 8-in. channel that starts at the firebox (drawing, facing page).

The parts that we bought from Barden included the cooking surface and its lids, the oven and its door, the firebox and ashpit doors mounted on a common frame, a grate with a frame, a slide damper and its frame, and two cleanout doors with frames.

The largest of the castings is the cooking surface. Its rigid frame is meant to sit above the masonry on a bead of mortar, which mates with a raised, ¼-in. bead on the casting. This design makes for a tight seal and guards against the top's warping. There are three lids—two normal-sized ones on the right and a big one—16 in. in dia.—over the firebox on the left. They are cast with deep waffling on their undersides to increase the area that can absorb and transmit heat to the top surface. The lids quickly get hot enough to boil water, and they draw off heat quickly enough so the rest of the stove doesn't overheat (tests show that mortar begins to disintegrate at temperatures above about 400°F).

The oven has shallow flutings that work like the waffle grids on the lids. It's small by American standards (13 in. by 10 in. by 20 in.), but it roasted a 16-lb. turkey last Thanksgiving and easily holds four to six loaves of bread. The oven's heavy, tight-fitting castings show evidence of fine workmanship.

The Finns have two notions about fireboxes that surprise most Americans. The first is that they should be small. Our firebox door is only 8½ in. by 6 in., and the firebox itself is just 21 in. deep. Wood burns more efficiently in a small combustion chamber, and the smaller fire is less likely to overheat the masonry. Besides that, it calls for small logs, so we tend to use less wood than we might in an American-style woodburner.

The second idea, which sounds almost like heresy, is that fireboxes should not be airtight. The Finns feel that woodstove efficiency depends on complete combustion, and on the route taken by the hot gases after they leave the firebox.

The positive-closing slide damper in the chimney flue up near the second floor gets shut tight when the fire has burned down to glowing red coals. This prevents room air from being drawn through the stove and leaving the house, setting up a convection current that would also cool the masonry from inside. With the damper shut, the house is heated all night by the stove top, the stove mass itself,

The stone cookstove designed and built by von Ranson weighs ¾ ton. He carted the stone from a nearby quarry and cut it to size in a sandbox on site. The firebox is small—only 21 in. deep—and accepts just four or five sticks at a time, but it's efficient enough to serve as the house's main heat source.

From *Fine Homebuilding* magazine (June 1984) 21:62-65

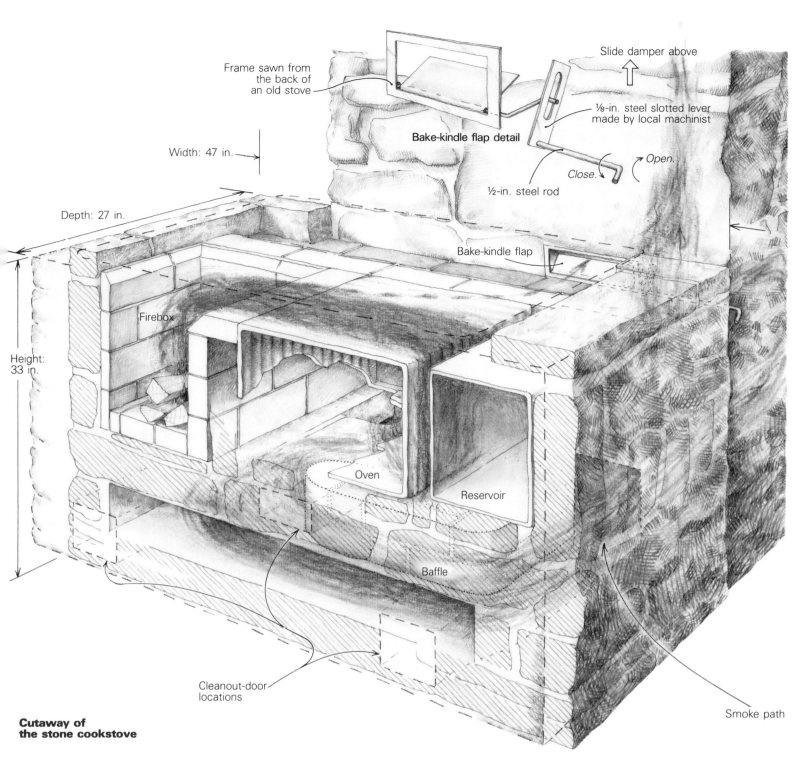

Frame sawn from
the back of
an old stove

Width: 47 in.

Slide damper above

⅛-in. steel slotted lever
made by local machinist

Bake-kindle flap detail

Open.

Close.

½-in. steel rod

Depth: 27 in.

Bake-kindle flap

Firebox

Height:
33 in.

Oven

Reservoir

Baffle

Smoke path

Cleanout-door
locations

**Cutaway of
the stone cookstove**

and the chimney base where the flue loops through, which I think of as our heat sink. They're all still warm to the touch when we come downstairs in the morning.

The ashpit door is beneath the loading door. The grate, with frame, sits between the two.

Both the visible and the hidden cast-iron parts have a spare, Scandinavian look. There was no thermometer in the oven door, so we bored a hole and installed one.

The design of our stove required some additional pieces, such as a small cleanout door that would let us get at the smoke path under the oven. We also got a welder to fabricate an 8-gal. stainless-steel hot-water reservoir with a scoop at the top for filling and a faucet at the bottom for drawing. Last, at Barden's suggestion, I made a flap that would allow the stove to operate in two modes: kindle, and bake-and-heat. The flap is open in the kindle mode, and smoke and heat go straight up the chim-

ney. This sets up a good draft, and gets the fire burning nicely. When the fire is well established, we close the flap, and the hot gases are routed through the entire masonry loop. This is the bake-and-heat mode.

We got the basic components of the bake-kindle flap secondhand at the Bryant Steel Works, a stove graveyard in Thorndike, Maine, and the under-oven cleanout door at the Good Times Stove Company in Goshen, Mass. (they have restored stoves, too).

The parts for the stove came to a little over $1,100, including everything: metal, cement, even travel. Most of this by far was for the Finnish castings. Don't take this route to save money on a cookstove purchase, unless your time counts for zero.

There are a few Finnish tricks we decided not to use. Some Finns sleep on top of their masonry heaters—a bunk is designed right into them. In another plan the smoke goes

through a masonry bench, which acts like a radiator and gives a toasty place to sit. They use flues like plumbing, piping the hot smoke to where it will give useful, even heat.

Another technique—short, hot burns—we don't employ out of protectiveness toward our stove. Theoretically, short burns are more efficient, since the fire burns only long enough to bring up the heat of the masonry. Also, combustion is more complete in a hot fire. But that technique is more appropriate to pure masonry heaters than to our masonry and cast-iron hybrid, in which the iron draws out enough of the heat to eliminate the need for fast, hot burns. Medium-length, medium-heat burns (just right for making pancakes or baking a casserole) seem to work best for us.

Stonework—With the metal parts in hand, it was time for us to decide on the material for the rest. Plans for brick faced with stone

looked cumbersome: too thick and too many mortar joints. All brick, then? No. Our house was made of stone, so we decided on stone for the stove.

Scandinavians almost invariably use brick, often surfaced with tile, for their *Kachelöfen*. For all we knew, we'd be building the only stone kitchen range on earth.

The smoke path is one of the things that makes our stove unique. As shown in the drawing on the previous page, there are three levels of it: under the stove top, under the oven (with a little eddy beneath the reservoir) and—here's the Finnishing touch—down in the stove's basement. On its way across the top level, the smoke heats the stovetop and the upper oven; descending, it heats the water reservoir and the side of the oven; once at the middle level it heats the bottom of the oven; then it's drawn down an opening into the "basement" where it leaves energy for space-heating the house.

I had four years' experience at masonry, but this stove was the most difficult thing I had ever undertaken. Such complex spatial relationships in three dimensions—such variables in mortar joints, expansion joints and missing parts—tax the ordinary mind. (Such heavy slabs of stone tax the ordinary back.) I had hardly anything in the way of plans. Barden had lent us a book, written in Finnish, with two or three plans for masonry cookstoves. From there on I was on my own. Over the next few months, the designs in the book—microscopically small, intended for brick—became clearer. Still, designing in a water reservoir and an optional smoke path produced powerful feelings of insecurity. The best advice available involved using three different mortars—the firebox looked too small to heat anything, let alone the whole house—and as I built, the stone had its own ideas about how things should happen.

I decided to use stones as close to 4½ in. thick as possible; that would be the thickness of the exterior walls and the horizontal divider, or baffles.

We began by hacking our way through a mile of forest to an abandoned quarry, and there I got my initiation in quarry-style stone-cutting. We picked out a 9-in. thick slab, tapped a line along the grain edgewise with a chisel, turning the stone several times, and it dramatically opened like a book, 1½ ft. wide and 7 ft. long. It took five trips with our 1947

The front stones (top left) were cantilevered to bring the heater out to full dimension and to create a kickspace. Combustion gases will be pulled into the crook of the brick baffle, and then directed around it and into the chimney. The frame for one of the two bottom-level cleanouts is in place. On the second level (center), three large slabs span from front to back. They will support the firebox, the oven and the hot-water reservoir. The hole at the back of the middle slab is the smoke passage. The photo at left shows the reservoir and oven in place, with the firebox almost completed. Smoke will curl first over the oven, down between it and the reservoir, then back under the oven before rising through the passage in the middle slab.

jeep to bring back a couple of tons of split stone, a stack maybe 2 ft. high.

I spent the next several weeks in the sandbox (the best place for stonecutting), creating particular shapes, fitting them together dry, and trying to fit them together wet in my mind. Steve Busch, a mason in South Paris, Maine, who had worked with Barden building several Russian fireplaces and was about to undertake a cookstove, suggested keeping the firebox isolated with an expansion space. But I saw no way to avoid cementing the firebox to the face of the stove, to maintain structural integrity and to control the fire. I didn't want cracks through which smoke and sparks could pass and short-circuit the carefully planned heat path. I was in uncharted regions of fire-taming and a little scared.

The ½-in. mortar joint I figured on turned out to be more like ⅞ in., a discovery that sent me back to the sand pile for more cutting, and eliminated almost an entire tier of stone and a couple of days of my life. Handcut stone tends to come out a little bigger than you mark it. Cutting the oven lintel took several tries, but finally one emerged from the sandbox, a virtual stone ruler about 1½ in. by 4½ in. by 17 in.

I laid the stone courses up in stages, allowing the bottom tier to set before going on to the next. First, I set four large stones on the slab for the floor of the stove. Once they set up, I built the walls, the front wall cantilevered forward 3 in.—more because the slab was that much too small than because we needed a place to put our toes. The overhang rested on 2x4s and shingles while the mortar set. A few red bricks were laid as a baffle around which the smoke would circle in the stove's basement (top photo, facing page).

Those brick were mortared with a special imported Finnish cement called Savi Uuni Laasti, which I also bought from Barden at Busch's recommendation, but I used a standard portland-base mortar for the rest of the stove (except for the firebox, which used furnace cement). The Finnish stuff—the mortar of choice for a brick stove—comes in a 75-lb. bag and costs $25. Not only is it expensive, but it isn't recommended for use with stone. So I stuck with a 1:1:6 proportion of portland, lime and fine sand from there on, even in the back where I laid a few more brick. This high-lime mortar is recommended for use in high-heat areas. It isn't quite as strong as ordinary mortar once cured, but it does stand up better to extreme heat. It sounds paradoxical, but it's the seismology principle: the weaker mortar will allow hairline cracks to form in joints without fatal resistance that might take the form of one large stovequake.

Putting things together called for a scalpel-and-tweezers type of masonry, especially at the front. I wanted to offset my vertical joints, but with five openings and several slender vertical stones between them, there wasn't often much bearing surface. Some of the stones didn't get tied in until the very top course, and I had to hold one of them in place with a pipe clamp until the mortar set.

Three horizontal 250-lb. stones capped the

basement smoke channel and spanned the cleanout doors (middle photo, facing page). It took two strong people to set these into place. Then the big day came when the large metal units—reservoir, oven and loading door—were set into place.

The method that day was to work from the right, setting the right wall stone, then the reservoir, then the 4½-in. deep stone between it and the oven, then the oven itself, then the next vertical stone. Doing it this way rather than dropping the metal units into ready-made stone openings allowed for a bed of mortar both under and alongside the metal pieces, sealing out drafts and sealing in smoke. It also reduced the likelihood of jostling the stonework after it had set up. I wrapped each metal unit with fiberglass insulation where it contacted the mortar to create about ¹⁄₁₆ in. of expansion room. Otherwise I didn't see how the stone could stay intact against the greater expansion rate of the metal. This seems to have paid off.

Next day the firebox fell together reasonably well thanks to a day spent earlier cutting the bricks with a carborundum blade in a circular saw. A thin tier of capstones went on the following day (bottom photo, facing page), as did a small section of brick to form the rear wall behind the oven.

The kindle-bake flap was next. It is a regular gravity-held flap for which I designed a lever of ½-in. steel rod that extends out the right hand side of the stove (drawing, p. 153). A twist of the lever and the flap opens and lets smoke go straight up the chimney; a reverse twist and it falls shut, sending the smoke on a 20-ft. journey around the oven before it goes up the chimney, perhaps 200°F cooler.

After everything had set up well, Susan and I laid a bead of cement around the top of the masonry and set the frame of the cooking surface into it, leveling it carefully. We meticulously pointed up the mortar so the iron would be able to expand horizontally without catching on a cement crumb. (All joints, incidentally, were pointed up both inside and out as soon as the mortar had set up.)

That finished the job. We were advised to wait a couple of months before using the stove to give the mortar time to cure.

Afterburn—We kept the first fires small, and felt the stones gradually heating up over a matter of days, not hours. The stove has been in regular use since last fall and has passed its first heating season in fine shape.

The only problems are that the oven gets a little hotter toward the rear, and creosote occasionally drips out of the slide-damper slot in the chimney. If I'd thought of this problem when I was building the chimney, it would have been easy to slope the damper inward a little so the creosote would run back into the

flue instead of out over the stonework. Now all we do is plug the gap. Actually, creosote isn't a big problem. We get none at all if we run the stove fairly hot on kindle mode for 10 minutes or so when we start a fire. We also try to avoid slow, smoldering burns, which let the chimney cool and creosote form.

To correct the uneven oven temperature, it will only be necessary to install a baffle of the kind found in old cookstoves to bring the smoke forward under the oven. This flat metal baffle (maybe ⅛ in. by 2 in. by 12 in.), which stands on edge catty-corner under the oven, would bring the smoke toward the front as it moved from right to left under the oven, (like herding cattle through a gate). Also, either firebrick expands more than I guessed or the iron cooktop didn't "flat" quite right, for a stone lifted slightly in the face above the firebox. This would be easy to repair, but the situation is stable and the stove is solid.

With a year of practice under her belt, Susan now calls the Finnish-style cookstove her favorite part of the house. The hours spent in all stages of anxiety and torment have been repaid—and still the payments continue. We casually stoke the front-loading firebox with the four or five small logs that it will accept at a time and heat the house while making breakfast and dinner. On grey, wintry days a longer fire is required. Our wood consumption is about 3½ cords a year—I like that. ☐

Homesteader Jonathan von Ranson writes and does stonemasonry in Wendell, Mass. Photos by the author.

K'ang
An under-the-floor masonry stove

by Jeffrey J. Smith

Karl Marcus, a stonemason in Missoula, Mont., was experienced in building Russian fireplaces, and he wanted to include one of these masonry stoves in his passive-solar addition, a south-facing greenhouse/office. But Marcus didn't want to sacrifice any of the floor space (170 sq. ft.), which was at a premium in his older home. The solution was to build a masonry stove beneath the addition's floor.

Marcus's idea wasn't a new one but rather an

adaptation of an ancient, multi-cultural tradition. In *The Book of Masonry Stoves* ($14.95 from Brick House Publishing Co., Inc., 3 Main St., Andover, Mass. 01810), David Lyle writes that the ancient Roman hypocaust and the Chinese

The k'ang firebox is small, 3.1 cu. ft., and designed to burn sticks rather than logs. But it must withstand temperatures of 1,100°F to 1,800°F. The firebox arches were cast in custom forms using refractory concrete.

k'ang, both subfloor heating systems, were used 2,000 years ago. An Afghan version, called the tawakhaneh, may have been around for as long as 4,000 years.

Marcus built his k'ang to wring every last Btu from his firewood. The stove does this by burning wood at high temperatures and by storing the tremendous heat that it generates. Marcus's k'ang won't need to burn a constant fire, either. He believes that an hour-long burn twice a day

Drawing: Chuck Lockhart

will be enough to heat the addition and contribute substantially to heating his entire house. There are 225,000 Btus in 30 lb. of firewood. If the k'ang is 90% efficient, it will generate and hold 202,500 Btus morning and night (see the sidebar on p. 159).

I had seen Marcus's work, and was intrigued by some other masonry stoves he'd built. I was especially impressed with the massive, under-the-floor stove in the conference center of the building where I worked. The heat it produced was very comfortable. So last spring, when Marcus mentioned that he had started to build his under-the-floor stove, I asked if I could watch the construction and write about the process.

Preparation—As soon as the ground thawed in the spring, Marcus began digging. At the center of the existing exterior wall, he excavated to a depth of 6 ft. below grade. He then laid in a concrete-block stairway, which began at the southeast corner of the house and dropped five steps to a 3-ft. square landing. All along the stairwell's interior wall, Marcus reinforced the existing foundation with concrete. As shown in the drawing, the stairway turned left at the landing and dropped two more steps to meet the existing 3-ft. wide basement corridor. Marcus strengthened his foundation at this entranceway using bond-beam construction with each course of blocks all the way to the floor. He used blocks that had knockout webs so that removing the knockouts in each block created a horizontal channel through the entire course. Marcus then filled this channel with concrete. Next, he tore out the rickety old wooden stairway leading to the living room upstairs.

The concrete stairwell divided the new addition in half, and also framed the west wall of the k'ang stove. Three courses above the stair's landing, Marcus placed the k'ang's 3.1-cu. ft. firebox. He and his wife would be able to stoke the stove while sitting on the stairs.

With the stairway and the exterior frost walls in place, Marcus removed the house's south wall. He rebuilt 12 ft. of flooring to fill in where the back entrance and basement stairway had been. Now he was ready to build the k'ang.

Construction—The stove consists of poured concrete slabs, concrete blocks and refractory brickwork (drawing, below). On the west, it is framed by the stairwell; on the east and south, it is bordered by the concrete-block foundation of the new addition. Marcus first laid 2-in. rigid fiberglass insulation on the packed, leveled earth where he would pour his foundation slabs. He then poured a 3½-in. by 40-in. by 48-in. foundation slab for the stove's firebox. When it cured, he poured another 3½-in. thick slab to support the body of the stove. This 9-ft. by 8-ft. main slab overlapped the firebox slab on one end. Both slabs were mixtures of 3 parts ¾-in. crushed stone to 2 parts sand to 1 part portland cement.

When the Anaconda Copper Mining Co. in Butte, Mont., shut down its smelting operation five years ago, Marcus bought a pallet of new smelter bricks from them. These were top-of-the-line, extremely high-temperature bricks. Except for 50 store-bought refractory bricks that he laid in the south wall of the stove (the wall that would be stressed the least by the k'ang's heat), all of the stove's bricks were smelter

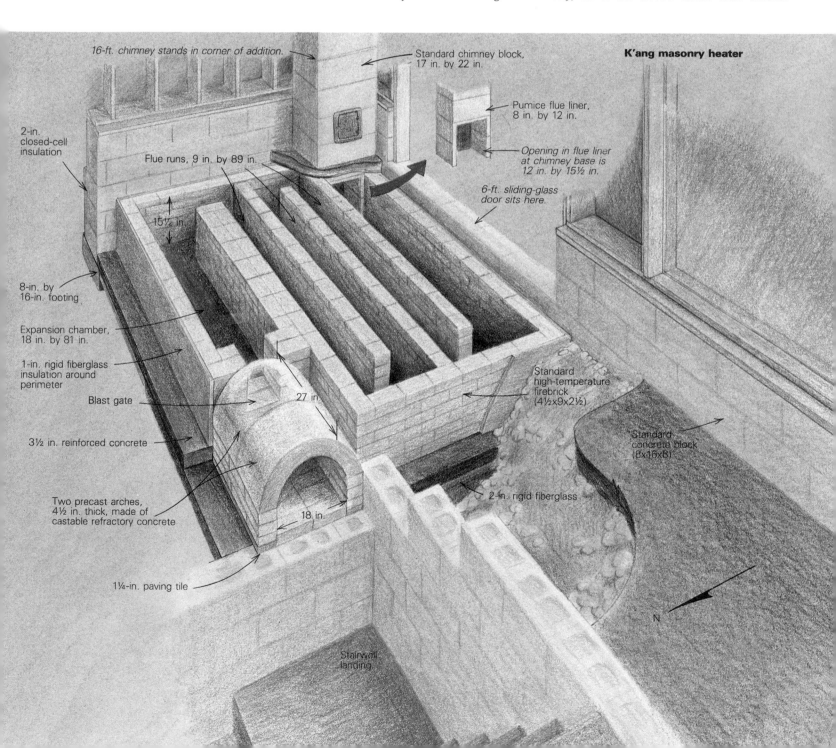

K'ang masonry heater

16-ft. chimney stands in corner of addition.

Standard chimney block, 17 in. by 22 in.

Pumice flue liner, 8 in. by 12 in.

Opening in flue liner at chimney base is 12 in. by 15½ in.

6-ft. sliding-glass door sits here.

2-in. closed-cell insulation

Flue runs, 9 in. by 89 in.

15½ in.

8-in. by 16-in. footing

Expansion chamber, 18 in. by 81 in.

1-in. rigid fiberglass insulation around perimeter

Blast gate

27 in.

Standard high-temperature firebrick (4½x9x2½)

Standard concrete block (8x16x8)

3½ in. reinforced concrete

Two precast arches, 4½ in. thick, made of castable refractory concrete

18 in.

2-in. rigid fiberglass

1¼-in. paving tile

Stairwell landing.

N

Like the firebox arches, the lids of the first three flue runs were also precast with refractory concrete. To allow for extra expansion and contraction due to the high temperatures, Marcus used nine separate sections to make the lid of the expansion chamber.

bricks. Marcus used Sairset refractory mortar manufactured by the A. P. Green Refractories Co. (Green Blvd., Mexico, Mo. 65265).

Flue runs and firebox—Next, Marcus laid out the elaborate interior flue structure, a labyrinth of vents and channels designed to absorb the firestorm created in the firebox. The fire pours through a 9-in. by 9-in. blast gate in the firebox's double-brick rear wall and enters the first flue run. The blast gate acts as a baffle, creating turbulence and burning up exhaust gases.

The first flue run must withstand great temperature fluctuations as well as occasional sustained, fiery blasts. It's called the expansion chamber because the gases and smoke are actively combusting, and therefore expanding, as they leave the firebox. Marcus made the chamber 18 in. wide—the same width as the firebox and is double the width of the succeeding flue runs. It is 15½ in. tall and runs 81 in. along the north wall, which is insulated with 1-in. rigid fiberglass. This is the only wall backed by packed earth. The interior wall is a double brick wall.

At the end of the expansion chamber, the hot blast collides with a brick wall one brick thick backed by 1-in. rigid fiberglass and the concrete block foundation of the room's east footing. The exhaust escapes through a 9-in. switchback into the next flue, which is also 9 in. wide. This second chamber is 89 in. long and ends in another switchback to the third flue run. The second, third and fourth chambers are identical in width, height and length. The final chamber is 13 in. wide and ends in a standard 17-in. by 22-in. pumice-lined chimney.

Marcus custom built the firebox's twin 4½-in. thick arches (photo, p. 156). The inner (bottom) forms are plywood and stretched galvanized steel. He put a piece of ³⁄₁₆-in. plastic on top of the curved form to keep the concrete from ad-

hering to the steel. The form's outer (top) piece was made by a local plastics manufacturing firm and was designed by Marcus to be held in place with clamps. He poured Refracrete (North American Refractories Co., 1500 Houser Way S., P.O. Box 975, Renton, Wash. 98055), a large-aggregate, castable refractory concrete, through an opening at the top of the arch, and the pieces cured in the same positions they would take above the firebox. The forms are reusable.

Marcus then built forms and poured a series of 2½-in. thick lids for each flue run. For the lids of the first three chambers, he used Refracrete. Marcus poured nine separate sections (9 in. by 27 in.) to cover the expansion chamber (photo above), thinking that the joints would allow for the expansion and contraction due to the high temperatures in this first flue run.

For the second and third chambers, he poured 13½-in. wide slabs that would cover all but 27 in. at the east wall. He needed cleanout ducts there, and he poured these duct lids separately to include 8-in. dia. holes that would accept airtight, sheet-metal lids (top left photo, facing page). The lids that covered the fourth and fifth flue runs Marcus cast from the 3-2-1 portland mixture he'd used for the slabs. He added a small amount of Mason's Blend, a fireclay made by North American Refractories, to the lids to increase their elasticity and heat resistance.

Final steps—When Marcus had mortared the seams in the flue-run lids and had filled in the sides of the arches with vermiculite, he laid a gridwork of rebar over the surface of the stove. He installed the two airtight cleanouts, one at the start of the second flue run and the other between the third and fourth chambers.

Marcus also installed a 3-in. dia. aluminum pipe across the top of the flue runs and insulated it with fiberglass (top left photo, facing page). He

was hoping that a 165-cfm fan would blow cold basement air through the pipe to two small ducts along the south wall, where the heated air would emerge and circulate through the house. This turned out to be ineffective. Marcus has since installed a small fan at floor level on the west wall; it simply blows air across the floor, up the wall and back into the house.

There is a carbon-monoxide danger associated with masonry stoves. Some European countries, most notably Switzerland and Austria, have a regulation that chimney dampers on masonry stoves cannot close more than 35%. Marcus is convinced that his k'ang will draft rapidly enough and is effectively sealed against the escape of exhaust gases. But as an extra precaution, he has no damper in his chimney.

When all this was done, Marcus poured a 4-in. concrete slab over the stove. It was made up of the 3-2-1 portland mixture with a small proportion of fireclay. He dyed it black for greater absorption of winter-time solar heat (top right photo, facing page).

The 16-ft. chimney was the last step, and it went up without a hitch. Marcus added an airtight, 8-in. by 8-in. cleanout door just above the floor. His business, Native Rock Design, got busy before he could build and install an airtight firebox door. But that didn't prevent him from christening the k'ang with its first roaring fire. There was no hesitation in the draft, and once established, a pocket of white-orange flames jumped through the blast gate to feed hungrily on exhaust gases to the end of the expansion chamber. Marcus brought out the Irish whiskey to celebrate, and I wished I'd saved one final frame of film to capture his satisfied grin. □

Jeffrey J. Smith is a Montana-based writer with a strong interest in natural-resource issues and energy efficiency.

From *Fine Homebuilding* magazine (December 1986) 36:30-33

After the flue-run lids were complete, Marcus laid in rebar and installed two 8-in. cleanout ducts with airtight lids (left). He also laid in a 3-in. aluminum pipe wrapped in fiberglass that was intended to heat cold air from the basement. Marcus added black dye to the concrete for the finished slab (above) in order to increase the absorption of winter-time solar heat. The k'ang, finished except for an airtight firebox door, is shown below. Photos above: Jeffrey Smith.

Metal stoves vs. masonry stoves

All masonry stoves are designed to have enough mass to store heat and deliver it slowly, over a long period of time. This contrasts sharply with the idea behind cast-iron woodstoves, which are now being used by more than 20 million Americans.

Most cast-iron, airtight woodstoves burn large loads of wood for quick response. Their thin airtight walls radiate the heat into your room. The warmth is stored in your walls, hearth, furniture, carpet, even in you if you linger near the stove. It is your room, then, that heats the air. Since most rooms lack masonry-mass heat-storage systems, you must maintain a fire. To keep the fire from burning too hot, you have to crank down the stove's damper, which deprives your fire of air and causes it to smolder.

Though it does the job for up to 12 hours without refueling, a cast-iron stove wastes much of the wood's energy. Researchers at Auburn University and at the New Mexico Energy Institute have found that one-half to two-thirds of the fuel value of seasoned firewood is in gases and volatile liquids.

The key to using those gases and volatile liquids is high combustion temperatures, 1,100°F to 1,800°F. But many cast-iron woodstoves actually begin to glow red at less than 900°F. Sadly, up to 50% of the fuel you've paid for will vanish up your chimney. And that's the same unburned fuel that is loaded with creosote and air pollutants.

Some cities, like Missoula, Mont., lose a half-dozen residences each winter because of chimney fires, and their officials are beginning to restrict wood burning. When the Missoula health department, for instance, finds more than 150 micrograms of respirable wood-smoke particulate per cubic meter of air, they require everyone to switch to fossil-fuel sources of heat. For 10 days last winter, Missoula's wood burners had to shut down their stoves because of air pollution.

Some countries have long traditions of clean-burning masonry heaters. Two-thirds of the new homes in Finland are built with masonry woodstoves. The Finnish government even gives tax breaks to encourage their use.

Also, though few tests of masonry-stove efficiency have been performed in the United States, European tests have placed masonry-stove efficiency at 70% to 90%. That means that almost all of the heat value of the wood (Btus) is used in a well-constructed masonry stove. The fires in these stoves reach 1,100°F in the first three minutes. After an hour, they reach 1,800°F.

—J. J. S.

The north wall of the building seen from the inside as it neared completion. Concrete from the upper courses dripping down the wall will be removed after the entire wall is constructed. The interior has a uniform and plumb appearance.

Form-Based Stone Masonry

A method for constructing cast-in-place stone walls

by Richard MacMath

The skills needed to build a stone wall are simple enough for self-taught masons to get beautiful results after a short learning period. Of course, construction must be done with care, and a few rules kept in mind.

The cast-in-place method we used to build the walls shown here is probably the easiest for the novice. We've had "first timers" who were able to learn the technique in one day. Many were producing beautiful results in a few days.

This method is best suited to construction with cobblestones—round-shaped stones in random sizes. Often these stones are found in fence rows between fields and in sandy or gravel soils. For such stones, using formwork is simpler than laying a freestanding wall, especially if one wants a

Richard MacMath is a partner in Sun Structures, Ann Arbor, Mich. The wall is part of Upland Hills Ecological Awareness Center in Oxford, Mich.

durable, true, residential wall. Using formwork as a guide also makes the work go faster. The stacking and leapfrogging of forms ensures plumb and level results.

The job's most difficult aspect, and one impossible to avoid, is getting the stones to the construction site. Once you have found an adequate supply, hauling the stone becomes a task for as many helpers as you can assemble. Although this seems like a troublesome job, remember that transporting free stone—even over a few miles—is inexpensive when compared to the cost of other building materials.

Even if you have stone and labor in abundance, it's important to be selective in the use of stone masonry. It is best suited for retaining walls and below-grade foundation walls with outside insulation covered with earth. Foundations, basements and earth-sheltered homes are appropriate applications. In such uses horizontal

and vertical steel reinforcing bar (rebar) is required to strengthen the wall against the extra pressure exerted by the earth mass. Stone is also a good choice for interiors, especially for thermal mass walls. A passive solar design rule-of-thumb calls for 1.0 to 1.5 cu. ft. of masonry exposed to direct sun for every 1.0 sq. ft. of south-facing glass.

As a note of caution, above-grade stone walls that require insulation are a problem. They must be insulated in one of two ways: either by constructing a cavity wall or by adding an insulating layer to one side of the finished solid wall. Cavity-wall construction is difficult, time consuming and not recommended for the novice. The other option—adding insulation to the interior or exterior face of a solid masonry wall—requires studs or furring strips, insulation and a finish surface (usually gypsum board) that will completely cover one side of the stone wall. In

From *Fine Homebuilding* magazine (June 1981) 3:28-32

both cases it would be easier to construct a well-insulated stud bearing wall first and add the stone veneer later.

Rules of stone masonry construction—The first rule of sound wall construction is to set the stones so that gravity rather than mortar holds them in place. Each stone must be set in a firm, relatively flat bed formed by the stone below and its covering layer of mortar. I have seen large stones fall out of the wall when formwork was removed because the novice stonemason relied too much on the concrete and formwork to hold them in place. This is the most common mistake in form-based stone-masonry. Place the stones as if you were laying a dry wall; use the concrete only as joint fill. If a level, flat bed cannot be prepared for a stone, then place it so gravity forces it inward. In this way, adequate beds can be made for angled and rounded stones.

The second rule is to place each stone so that its weight is distributed over at least two other stones below. This is the principle of crossing joints for maximum strength. Crossing joints ensures that the wall will work as a unit rather than as individual parts. This rule doesn't apply when setting small stones on top of large ones, but it's best to avoid a continuous joint from top to bottom in a wall. Otherwise you invite cracks, and the finished appearance lacks the random overlapping pattern that makes stone masonry so attractive. In long walls, however, a continuous joint is constructed intentionally, every 30 ft. or so. This is called a control joint, and it allows for expansion and contraction of the wall under changing thermal conditions.

The third rule concerns both structure and aesthetics. The mason must not only look at each stone carefully, but also at how each stone fits into place and contributes to the structural integrity of the wall as a whole. For example, a particular stone may have one face that is flat

"Masonry is as fundamental as gravity, nothing more. Once you are aware that a stone's tendency is to fall straight down, you will know how to build a strong wall."
—Ken Kern

and colorful. The beginner's impulse is to expose this face on the finish side of the wall. Unfortunately, its best use may be as a flat surface, providing a firm bed for succeeding stones. In this case, aesthetics must yield to the structural requirements of the wall.

Remember that these rules are easy to overlook when employing this cast-in-place method because the formwork holds the wall in place while the concrete is setting, hiding the finished surface from view.

Designing the wall—If the stone wall is going to be a bearing wall or a retaining wall, then the structural loads must be calculated. These loads will determine wall thickness, footing size and the amount of steel reinforcement required. Fortunately, we have engineering training so we perform the necessary calculations ourselves. Also, construction details must be figured out and drawn to scale to illustrate connections to other structural elements. For example, anchoring the roof to the top of the wall requires proper planning. Our design called for 12-in. long anchor bolts set into the top of the wall on 4-ft. centers. These secured a top plate that served as a nailing surface for 2x12 roof rafters. Since we were constructing a sod roof, we had to continue

the waterproof membrane down the outside of the wall and over a layer of rigid insulation.

In our building the north enclosing wall is a below-grade bearing wall that supports both roof and earth loads. Vertical loading from floors and roof is easily transferred to the footings because a masonry wall is very strong in compression. However, the major structural load below grade is the lateral force of the earth pushing on one side of the wall. These lateral loads force the wall to react in tension on the interior side. Pure masonry walls have almost no tensile strength, so steel reinforcement must be added. As a rule-of-thumb, earth backfilled up to 6 ft. high against a 12-in. thick wall requires no vertical reinforcing. We designed our north wall to support the lateral load of almost 10 ft. of earth. The vertical loads from the roof and the weight of the wall itself provide some resistance to the force of the earth, but we still had to add substantial reinforcement. Note that the rebar is placed on the tension side of the wall—in this case the inside—and additional vertical reinforcing was added to the bottom half of the wall where the earth loads are greatest. Horizontal rebar is laid down between stone courses at 2-ft. intervals during wall construction. The amount of reinforcement is determined by structural calculations.

Footings are generally twice as wide as the wall is thick and equal in depth to the thickness. Here, in a retaining wall situation, the wall is placed close to the interior side of the footing so that the weight of the earth on the outside helps to keep the wall from overturning.

There is one place in our building where the wall changes from a retaining wall enclosing the building to a freestanding exterior wall. Because these two sections of wall experience different loading and temperature conditions, we built a continuous control joint in between them. At the control joint the two walls are tied together with horizontal rebars wrapped with felt paper. The

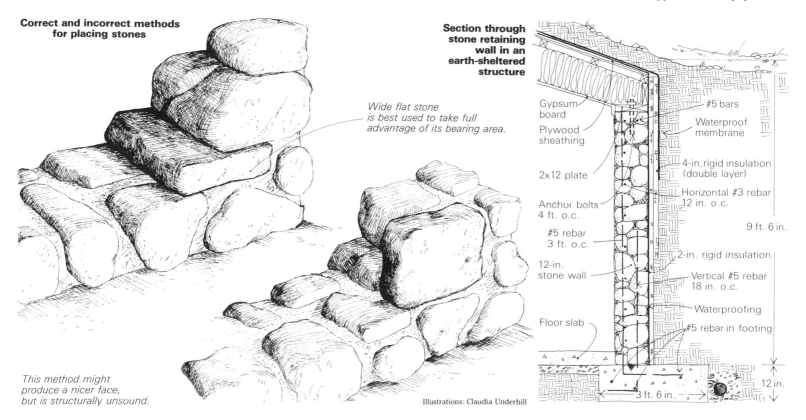

Correct and incorrect methods for placing stones

This method might produce a nicer face, but is structurally unsound.

Section through stone retaining wall in an earth-sheltered structure

Wide flat stone is best used to take full advantage of its bearing area.

Gypsum board

Plywood sheathing

2x12 plate

Anchor bolts 4 ft. o.c.

#5 rebar 3 ft. o.c.

12-in. stone wall

Floor slab

#5 bars

Waterproof membrane

4-in. rigid insulation (double layer)

Horizontal #3 rebar 12 in. o.c.

9 ft. 6 in.

2-in. rigid insulation

Vertical #5 rebar 18 in. o.c.

Waterproofing

#5 rebar in footing

12 in.

3 ft. 6 in.

Illustrations: Claudia Underhill

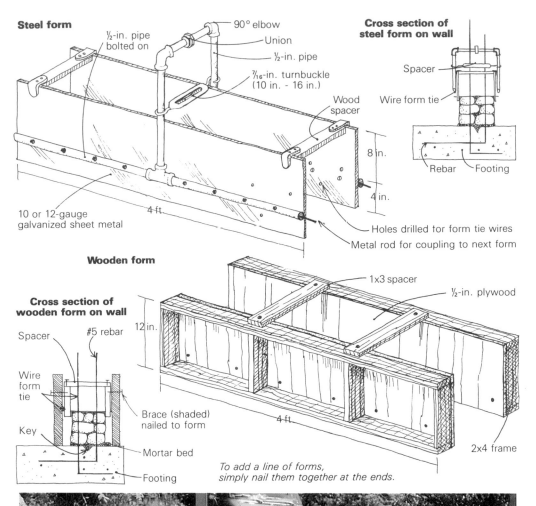

Steel form

½-in. pipe bolted on
90° elbow
Union
½-in. pipe
⁷⁄₁₆-in. turnbuckle (10 in. - 16 in.)
Wood spacer
10 or 12-gauge galvanized sheet metal
4 ft.
Holes drilled for form tie wires
Metal rod for coupling to next form

Cross section of steel form on wall

Spacer
Wire form tie
8 in.
4 in.
Rebar
Footing

Wooden form

1x3 spacer
½-in. plywood
12 in.
4 ft.
2x4 frame

Cross section of wooden form on wall

Spacer
#5 rebar
Wire form tie
Key
Brace (shaded) nailed to form
Mortar bed
Footing

To add a line of forms, simply nail them together at the ends.

Wood forms work just as well as metal ones. Concrete, stones and water must always be on hand. Working with at least three or four people at a time keeps the work progressing quickly. Here two people are setting stones in a bed of concrete, layer by layer. The wood forms shown are heavier than metal ones, but last longer and are easier to clean.

rebar holds the retaining wall in place laterally while allowing for expansion and contraction in a direction parallel to the wall.

Formwork—We used both metal and wood forms, both serving the same purpose. When set on the footing or clamped to the top course of the wall, the forms held the concrete and stone in place while drying. Metal forms were lighter and easier to assemble, but sometimes bulged at the bottom as they were being filled. Wood forms, though heavier and clumsier, better resisted the weight of concrete and stone. Both types were coated inside with used motor oil to keep them from sticking to the concrete.

Design for the metal forms came from Ken Kern's book, *The Owner Builder's Guide to Stone Masonry* (Owner Builder Publications, Box 550, Oakhurst, Calif. 93644). Kern uses galvanized pipe, pipe fittings and sheet metal, together with a metal turnbuckle and wooden spacers. The forms are 12 in. high and 4 ft. long and can be interlocked at each end with other forms. We assembled 4 forms, each weighing 20 lb., but they gave us some problems. First of all, we had difficulty keeping them clean. Concrete was hard to remove from the threads on the turnbuckle and from the pipe fittings. Also, tightening the turnbuckle sometimes bent the pipe.

Generally we preferred the wooden forms even though they were heavier by about 10 lb. They were made in two parts and connected when clamped to the wall. One advantage of using wood was that the forms could be nailed to stakes and other forms for alignment and stability. We always used the wood forms when building the curved north wall. Bending pipe to the proper radius seemed difficult to us. Spacers and wire were used with these forms as with the metal ones.

Required materials and tools—Besides the stones and forms, you'll need the following materials and tools to build a form-based stone wall:

Sand, gravel, portland cement.	Mortar boards
Water	Hose, containers of water (for soaking stones)
Reinforcing steel (rebar)	
Waterproof coating (asphalt or cement base)	Wood stakes (for layout work)
Polyethylene sheets	String
Used motor oil	Gloves (lots of pairs)
Cement mixer or deep wheelbarrow and mason's hoe (has holes in it)	Wire
	Hacksaw
	Level (preferably 4-ft. length)
Wheelbarrows (for hauling stone)	Chalk line
	Paintbrush
Shovels	Wire brush
Buckets or pails	Plumb bob
Mason's trowels: large size (for filling forms with concrete), small size (called pointing trowels, for finish work)	Nails
	Hammers
	Sledgehammer
	Mason's hammer and chisels (if you want to shape stone)

We used a standard 3-2-1 concrete mix: 3 parts gravel, 2 parts sand and 1 part portland cement. To estimate the amount, assume that the concrete mix makes up approximately one third of the wall's volume. Whenever we were asked how much water to add to the dry concrete mix, my friend Wayne would reply, "to taste." There are so many variables—weather, wet or dry sand, gravel size—that the right amount of water will often vary from day to day. We preferred to use a mix that was dry enough to stand on a trowel, but wet enough to fill all the gaps between the stones. A dry mix can be held in by the form, but must be scooped out of the wheelbarrow with a trowel or shovel.

Reinforcing steel (rebar) is available in diameters from $\frac{1}{4}$ in. (#2) to $1\frac{1}{4}$ in. (#10). After calculating the amount required, you have a choice of using a small number of large diameter rods or large number of small-diameter rods. Small-diameter rebar is easier to bend and cut to the required length. For example, we often substituted three #3 bars ($\frac{3}{8}$ in. diameter) for one #5 bar ($\frac{5}{8}$ in. diameter). The cross-sectional area is the same, but the #3 bars are easier to bend into the footing and cut with a hacksaw. Since rebar is sold by the pound, there is no additional cost in using bars of smaller diameter.

Many different products are available for waterproofing below-grade walls. The standard products are asphalt-base and cement-base coatings, either of which can be applied with a brush or trowel. We brushed on an asphalt base coating and added a sheet of 6-mil polyethylene over that. Polyethylene sheets can also be used to protect the masonry during cold weather and to keep various floor and wall surfaces clean during construction.

If you're using a cement mixer (ours was driven off a wind-powered electric system), then have a few large buckets on hand for filling it quickly and easily. Using a wheelbarrow and mixing by hand requires measuring by the shovelful—a slow process. Add water by the bucketful or with a hose.

Be sure to have some large containers on hand for soaking the stones. We used old 55-gal. drums. The wall cures more uniformly when the stones are holding water as they are set into the wet concrete. Concrete and stones then dry together, and a better bond results.

Always have plenty of work gloves on hand. After the fabric of the glove is worn away, (sometimes in one day), handling the stones removes skin from your fingertips. It hurts just thinking about it. Prepare yourself for seeing a lot of useless gloves lying around with holes in the fingers.

Spend the money on a 4-ft., 3-tube level; the time saved and accuracy provided more than justify the expense. Wall surfaces will never be perfectly flat, and when you check for level and plumb, the level must span four or five stones. This isn't possible with a shorter level.

The construction sequence—Building the wall requires proper design and construction of the footings. Since we had continuous vertical reinforcement from the footings up into the wall, we placed the rebars before pouring the footings. As the concrete began to set, we chiseled a key way—a groove 1 in. wide by $1\frac{1}{2}$ in. deep—down the center of the footing. This provided a better bond between the footing and the first course of stone.

As the wall begins to go up, make sure there are enough stones stacked within reach of the

With the first course of stone and concrete in place, top, the forms have been removed. Note the width and depth of the concrete footing and the size and spacing of the reinforcing steel. The string line is a horizontal reference 4 ft. above the top of the footing supported by stakes 8 ft. high. A plumb bob is dropped from this string to plumb the wall. For a proper structural connection between the two poured layers, the top of each layer is left rough, with some stones and reinforcing steel projecting up. Each stone must be placed so that a relatively flat bed is created for the stones above. After the second course is laid, above, the metal forms are still in place on the projecting retaining wall to the left rear. The concrete has already been scraped away from these stones with a trowel. Each stone is set so that its weight is distributed over at least two stones below. Unless a control joint is required, joints are never placed directly above each other because this leads to cracks.

55-gal. drums, which should be convenient to the section of wall being worked on. One or two people should mix concrete while others fill the drums with water and stones. At the end of the day, we refill the drums for the next day.

We often worked on two different parts of the wall at once, a team completing one course while another started the next. We find the optimum team to be two people filling the forms, with one strong assistant to lug the stones, mix and haul concrete, keep the drums filled, and point the finish side of the wall once forms are removed and shifted to another section.

After the footing has cured, use a chalk line to mark the inside and outside surfaces of the wall. The first row of forms is set on these lines. We used 8-ft. 2x4 stakes with string running the entire length of the wall as a plumb guide. Adjust the stakes so that a plumb bob held from the string touches the outside form. Using 8-ft. stakes allows you to move the string up the wall as construction progresses.

The curved part of our wall was more difficult to lay out. When we were drawing the building floor plan, we calculated the radius of the inside

of the wall. After locating the center point of this curve on the ground, we marked the footing with string and chalk. We relied on the level for keeping this part of the wall plumb.

Whether you are starting at the footing or on top of a previously completed course, you should begin by hosing down the surface and removing any loose material. This ensures a strong bond. When the surface is clean and wet, set the forms in place. For the first course, this simply means setting them down on the footing with the wooden spacers in place to make sure the form is tightened to the proper dimension. For subsequent courses, the forms have to be clamped in place.

Metal forms can be erected quickly. Hold the form in place, and tighten down the turnbuckle that is just above the metal sides. Wooden spacers help prevent over-tightening. After many setups, our forms began to bend at the vertical pipe, causing the bottom to bulge out slightly. To prevent this we drilled holes in the sheet metal and improvised wire form ties that were tightened by twisting the ends around nails on the outside of the forms. Thicker sheet-metal

The exterior of a masonry wall must be kept as smooth and clean as possible for waterproofing later. In the foreground, mortar is being applied for a smooth finish. In the background, wood forms are in place on the curved part of the wall. Temporary wood forms made of 2x12s can be used when a lot of people are willing to help and additional formwork is needed.

first if the mix is wet, but as you pack stones and more concrete firmly against the form this waste will be minimized. Pack concrete firmly around all of the rebar, both vertical and horizontal, to bond the masonry and steel.

Fill the entire form in this fashion until the stone projects a few inches above it. This provides a better joint with the next course and makes it easier to clamp the form to the protruding rocks when it is moved. When proceeding smoothly, three people were able to fill a form 12 in. high and 4 ft. long in about half an hour.

If you build a below-grade wall as we did, each side of the form should be packed differently. For the earth side of the wall, try to achieve a smooth finish for a waterproof coating by packing the concrete in tightly. On the finish side, consider final appearance as you select and pack stones. Packing the concrete is not as critical on this side since the wall will be pointed after the forms are removed. However, you won't be able to see how the faces of the stone fit together on the finish side of the wall until the forms are removed. As part of the learning process you should start the wall at an end that will not be highly visible, because until you've gained some experience you won't be able to visualize the fit of the stones.

How long the forms must be kept in place varies with the weather and the concrete mix, but they must be removed soon enough to tool the joints to a finished state. We never keep our forms in place longer than two hours, but you will have to test this timing for yourself, since the variables of mix, weather and stone may differ. Detach the metal forms by loosening the turnbuckle and removing the nails from the wire end. Pull the wire through the wall and lift off the form. Removing the wood forms requires only pulling out the wire at the bottom, since we keep the top wood strips in place. Scrape the form clean, give it another coat of oil, and set it again atop the wall. The forms need be hosed down only at the end of the day.

Once the wall surface is exposed, it begins to dry quickly, so the joints should be tooled tight to the stone immediately. On the finish side, remove concrete around each stone with a small pointing trowel. Recessing the joints in this way exposes more of each stone, making them the prominent elements in the wall and giving it a laid masonry appearance. On hot days we keep the wall wet by hosing it down with a fine spray as we work and at the end of each day. The slower the concrete dries, the stronger it cures.

To give an idea of required construction time, it took us six weeks to complete a wall 60 ft. long and 8 ft. high. The wall cannot be thoroughly cleaned until it is completed, because concrete continues to spill down the sides during construction. We waited until the building was enclosed before sandblasting the inside of the wall. This is a quick method, but you must be equipped with gloves, goggles, proper clothing, a compressor and fine sand. Scrubbing with muriatic acid and a stiff brush is another method of cleaning that removes most of the surface dirt and excess mortar. Once the wall is clean, everyone is surprised by the beauty of the color and texture in the stone. □

forms, 10 or 12 gauge instead of 14 gauge, would also prevent bulging.

Wood forms take a little more time to set up. The two sides are connected by furring strips nailed into the top and by wires threaded through the form at the bottom.

After the forms are in place, use a trowel or shovel (trowels are easier to handle) to lay approximately 1 in. of concrete on the footing or previous course. Begin a layer of stones by packing them against both sides of the form and then filling in between with small stones and rubble. Make sure each stone has a face set into the bed of concrete. Some stones may be large enough to touch both sides of the form so that center-

filling isn't necessary. Pack the stones as tightly as possible, allowing minimal space for concrete. This is the best way to save money, since cement is your most expensive construction ingredient. Once a stone is laid in place, do not move it. Movement weakens the bond. Always have stones of various sizes and shapes on hand, so that placing stone around the vertical rebar is no problem.

Once the first course of stone is complete, add another layer of concrete. We often shoveled concrete into the form, and then agitated it with a small trowel, to fill gaps between the stone. Don't worry about concrete running out of gaps at the bottom of the form. This may happen at

The Structural Stone Wall

Save your stones for pointing, bed your stones in concrete, and remember that there's no substitute for gravity

by Stephen Kennedy

Adams County, in south-central Pennsylvania, has hundreds of stone buildings, most of them built before the days of portland cement. I got started in stonemasonry by pointing up some of these old beauties, and I've always been impressed with the soundness of their stonework. After nearly two centuries, these buildings are still young. The masons who built them couldn't rely on their crude lime mortar to hold stones together. Instead, they used a far stronger glue— gravity. The stonemasonry I do today follows this old-fashioned philosophy, but fortunately I'm able to take advantage of some new materials that weren't available 150 years ago—concrete and pointing mortar.

The trademark of my stonework is spotless pointing (bottom photos, p. 166). A good pointing job keeps water from getting inside the wall and also provides a consistent background that can really show off the variety of textures, colors and shapes in a rock wall.

High-quality pointing mortar isn't cheap, but I don't use a lot of it. The jointwork extends only an inch or two into the wall. The rest—the part you can't see—is just stone and concrete. There are a number of advantages to using concrete instead of mortar to lay up a stone wall. The most obvious is cost: mortar mix is a lot more expensive than concrete, and you'd need quite a few more bags to complete a comparably sized wall. This is because the gravel (technically known as *coarse aggregate*) acts as a filler in the mix, making it go twice as far without sacrificing strength. The large stones still rely on gravity to hold their position in the wall, and the concrete ensures that there will be enough space between stones for me to point later.

With a 5-gal. bucket as a measuring unit, I make and mix the concrete on site. Two buckets of sand, three buckets of gravel, ⅔ bucket of portland cement and about ½ bucket of water (depending on the moisture content of the sand) yields a large wheelbarrow load of concrete. I mix everything together at once except

the gravel, which gets added last. The final bucket of gravel really dries up the mix, and then it's ready to use.

Concrete also gives me greater flexibility in laying up the wall than mortar would. Without the added gravel to hold adjacent stones apart, mortar tends to squish out between joints. With a stiff mix of concrete, I can keep the mix well away from the exposed face of the wall. This leaves room in the joints for my pointing mortar and cuts down the risk of messing up the faces of the stones with squeeze-out. The concrete won't compress with added weight even if it hasn't set, and this allows me to work vertically without worrying about the joints collapsing. Regular mortar mix can't stand up to much compression before it sets, so you have to work horizontally, which isn't always convenient.

The search for stones—To build a structural stone wall you need plenty of stones—about 30 tons for a 2-ft. thick wall 10 ft. high and 10 ft. long. Finding the rocks and getting them to the site is at least half the work. Both your back and your pickup truck will probably suffer for it, too. Fortunately, this is stone country. Mortarless stone walls hastily piled up by earlier generations of farmers crisscross the landscape, so few house sites are bereft of material. This isn't usually enough, though, so I end up looking on mountaintops and through dry washes and stretches of woods for the many elusive "ideal" stones that almost every job demands. I sometimes take out permits to get rocks out of state forests. In fact, when I'm in the middle of a job it's hard for me to drive down the road without scanning the countryside for rocks.

With the help of fellow mason Paul Qually, the author built the walls of this small house (above) in 1977 using locally gathered stone. Large, square-edged stones that span the full thickness of the wall were saved to build corners that look and work like rough, massive finger joints.

I hardly ever cut or dress my stones because this takes lots of extra time and because it alters the rock's naturally weathered surface. If I find a nice stone that's covered with lichen, I'll often leave the lichen exposed in the finished wall.

I look for stones that are shaped like large boxes, books, bricks and milk cartons. These and other cubic forms are far easier to stack than bowling balls, footballs, turtles and sausages. On every job I always have a few unusual stones that I want to fit into the wall somehow, but the heart of the wall should be made up of parallel-sided rocks.

Wall construction—The size and construction of the footing, or footer, for a stone wall depends a lot on wall dimensions and local soil conditions. I've dismantled old walls 2 ft. thick that had no footing to speak of. I prefer to foot a stone wall with a stone base, rather than use rebar and poured concrete. This is another method used by early stonemasons, and it's stood the test of time well.

I usually start by digging a trench about 3 ft. wide and 3 ft. deep, and fill it with large rock rubble up to just below grade level. Then I pour a broad concrete cap about 3 in. or 4 in. thick. This is where the visible stonework begins (drawing, p. 167).

Keeping a stone wall plumb is always a challenge. Inevitably, some stones will protrude beyond the plumb line while others will fall slightly short of it. If the average between the proud and shy stones is close to plumb, the wall should be plumb (and look plumb) overall.

I use a plumb bob whenever I can, but sometimes there's no place to hang the bob. An alternative to the plumb line that works well for me is a 6-ft. level upended in a bucket of sand. I often use two or more of these guide sticks, positioning them as close as possible to the work in progress. I usually align the levels with the chalklines on the footer that describe the average wall width I'm aiming for. Eyeballing my stonework off these verticals is fairly easy, and they're easy to reposition.

The stones for the first course sit in a bed of concrete laid directly on the footer (drawing, facing page). Their exposed faces should fit together with a fairly even space around them. Angular stones should be positioned so the faces slope in, toward the center of the wall. This sometimes creates depressions inside the wall, which I fill with concrete or small rocks. I often have to test-fit a rock, looking under it or lifting it up to see how solid a "print" it makes on its rock and concrete base; then I adjust with more or less concrete and reposition the rock.

I hold the concrete back from the face of the wall, so that there's room between rocks for pointing. The real structural bearing starts an inch or more back from the wall face.

Vertical joints should be staggered. And it's important to use large stones here and there that span the full thickness of the wall. Large stones are essential structural and visual elements, especially at corners.

I avoid standing narrow rocks on edge in a wall. This often sacrifices the look of a really big stone, or even hides an attractive rock face. But anti-gravity stunts don't hold up over time, even if you glue a shaky stone in place with mortar. The challenge in my kind of work is to create a nice-looking wall that will last for generations.

Pointing—Before I start pointing up the joints between rocks, I hose down the wall thoroughly. This removes loose grit and dust that would prevent the pointing mortar from adhering strongly. Once the wall has dried so that it's damp rather than wet, I can start pointing. I usually buy the best cement and the cleanest, finest sand I can find. This sounds extravagant, but it isn't, because a little pointing mortar goes a long way in my walls—the pointed joint is only one or two inches thick. I use either black or white pointing mortar. For a really white mix, you have to use white sand. A grey mix can be darkened by adding black pigment. On a Trombe wall that I built recently, the north wall face is

The heart of the wall. The ingredients are stones and a very dry, stiff mix of concrete. The stones are gravity fit with overlapping joints. Concrete and small stones are used to fill the voids between larger stones. As the wall is built, the concrete is kept back at least an inch from the face of the wall. This leaves enough space between stones for a thin, strong mix of pointing mortar.

Pointing. A rich, buttery mix of pointing mortar made with fine sand sits well on the trowel and is easy to work. At left, Kennedy uses a pointing trowel to work the mortar into joints between stones. When the mortar starts to dry, he goes over it with a stiff, dry brush, right, smoothing the joint tight against the stones and brushing away small splatters and crumbs. Pointing mortar should be kept damp to prolong its curing time.

Photos: Dannie Kennedy

pointed with white mortar, while the south-facing joints (photo center right) are black for better solar absorption.

I always use a rich pointing mix: 2 parts fine sand, 1 part cement. This keeps the mortar buttery and generally easy to work; it won't slide off your trowel or out of the joint as easily as a 4-to-1 mix will.

You've also got to keep the mix as dry as possible. I add just enough water to get the mortar past the crumbly stage. This way, I can pack the joint well without having the mortar run down the face of the wall. I load a triangular mason's trowel with a fist-sized blob of mortar, flatten it and then pack it into the joint with any one of several thin pointing trowels (photo facing page, bottom left). The wider the pointing trowel, the better it will hold mortar, but for thinner joints you need skinnier trowels.

The more pointing you do, the less you'll tend to lose mortar off your trowel. A few drips are inevitable, though, and if these land on exposed rock faces, just leave them be. If you can resist the urge to clean up these splatters immediately, you'll avoid staining the face of the wall. Let the misplaced mortar stand until it's very dry, but not hard; then scrape it off with a trowel or stiff-bristled brush. It should fall off like dust.

Every so often there's an especially large or broad joint that looks out of place among its narrower neighbors. When this happens, I pack pointing mortar into the space and then push a small face stone into the mud. This fills in the space nicely and eliminates unsightly fat joints in the finished wall. Whenever I insert such non-structural stones, though, I make sure that they penetrate the full depth of the pointing mortar. Otherwise, they're liable to fall out sometime in the future.

Another way to fill fat joints is with small ornaments. Not all clients like this kind of thing, but a sculptor I built for recently supplied me with quite a nice variety of inserts, including a brass horse, a steel toy truck (photo bottom right) and a small, cast-bronze Mickey Mouse.

Once the mortar is worked into the joint, I don't fuss with its rough texture for about an hour or so. This gives the mud time to harden up slightly. I then scrape it back and smooth it with a pointing trowel. After troweling the mud smooth, I go over each joint with a small (2-in.), dry paintbrush (photo facing page, bottom right). This really tightens the joint against the edges of the stones, and it removes any remaining grit from the mortar.

The next day after pointing, when the mortar is very firm, I give the wall a thorough hosing down. This helps the mortar to cure better by prolonging its drying time. With hose in hand, you can also go over the wall and wire-brush away any stray mortar globs that might have escaped your scrutiny earlier. The second day after pointing, I wet the wall again, this time setting the nozzle to its hardest spray. This is the last chance to brush off excess mortar without having to resort to muriatic acid or sandblasting. Once the pointing mix has cured, what's on the rock will pretty much stay there. □

Stephen Kennedy lives in Orrtanna, Pa.

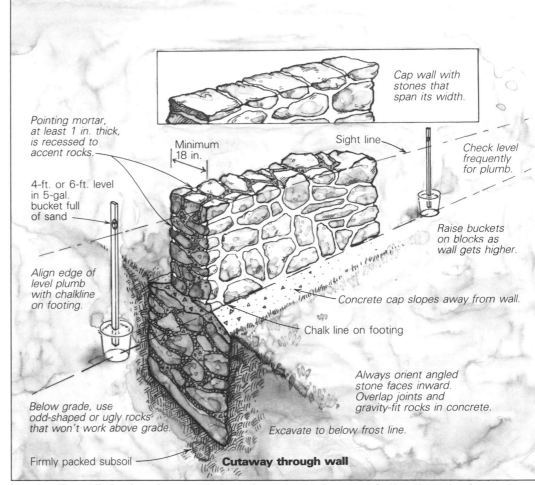

Cap wall with stones that span its width.

Pointing mortar, at least 1 in. thick, is recessed to accent rocks.

Minimum 18 in.

Sight line

Check level frequently for plumb.

4-ft. or 6-ft. level in 5-gal. bucket full of sand

Raise buckets on blocks as wall gets higher.

Align edge of level plumb with chalkline on footing.

Concrete cap slopes away from wall.

Chalk line on footing

Always orient angled stone faces inward. Overlap joints and gravity-fit rocks in concrete.

Below grade, use odd-shaped or ugly rocks that won't work above grade.

Excavate to below frost line.

Firmly packed subsoil

Cutaway through wall

Keeping the wall plumb and strong. *Instead of a conventional concrete footing, the author frequently fills a trench with a rubble of rock, gravel and concrete. The trench should be excavated twice as wide as the wall to a packed base below the frost line. The rubble footing is topped with a concrete cap 4 in. to 6 in. thick that slopes slightly away from the wall. Stones should be gravity fit with overlapping joints. Some stones will have to protrude slightly beyond the intended width of the wall, while others will set just shy of it. If plumb lines can't be set conveniently, you can work off sightlines established by two or more levels upended in buckets of sand. The levels should be positioned at opposite ends of the wall, vertically aligned with the layout line chalked on the footing.*

Skill, patience and a varied selection of shapes and sizes make it possible to accomplish intricate stone joinery without cutting any stones to size. Kennedy built the wall shown above in his own house and pointed the joints with dark mortar. Several stones have a natural lichen mantle that Kennedy decided not to disturb. At left, a toy steel pickup truck is used to fill a large joint that would otherwise stand out badly in the finished wall. Small, non-structural stones can also be used in this way, but they must be pressed firmly in the pointing mortar before it sets.

Stoneoak

Massachusetts homesteaders build a cottage of rock and timber

by Jonathan von Ranson

Our house was a family project that took four years of intense construction, plus three years of sporadic finish work. Susan DunLany, the children and I found, hauled, washed and laid stone, and cut, hauled and hewed timbers until the structure was finished and insulated. Then, living inside, Susan and I outfitted and paneled it as time permitted.

Planning and design decisions were made and remade throughout the project. But the essentials were constant: it was to be a house for people who would spend a lot of time in and around it, who were living self-sufficiently, agriculturally. In other words, a house for people who didn't want bills to pay. This meant a small, wood-and-solar-heated house with plenty of natural light. It could involve a high initial investment of our labor, but little of money, and should require little maintenance thereafter. In keeping with that goal, we gravitated toward simple, compact design with as few seams, stays and moving parts as possible. This suited our moderate level of construction expertise.

The 20½-ft. by 22-ft. house sits in a crease alongside a strewn-out jumble of boulders, between a small pond and a cliff, near the top of an otherwise uninhabited mountain. Good sun and the proximity of water governed the location of the site. Nearby sits a cluster of weathered outbuildings. We built the first of these, and lived in it as we were building the house.

The homestead is tucked away in western Massachusetts, two miles from neighbors, maintained roads, electricity and telephone. To outsiders, Stoneoak probably looks as if a medieval artisan met a New England pioneer and found Thermopane. The factory-made doors, the large windows and slate roof with skylight look frankly contemporary, while the fieldstone walls draw the house back into the rocky, post-glacial landscape and into an earlier age. As it sits on the site, the house is more assertive than I'd hoped. The final effect awaits a back porch, a retaining wall to bring the landscape up a little higher around the house, and some vegetation.

Both upstairs and downstairs are essentially single rooms. Downstairs, three walls are paneled and only one wall—the south—shows stone (photos facing page). The flooring is T&G strip oak. Upstairs, the space is occupied by the bed at one end and a desk at the other.

Post-and-beam construction in a stone house may seem redundant. But it made sense here because, wherever possible, we preferred to use available materials rather than earn money to buy materials. Oak was plentiful—the 1978-79 gypsy-moth pestilence had left some trees dying or dead. Cutting and hewing 16 logs between 8 ft. and 20 ft. long gave us the supports for walls and floors. Because of free spans, beams

From *Fine Homebuilding* magazine (April 1987) 39:56-60

Sheltered on the north by a cliff, the house (facing page) is two miles from neighbors and paved roads. Its stone walls and slate roof enclose an oak timber frame. Inside, exposed beams and hand-planed oak paneling lend warmth (right); fixed Thermopane units bring in light. To the left of these windows, the stonework is exposed (bottom right).

were needed to carry the second floor anyway, with the help of 4x4 oak joists on 24-in. centers. This type of system looks good exposed, and it eliminated the need for a finished ceiling.

The timber frame had another function to perform during stoneworking. It was erected after the foundation and deck were in. It held the roof over our workspace, and best of all, braced the forms against which I laid up stone.

In addition to timbers, the site offered abundant stone. Here, digging the garden served a double function—building stones for the house were actually the first crop out. Others were pried out of the duff, or found lying on the surface alongside the road. I estimate 100 tons or so were gathered, mostly by sons Erik and Dean. Nearly all the stones had to be washed and scrubbed with a wire brush to remove clay and lichens. Anything loose on a stone can interfere with the bond with the mortar.

Stone construction—We had a backhoe dig the hole for our foundation and root cellar. For the foundation, stone was laid without forms. The 18-in. thick walls were laid by using two parallel strings, one directly above the other, and sighting down to find the outer surface. Both strings were kept above the height at which I was working, to be out of the way. The inner surface was located by measuring 18 in. in. In laying up the stone, I tried to break not only vertical face joints, but also the vertical interior joints where two opposing face-stones backed up to each other, for increased strength.

The mortar for this and the entire job was 9 parts sand to 2 parts portland cement to 1 part lime. This is a fairly stiff, quick-setting mixture that handles better than mortar made of only sand and cement. It is less plastic than a standard brick and block mixture, which includes at least an equal ratio of lime to cement, so it doesn't squish out when heavy stones are laid.

Mortar joints were kept between ½ in. and ¾ in. wide whenever possible. In the foundation they were pointed up both inside and out. Susan pointed up most of our stonework (top photo, next page), moving in with her pointing trowel as the mortar set up. Where the walls were to be underground, she took special care—we wanted to avoid frost damage, as well as a wet cellar.

Just outside the wall, we laid a foundation drain of 4-in. perforated plastic pipe covered with ¾-in. stone. Together with the pointing, this precaution has left a dry cellar. A dirt and ledge floor makes it sufficiently humid to be an excellent root cellar, though. We regularly eat carrots and apples from it all winter and into April and May—seven months after harvest.

Once the foundation had reached the height of the basement ceiling, the interior face of the wall was stepped 6 in. back to form a shelf for

Meticulous pointing (top) has kept the cellar dry and enhances the look of the masonry. Stone 'arches,' laid up against a board form, span the operable windows (middle). Rough window and door frames were braced in position (above) and the stone was built up around them.

the post-and-beam sill. From here on, the masonry was done from the outside, laying up the wall—now 1 ft. thick—against movable 2-ft. by 8-ft. wooden forms. A good day's work for me was doing the 16 sq. ft. of wall represented by one of these forms. In cooler weather, when the mortar wasn't setting up, I'd alternate between a couple of forms already braced in position.

The next day, when the wall was sufficiently hardened, I'd move the form to the next position—often just slide it 2 ft. higher. The forms were anchored to the post-and-beam structure itself, or by long braces to a cleat on the floor.

We got a good buy on eight thermal sliding-glass-door inserts, which we installed as fixed windows. The four that were installed horizontally were set at a height that permitted a tall person to see the surrounding mountain ridges while standing up.

Downstairs we ventilate by three double-hung windows and two doors. Upstairs ventilation is provided by two 18-in. by 36-in. north windows, which hinge out under the eaves, and two 6-in. by 46-in. vent flaps over the gable windows.

Window openings were created by assembling permanent boxes out of 2x10s the size of the rough opening and tacking them into position on braces. Stone was built up to and around them (photo bottom left). A 1x1 strip nailed around the outside of the boxes served as a barrier to air infiltration between shrinking wood and masonry. Threaded anchors set into the wall were used to prevent movement of the 2x10s and allow them to be replaced later if it ever becomes necessary.

Visually, the board-and-batten panel in the gables relieves and sets off the stone (photo facing page), but it was essentially a way to avoid the difficulty of spanning the 76-in. wide window openings with stone and the need to hoist stone all the way to the peak. Without stone, the 12 in. of gained interior space above the windows permits a bookshelf downstairs and a shallow alcove upstairs. For the same reasons, the triple window unit in the south wall has no stone above; it is fixed directly to the top plate.

Over the smaller, double-hung windows and the front and back doors, however, there is stone. I used an "arched" type of construction: tapered stones placed fanwise with a keystone similar in size and shape to the others in the center (photo middle left). They were given extra temporary support while the work set up.

About half of the stones used in the house needed some shaping. For shaping and cutting, I sometimes used a hammer and chisel, sometimes a stone hammer alone. To split a large stone (with the grain), I'd put it in the sandbox and hit it with the wedge end of a sledgehammer. To cut it (across the grain), I'd support the two ends on sand, hollowing out the sand between, and strike it with the flat end of the sledge, in a line where it was supposed to break. In both splitting and cutting, I'd work close to the middle of the stone—where reflected shock waves collide to help create the fault.

The chisel is the tool of choice for fussier stonecutting. It can be used to score a stone and split or break it other than in the middle. Another technique I developed was using the

Photos this page: Jonathan von Ranson and Susan DunLany

Board-and-batten siding on the gable ends eliminated the need to span the wide openings above the fixed windows with stone.

edge of a heavy stone or ledge as a backup. That way it was possible to control accurately where a stone would break.

The summertime presence of our teenage children left its mark in the ambitiousness of the undertaking and the pleasure of the accomplishment. Everyone in the family had a main responsibility. Joel, at 12 the youngest, kept the mason supplied with clean stones. Erik, 15 (these ages on the particular summer I'm remembering), drove the Jeep and trailer out for stones, often helped by Dean, 13. Kristin, 14, mixed the mortar in the wheelbarrow—hundreds of batches at the rate of six or seven a day.

The timber frame—After the foundation was finished, we turned to hewing and joinery. The beams for the first floor—sill, girt and sawn joists—were installed first and the subfloor laid. Then preparations were made for the second floor and roof members.

Layouts and measurements reflected the fact that the posts and plates would be set 1 in. away from the stone walls. The space was intended to allow air circulation and lessen moisture "wicking" to the timbers from the masonry. The only connection between the stone and timber frame above the sill was to be the roof rafters.

To hew the oak logs into 8-in. posts and beams, we placed them on a set of skids at a

little above knee height. An 8-in. by 8-in. square was drawn on each end of the log using a graduated level (it was important to orient the squares identically to avoid hewing a twist into the beam). These lines were projected out to the surface of the bark. A chalkline was stretched between two of these points on the log's surface and pulled on the plane of the slab that was to be hewn off.

The step was repeated on the other side of the slab to be removed. Then, with an ordinary ax, I chopped to the line, swinging at an angle for easier chopping. Next I hewed off the chips with a broadax; its silvery ring filled the clearing for about three weeks while the beams were being prepared.

Cutting the mortises and tenons took another week or more. Much of the work was done with an old hand-powered boring mill with a 2-in. bit. A joiner's slick—similar to a large chisel, this one just under 2 in.—removed the remainder of the wood from most of the mortises. A 1-in. bit in the mill made the holes for pegs. Tenons were cut with standard crosscut and rip handsaws.

The beams were worked green, at the recommendation of an experienced joiner, who said red oak would be too hard to work efficiently when dry. (I confirmed that myself on a couple of the drier logs.) The beams have not experienced serious twisting or splitting as they dried

in position. However, we painted all ends before assembly to control uneven drying.

With rough-hewn beams, cutting joints can be tricky, since the surfaces are neither regular nor square in respect to another surface. I tacked a length of 1x8 against the top surface of a beam (and the outside surface of a post) and measured and squared from it. Having rationalized one dimension—height—I was content to custom-cut mortises by measuring for and matching widths. All joints were lettered with red grease pencil that would survive outdoor weather.

The beams that were hewn first, in the fall, turned an attractive silver as they dried. Those hewed closest to installation time are more pink-cheeked, but they show the grain nicely. There is a gradation in tone from the east end of the house to the west that allows the two looks to coexist comfortably, but ideally I would have let all of them weather over the winter.

Neighbors joined us for a beam-raising—surely one of the few ever done for a stone house. With ropes and come-along and a long line of willing lifters, the 13-ft. high north bent was raised as a unit. Everything was assembled that day except the 4x4 oak joists. During the following weeks we framed and sheathed the roof with rough-cut boards, covering it with 30-lb. felt paper while we looked for some slate. Meanwhile, stonework resumed, and all the masonry above the foundation was completed in one arduous four-month push.

The rafters, extending over the stone wall, were ready with bird's-mouth cuts and double 2x10 plates with anchor bolts dangling out of them to be caught by the rising walls. The roof, which until this point had been supported only by the timber framing, now was supported also by stone. Similar anchor-bolted 2x8s nailed to the end rafters were used at the gables for the interface with stone.

Roofwork and entry—After the basic structure was done, outside finish work included roofing and building a front entrance. The front stoop and steps were laid, starting with a 4-ft. deep excavation. This job gobbled up an amazing amount of stone, including an ancient 3-ft. square stone inscribed with a perfect circle and a line, a little like a giant Q. Next to this Druidic marvel, which we later learned might have been the bottom of a large cider press, we installed a mundane piece of 2-in. Styrofoam insulation to try to keep the steps from heating and cooling the cellar too much with their exposed mass. A small dormer-style roof, which was prefabricated in the workshop and lag-bolted into several rafters, protects the entry from rain and snow.

We found and bought some used black Guilford slate from the owner of a mansion near Boston. At $1,200, this was the largest expense of the entire project. The slate came up a rope and pulley fastened to a white birch tree, or a tripod on the other side. About three tons of slate went up that way, with Susan customcutting end pieces as we needed them.

The skylight on the south roof was installed flush with the top of the slate, the large thermal panel set in a rabbeted 2x12 frame. The glass is embedded with the flashing in a bed of clear, high-quality caulk. The flashing laps out every three or four courses.

Ice occasionally builds up on the roof, and after a little melting, lets go in a single sheet. These hair-raising avalanches slide right over the skylight. If it were raised, the unit might trap water or be subject to damage. Aesthetics more than anything influenced the design, though.

Framing, insulation and hearth—All of our framing lumber was roughcut. Two-by-four studs were installed 2 ft. o. c. between single 2x4 plates and sills nailed to the oak plates and sills. All work was still kept 1 in. away from the stone.

In the ceiling, 9-in. R-30 fiberglass insulation was installed. The walls and the first floor received 3½ in. We used a 6-mil plastic vapor barrier toward the living space except in the cellar ceiling, where it seemed the worse moisture problem would be from below.

We didn't use foundation insulation because our cellar is for storing vegetables and fruit, and we wanted it to remain cool. Firewood grows handily for us, and the energy cost-benefit ratio didn't justify it either.

With the house closed in, we began seriously to weigh our preferences in the general matter of cooking and heating. We had considered building a huge central masonry core that would have included a fireplace, a stair to the second floor and a flue for a cast-iron cookstove, but meeting Albie Barden of Norridgewock, Maine, changed our plans. He encouraged us to build a masonry cookstove-heater. This we finally did (see article, pp. 152-155), and luckily, by scaling down the fireplace and moving the stair

elsewhere, we obtained enough room on the chimney slab to accommodate the stone stove.

The masonry core is the functional and visual focus of the house. The 26½-in. by 48-in. stove, 33½ in. high, sits on the chimney's east side. The Rumford fireplace is used mostly for small fires in the evening. Its shallow design radiates and reflects into the room heat that would be sent up the chimney in a boxier design.

Once the stove was built, we moved in, and turned our attention to the kitchen, which features a counter/cabinet unit of cherry, with a slate sink and a hand pump (photo below left). The wood was salvaged after a 1977 tornado in Connecticut, and the work of making the butcher-block counter and frame from the thoroughly dried lumber was bartered (for cordwood) from cabinetmaker Adin Gilman. We installed it and built the doors and drawers.

The short section of south wall that is not window, door or closet was left with the stone revealed (bottom photo, p. 169). It was built without the inner form, and stones were laid and pointed up inside as well as out. Its 12-in. thickness made it too thin to consider breaking it with internal insulation. We live with the small heat loss in exchange for the reminder that this is a stone house.

The stairs to the second floor are more like a ladder—1x6 treads let into 5/4 by 8-in. stringers with mallet and chisel. There is a 12-in. rise between steps, starting with the bottom "landing" (which has a hinged lid on top and doubles as a kindling box). Under the stairs is the main woodbox covered with an oak counter. The need to arrange for multiple uses and to think like a yacht-builder is always present in a small space like this.

The house is entirely paneled, downstairs with oak and upstairs with birch. The oak paneling was planed and tongued-and-grooved in my workshop using hand tools. The 60 ft. of wall involved roughly 80 or 100 hours of aerobic work, including installation.

The living experience—We've lived in our house five years now, and we use less than four cords of wood a year for heating and cooking. We are comfortable with that and with the air exchange we've purposely allowed through imperfectly sealed openings such as door sills and the trapdoor into the attic. Particularly since we light with kerosene, we feel a supertight house would conflict with the need for clean, fresh air.

Because the house has limited luxuries, Susan and I spend more time on the normal rituals of living. The lamps have to be filled and cleaned, the ice in the icebox replenished, the outhouse visited, sponge baths taken. But since we don't need to pay the bills for their mechanized counterparts or do as much repair and maintenance, we save time in comparison with other home owners. This time is mostly spent at the homestead, some of it writing, reading or shelling beans. The house, I am happy to say, is sufficiently pleasurable to live in that the time we spend at home is its own reward. □

Cherry cabinets and counters set off the slate sink in the kitchen. A hand pump brings in water from a nearby spring. The towel peg holds up the trap door to the root cellar when it's open.

Homesteader Jonathan von Ranson writes and does stonemasonry in Wendell, Mass.

From Boulders to Building Blocks

How a traditional stonemason quarries and dresses sandstone

by Charles Miller

Benny Soto doesn't have to look at the chisel anymore when he dresses a block of sandstone. His hammerhead instinctively finds the butt of the chisel, sending a steady clink, clink, clink ringing around the building site. The stone chips fly about, as he transforms another ordinary rock into a hand-tooled flagstone.

It wasn't always this easy for Soto to hit the chisel butt dead center. Sixty years ago, when he moved to Santa Barbara, Calif., from his native Guadalajara, Soto started his masonry career by lugging stones and digging ditches for a group of Italian stonemasons. Sensing that he had more to offer than a strong back, his boss urged him to learn the stonemason's trade. Soto agreed to give it a try—anything had to be better than lifting and toting rocks about all day, broken only by bouts of ditch-digging.

But the shift from hard labor to skilled craft wasn't without difficulty. Many of Soto's unpracticed mallet strokes hit the chisel butt slightly off center. The hammerhead would glance to the side, and the big knuckle on his left hand would take most of the shot. He hit his hand so many times that he developed blood poisoning, and he nearly lost his resolve to learn the trade during the two weeks that it took him to recover. But the thought of going back to the ditches was a powerful incentive, and Soto stuck with it.

The on-site quarry—Over the last 60 years, Soto has built walls of random rock, flagstone patios, fireplaces with squared-off sandstone blocks and baronial entryways topped with S-curved capstones. He quarries the stones himself, and given Santa Barbara's notoriously rocky soil, he usually needs to go no farther for raw materials than any nearby foundation trench. Some of these virgin stones are the size of beach balls, others are as big as hippos. The big boulders are easiest to work, for the same reason that you get more uniform slices from a loaf of bread than you would from a biscuit.

Although electric and pneumatic drills and chisels are now available, Soto relies on the kinds of tools that stonemasons have used for

Reducing a boulder into building blocks begins with cutting it in half, then dividing the sections into ever smaller pieces. In the photo at right, Soto uses a lifter to start holes for the wedges that are used to split the stone. Once the wedges are in place, he drives them into the stone with a sledgehammer. The wedges have to be hit alternately to ensure a smooth cut.

From *Fine Homebuilding* magazine (October 1986) 35:35-37

centuries. He thinks that handmade work should look handmade, and power tools (besides being too noisy) take away some of the artisan's control. Soto's tool bucket contains steel wedges, cold chisels and a 4-lb. hammer (photo right). The hammer has a hickory handle, which absorbs some of the shock of hitting the chisels. If the handle gets slippery, Soto roughens up the hickory on the edge of a stone. If a handle breaks, he shapes a new one to the right contours, using a piece of broken bottle as a scraper.

The chisels are of four varieties: pitching tools, points, lifters and toothed chisels. A pitching tool looks a little like a brick chisel, but its cutting edge is blunt. It's used to whack off pieces of stone near the edge of a block. A point is a cold chisel with a tip that's about as sharp as a railroad spike. It's used to excavate the holes needed to split the stones, and to dress the stone. A lifter resembles a point with a blunt tip, and Soto uses it primarily to begin the slots in the stone that will accept the wedges. The toothed chisel creates a texture on the stone's surface that resembles cross-hatching.

Most of the tools that Soto uses are available commercially (The Bicknel Co., P.O. Box 627, Rockland, Maine 04841, and Trow & Holden, P.O. Box 475, Barre, Vt. 05641 are two sources). He has his wedges made by a blacksmith.

Sandstone—Like limestone and shale, sandstone is a sedimentary rock. Sandstones are held together by various kinds of naturally occurring cements. The yellow and reddish versions indicate iron-oxide cement. Other types can be white, black, cream-colored or even green. When sandstone breaks, the fissure usually opens through the cement, rather than through the grains of sand. This property makes sandstone relatively easy to shape.

When Soto sizes up a rock that he is about to break into building blocks, he thinks about waste. How can he best use the rock with as little waste as possible? Soto is adamant on this point, and tries to put every offcut to use. He won't, however, reuse stones that have previously been in contact with mortar. Elements in the mortar evidently leach into the sandstone, making the stone brittle and unpredictable to cut. Soto says such stones are dead.

Quarrying a sandstone boulder is a matter of reduction. A large rock is cut into ever smaller pieces and eventually into usable blocks. Some sandstones have a grain to them, and the first cut should follow it. Typically, a boulder will be quartered (drawing, right) and the dimension of the slices taken off the quarter-sections will be determined by the task at hand. If, for instance, Soto is making fireplace veneer blocks, which are about 18 in. long, 9 in. high, and 5 in. deep, he will make sure there is a usable 20-in. thick portion in the next slice he takes off the boulder. The excess "meat" is an allowance for a slightly erratic cut—it can be easily trimmed away when the stones are dressed. If the cut goes radically awry, chances are he will still end up with a piece of stone that has usable dimensions. If he tried to carve off a 5-in. thick piece and failed, it's likely that little of the material would be salvageable. Also, it's easier to get

straight cuts when there are roughly equal amounts of stone on both sides of the cutline. Much of this quarrying process is guided by an intuition that comes only with experience.

Once he has the 20-in. thick piece lopped off, he cuts it in half again. If the stone co-operates, he may now trim off the waste portion near the curved edge. If he's in doubt about the accuracy of this cut, he will split out the blocks and trim them individually. In this manner, large boulders are cleaved until they are reduced to blocks that are about 20 in. by 10 in. by 6 in.

Making the cuts—Soto begins a cut by using a point or a fat, soft pencil to mark a line on the stone. If the stone is still in the round, the line he makes is across the top of the stone, and it is straight in plan. Soto's straightedge is an ancient length of 2x2, and if he needs to square it with another line he relies on his eye.

He uses a lifter to begin a series of wedge slots along the cut line, as shown in the photo on the previous page. The slots are 3 in. to 4 in. on center, and never closer than 2 in. to the edge of a stone. When each slot is about ½ in. deep,

Clockwise from the top, Soto's hammer, a point, two pitching tools, a toothed chisel and a lifter. The three wedges were custom made by a blacksmith.

Quarrying a boulder
Large boulders yield the best blocks with the least waste. They are typically cut in half, and a section is levered onto its side and again halved. Then slabs of the appropriate dimension are cut away from the quarter sections and reduced, roughly by halves, to the desired blocks.

Drawings: Chuck Lockhart

The weighty presence of hand-wrought stonework is entirely in keeping with the sturdy detail in this Spanish Colonial Revival style home. Soto used a plywood template to regulate the curvature on the bottoms of the corbels that support the mantel, and he shaped them with a pitching chisel.

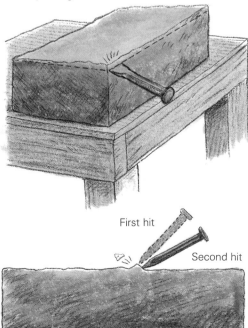

If opposite faces are not in the same plane, mark sides with parallel lines and remove excess stone with a pitching tool.

First hit

Second hit

To remove surface projections, use a point held at about 45°. Lower the angle for stubborn bumps.

Soto switches to the point and excavates the slot another ½ in.

Now the wedges are inserted into the slots, and Soto methodically drives them, alternating from one wedge to the next, with blows from a 16-lb. sledge. Soon a fissure opens, and the stone fractures in half. If the stone is a big one, he uses a long prybar to lever one of the halves on its back.

Stone dressing—Once he has a pile of rough-cut blocks on hand, Soto takes them one by one to his work table, a sturdy platform made of 2x6 braces, 4x4 legs and a ¾-in. plywood top. It measures about 3 ft. square, and its height is about 6 in. below Soto's beltline. With a stone on the table, he can hold his tools at a comfortable, waist-level height without having to bend.

If he's making blocks that need regular dimensions, Soto will check the block for twists or out-of-square corners. A straight 2x2 is used for the twist test, a framing square for the corners. If a block needs trimming to bring opposite faces into the same plane or to straighten an edge, Soto marks the stone accordingly. Removing this unwanted material is the pitching tool's job. With the curved back of the tool on the side opposite the workpiece, Soto cleaves away unwanted stone with sharp raps from the hammer (photo below left). It is the pitching tool that gives the edges of the blocks the broad facets that make handhewn stone so attractive.

Any bumps and projections on the face of a block are removed with the point (drawing, below left). Soto makes this work look effortless, with the tip of the point finding the base of a projection a millisecond before the hammer strikes the butt of the chisel. Stubborn bumps get two or more hits, the first with the point held at about a 45° angle, and subsequent shots with the angle approaching 30°.

If he's making flagstones, Soto doesn't have to worry about square corners and parallel edges. Instead, the task is to make the stones as flat on one side as possible, and then finish them with a pleasing texture that won't get slippery after years of use. For this he uses a tool called a bush hammer, which looks something like a meat tenderizer. Soto's bush hammer weighs 5 lb. and has 36 teeth on its face in six parallel rows. Soto lets a helper use the bush hammer, which has to be pounded over the entire surface of the stone within about 2 in. of the edges—any closer and the stone is liable to break. This work is for bruisers—15 minutes on the bush hammer will make your forearms blow up like Zeppelins. After hammering, the stone is swept clean with a stiff bristle brush to reveal a pleasing stippled texture.

Many of Soto's clients hire him to craft fireplace surrounds, hearths and mantles (photo above left). Whenever he does a fireplace, he cautions the mason who installs the pieces to use a stiff mortar mix, and thereby avoid messy drips that could discolor the stone. If some mortar does get on a stone, he recommends cleaning it with a stiff bristle brush. Dip the brush in water, shake off the excess and run it over the mortar stain, but in only one direction. Back and forth will drive the stain deeper. □

Sticks and Stones

A house handbuilt from red cedar and basalt in the Pacific Northwest

by Sebastian Eggert

The voice on the telephone inquired, "Would you like to help build a stone house on the Olympic Peninsula? You know, in Washington State?" It was Rick Hayton, a designer and builder who lives in Brooklyn, N.Y. I had worked on and off with Rick for about six years, and knew him well enough to expect weird phone calls and unusual requests. I quickly said yes, but didn't know at the time just how challenging the project would be. When I saw his drawings, I wished I'd said no.

Rick had designed a compact house with a truncated rectangular floor plan (drawing, below), to be notched into a south-facing hillside. The north and west walls were to be a curving mass of rock 2 ft. thick and 22 ft. high at the peak. The south and east walls would be almost entirely glass, inviting the sun's rays deep into the interior of the house.

Inside, a spiral staircase and a cylindrical stone chimney would be strong points of visual interest and essential structural elements as well. Each would carry heavy steel brackets to secure the fan-shaped patterns of the rafters. The layout would require precision work with wood, steel and 200 tons of irregular stone. I was more than a little apprehensive, but Rick was convinced we could do it.

The Olympic Peninsula is a region known for its frequent seismic shakings. A call to the local building department gave us two particularly important guidelines for rock-wall construction in this earthquake-prone area. First, the foundation had to be connected to the rock wall with sections of rebar, 3 ft. on center and projecting at least 3 ft. into the stonework. Second, the wall had to have bondstones. At 140 lb. per cu. ft., a manageable rock is roughly 1 ft. square by 3 in. or 4 in. deep. Consequently, a wall; as thick as ours would actually be two walls built side by side. The bondstones are large enough to span the full thickness of the wall; they work as ties to keep the two walls from pulling apart. The building code requires that bondstones be placed every 3 ft. on center, both horizontally and vertically.

Another important consideration was the weight of the structure. At the highest part of the wall, we calculated the combined load of the walls, roof, and floors to be close to three tons per linear foot. This is a staggering load, and the soil survey told us that for most of the foundation, footings would have to be 3 ft. wide and 2 ft. thick, reinforced with three #4 rebars, two at the bottom and one at the top. A foundation width of 3 ft. would allow us room for mud sills, and the inevitable variations in the rock wall. The cylindrical chimney would be built on

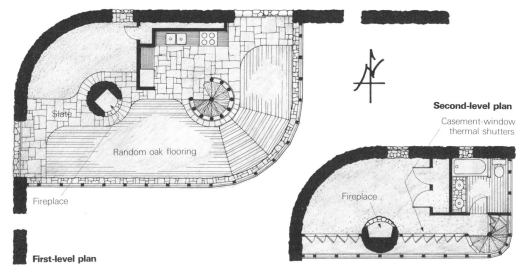

Slate

Random oak flooring

Fireplace

First-level plan

Second-level plan
Casement-window thermal shutters

Fireplace

a monolithic concrete pad, bristling with vertical rebar for a solid connection with the cinder-block and concrete core of the chimney.

Building with basalt—Efficient stonelaying requires rocks in many different shapes and sizes. Squarish blocks for framing window and door openings are essential, and if they aren't included in the random selection delivered to the site, they have to be hand-picked at the quarry. We were lucky to be just three miles from a local quarry, and spent many hours there selecting the square-cornered blocks we needed. Although the quarry's shaker mill separated the rock into different sizes, the "one-man rock" we ordered came to us as anything from small chunks to boulders that three of us together could barely move.

We discovered immediately how different the local blue-black chunk basalt was from the Catskill Mountain bluestone we were used to. Sedimentary rock can be quarried in such unvarying units that it requires little mortar to set in place. The Olympic Peninsula igneous basalt was so irregular that it required a surprising amount of mortar to pack the jagged voids between the rocks.

The same overlapping principle used in bricklaying applies to building with stone. The joints between the rocks should never occur directly over one another, and the stones should be arranged in interlocking patterns that take advantage of their natural hollows and projections. The mortar should be thought of as padding, not glue, and small chunks of rock should fill the frequent odd spaces. Gravity is the ultimate test, and the stones should be able to stand as a unit—with or without mortar.

We used a mortar recipe of one part lime, three parts portland cement, nine parts washed masonry sand and plenty of clean water. The lime gives the mortar a plastic quality, allowing it to hold its shape, yet still stick to vertical surfaces. Like the perfect martini, the perfect mortar is an elusive mix: It must be able to hold its shape with a minimum of slump, yet elastic enough to be squeezed into the irregular cavities between the rocks without crumbling or dribbling out. The volume of water in our mix varied from day to day, depending on the humidity, temperature and moisture content of the sand, variables we just couldn't control.

We mixed our mortar in a 2-cu. ft. cement mixer, and passed the prepared mud to the scaffold in 5-gal. joint-compound buckets. There we dumped the bucket loads onto plywood mortar boards. We liked 12-in. bricklaying trowels best for moving and spreading large quantities of mortar. Their wide blades carry plenty of mortar, and their pointed tips help to work it into the irregular surfaces. For applying small amounts of mortar, we used 5½-in. pointing trowels, which are just smaller versions of the bricklaying trowels.

Laying up the wall, we chose rocks that fit well with the rocks under them, and tested them in place without mortar. Slight projections that prevented a close fit were knocked off with the chisel edge of a brick hammer. Sometimes the rock broke where it was hit. Other times it shattered along unseen fissures, leaving a pile of chunks to use somewhere else.

Quarried rock has been blasted, sorted, stored and transported, all of which contribute to a thick layer of dust and sometimes caked-on mud. Mortar won't stick to dirty rocks, so we soaked each rock in a bucket of water and hosed and brushed it off before it went into the wall. We removed dried mud with long-bristled wire brushes. Then we laid the clean rock in position on a mortar bed thick enough to ooze out slightly around the edges when the rock was pushed into place. If it was a hot day, we'd spray the wall occasionally with a fine mist to keep the mortar from drying out too quickly.

When it had begun to set up (usually in about four hours), we pointed the edges with a ½-in. caulking trowel. In general, if the mortar wants to fall back out of the joint while pointing, it hasn't set up enough. If it's hard to pack and tends to crumble, it's too late.

Pointing cleans up excess mortar and also forces mortar back into any voids between the rocks. We wire-brushed the previous day's jointwork each morning while it was still green, before we fired up the mixing machine. The brushing gave the joints a pleasing, consistent texture, and removed any misplaced mortar.

We laid up stone from the footing to as high as we could comfortably reach, checking the wall for plumb by dripping water from a sponge adjacent to the wall on windless days. We erected the scaffolding on the inside of the wall, where the appearance of the stonework was most important. We doubled up 2x4s for uprights, and used 2x6s for joists and 2x12s for planking. We cross-braced the scaffolding with 1x4s, and added plywood sides to the planking to keep rocks from falling from the top. Small rocks were passed up in buckets, and large rocks were handed up one at a time. We were all aware of the danger of tools and materials falling as the wall went up, so we instituted a fifty-cent fine for anyone who dropped anything from the scaffolding. The money went to the refreshment fund. When we heard anything drop, we'd yell "fifty cents!" The guilty party had to cough up the change on the spot.

While the wall builders struggled with their burden, the chimney crew began building the stone cylinder. In addition to housing the open fireplace on the first floor, the cylinder was to contain a bedroom fireplace on the second floor. Structurally it would serve to support the roof and second-floor deck; it was fitted with numerous beam and rafter brackets. To keep the weight down, we used lightweight cinder blocks for the chimney's structural core (drawing, facing page). In the mortar between each block, we placed masonry ties to connect the outer layer of rock to the chimney. The ties are corrugated metal strips, ⅞ in. wide and 4 in. long, which project into the mortar joints of the rock veneer. Like the bondstones, the ties prevent two adjacent walls from pulling apart.

We laid rock eight to ten hours a day for ten weeks, and escaped without permanent physical injury. Caring for your body is an important part of this kind of heavy work. A bad back can spoil all the fun, so we always lifted the heavy stones with our legs bent and our backs straight. We wore loose-fitting jeans, which allowed easy squatting without splitting seams. Our hands suffered considerable abrasion, so we used skin lotion daily, and avoided handling mortar like the plague. When moving rocks up to the scaffolding we wore gloves, but they were too clumsy to wear while laying stone.

For windows and doors, we built simple wood forms to keep the rough openings within

The evolution of a structure: Below left, the perimeter foundation and chimney footing have been poured. At the northeast corner of the house, a concrete-block basement is under construction. Center, the chimney has been built, and the floor and rafter brackets can be seen at the middle and upper levels. The single pole at the right is the center of the spiral staircase. Right, the last of the red cedar beams has been bolted into the brackets in the masonry wall and chimney cylinder, and the roof is ready to be sheathed. Photos: Sebastian Eggert.

Concrete chimney cap

Flue tile

Plywood form (left in place)

Rafter-bracket ring

Second-floor fireplace

To build a chimney that supports the roof and second story, a block foundation was laid on a concrete footing. Upper and lower fire boxes are encased in concrete, and circular courses of block form the core of the cylinder, which is tied top to bottom with #4 rebar. Stone veneer was laid up and attached to the block with masonry ties.

Curved steel lintel

Steel-cone damper extension

Poured concrete

Firebrick firebox

Rebar from foundation

Concrete footing

Structural chimney cylinder

Illustrations: Frances Boynton

Designing around the site

When we first visited our client's property, we were struck by the lush natural beauty of the area. This portion of the Pacific Northwest has a climate that comes as close to a temperate rain forest as any region in the Northern Hemisphere. The dense fog that often blankets the land creates an aura of mystery. In designing the stonework for our client's house, we tried to make shapes that would evoke this mood of brooding uncertainty and heighten the setting's natural sense of drama.

Given our background in landscape design, we conceived a plan for the property, within which the house would be integrated as a landscape element. Thus the building's major features were established on our site plan well before the floor plan and functional details took shape.

We decided early on to have two stone walls running through the property, roughly perpendicular to each other. From the east, one wall would border a scenic approach path. From the south, the other would edge an access road routed inconspicuously to the rear of the site. The stone arms would converge on the site, rising and assuming the hill's pitch as they ascended the slope. Just below the crest of the hill, having reached a height of 22 ft., the walls would join in a 12-ft. radius arc and become the north and west walls of the house. Broad-leaved evergreens planted at the base of the walls would soften the abruptness of the stone as it rose from the ground.

To the south and east, the house looks out over meadows that slope down to a natural depression, which we planned to make into a pond. A section of the low western stone wall running alongside the driveway could expand into a bridge and dam for the pond. The upper expanse of the meadow would be seeded with wild flowers, willows would be planted by the pond, and crabapple and cherry trees would complete the major landscape additions. To date, however, only the house has been done, but the owners hope to complete the entire landscape plan some day.

Cradled by the two long stone arms where forest and meadow meet, the house has its masonry back to the woods and to bad weather. In good weather, the large windows, which face south and east, catch the sunlight and let the stone wall function as a heat sink for solar radiation.

We chose local red cedar as the second structural material, both for its durability on the outside and for the warm earthy tones it takes on when oiled. We decided that all the structural woodwork would be exposed on the interior. This meant joining beams and headers without evidence of fasteners (except for those held by steel brackets), and sanding and oiling all wood surfaces. For convenience in construction and for a more substantial look, we decided that all wood should be milled to full (not nominal) inch dimensions.

For the grace it lends to the overall scheme, quite often we like to use curves where a plan might ordinarily have a right-angle corner. In this case, we rounded the opposing northwest and southeast corners and decided to place a major structural element concentric with each—the stone arc has a cylindrical stone chimney, and the glass wall, a cedar-pole spiral stair. The whole plan had the look of a one-celled animal that just had undergone division of its nucleus. We tried to work complementary curves into other aspects of the design. The boundary between the oak and slate floors, for example, curves around the chimney and the spiral, then sweeps into the kitchen and past a sliding glass door in the north wall.

Given the smallness of the house, an open plan seemed the best arrangement for the ground floor. The bedroom and bathroom upstairs have a full clerestory exposure. The bedroom fireplace has a raised hearth. The area directly below the bedroom, with its lower ceiling, carpeted floor and pot-belly stove, is a cozy sitting room, a nice contrast to the openness of the front area.
—*Richard Hayton and David Zatz*

Rafter-bracket layout

The roof framing plan called for 4x6 cedar rafters radiating from the chimney to the stone wall (drawing, below). Rafter #1 is perpendicular to the wall and slopes down to the chimney cylinder at a 4:12 pitch. Subsequent rafters change angle and height as they follow around the arc.

On the chimney cylinder these beams were held by brackets welded onto an elliptical steel band, which we embedded in the chimney's stone veneer so that bracket #1 projected at a 4:12 pitch. The connections between the rafters and the stone wall were made by individually placed brackets. To lay out this puzzle, we turned to string lines.

Before building the wall, we laid out reference marks on the foundation and plumbed up from them to locate the position of the chimney brackets, and wall brackets #1 and #7. The height of bracket #1 was determined by the rafter's 4:12 pitch. With brackets #1 and #7 in place, we ran a taut string between the bottom of bracket #1 and the point at which the rafter plane intercepts the wall at bracket #7, establishing the plane on which all the rafters would lie.

I found the midpoint of the string, where rafter #4 crosses it, and placed the end of a straight, rafter-length 2x4 in the #4 chimney bracket. Carefully holding this jig stick above the string at the center mark, we jockeyed wall bracket #4 into position following the angle and pitch of the stick. As soon as the #4 bracket was secure and the mortar dry, I reset the string to connect the three installed wall brackets. Again dividing the span of the string, this time in thirds, I set the remaining four brackets in the wall using the jig stick as a guide. —S.E.

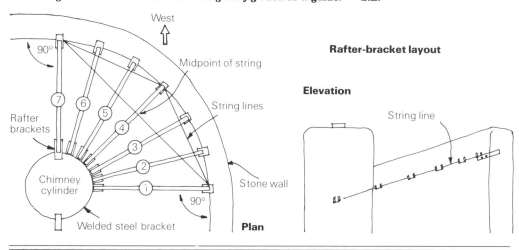

Rafter-bracket layout

Elevation

Plan

prescribed tolerances. Two lengths of ³⁄₁₆-in. by 6-in. angle iron, welded back to back, served as headers (detail drawing, below). The headers rest on leveled mortar pads, which were allowed to set up before the steel was put in place. Laying stone over the headers required a run of fairly symmetrical rocks and some delicate balancing. Rocks were set in mortar on opposite sides of the welded steel flanges, and then weighted down with loose stone. When the mortar had hardened, we expanded the wall back to its standard thickness.

The steel brackets that join the structural members to the stonework are an important element of the overall design. Mounting the beams in recessed pockets in the stonework would have been adequate, but the brackets give an added measure of safety during earthquakes. These brackets also work as decorative elements, accenting junctures between wood and stone. We kept them simple, with a generous saddle for the beam, and length enough to secure them about halfway into the stonework. Short lengths of ½-in. rebar, welded horizontally on the hidden ends, secure the brackets during an earthquake (drawing, bottom).

Before we began construction, we plotted the exact location of the beam, rafter and joist brackets. When the wall reached the proper height above the footing, we located the reference points on the foundation, plumbed up to the correct height and set the bracket in the stonework. The beam brackets were easy to locate, but the chimney rafter brackets were a different story, explained in the box above left.

Finally we could see the end of the stonework. The wall was 22 ft. high, and acrophobia was setting in when we leaned over the wall to point the outside joints. After the last rock was in place, we smoothed off the top of the wall with a crowned mortar cap to prevent water from collecting on the top of the wall. Mortar caps will crack if they dry too quickly, so we kept ours covered with wet burlap and plastic sheeting until it set up.

When the last of the mortar had set, we gave the whole business an acid wash. We suited up in rubber boots, heavy rain gear, filter masks, gloves and goggles, and equipped ourselves with long-handled brushes. After soaking the stonework with plenty of water, we brushed a dilute solution of hydrochloric acid and water over the entire wall. The mixture smoked and fizzed furiously, bubbling greenish foam until the acid was neutralized by the mortar. The acid bath cleans the mortar smears off the rocks (heavily caked mortar has to be chipped off) and leaves the stone its natural color. It also helps even out the texture of the mortar joints.

As we took down the scaffolding, we caught an unobstructed view of the stonework. With the sun low in the sky, it looked more like an ancient ruin than a house under construction.

Closing in—The autumn rains weren't far off when the stonework was completed, so we quickly turned to the framing. We had red cedar beams, posts and rafters rough-cut by a one-man mill for $200 per 1,000 bd. ft. Even in 1977, this was a good price for structural

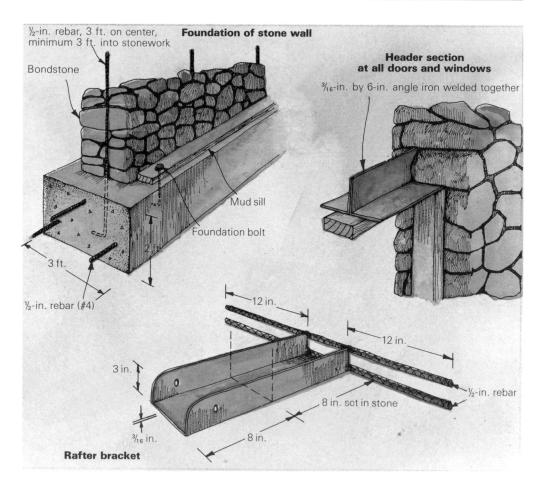

Foundation of stone wall

½-in. rebar, 3 ft. on center, minimum 3 ft. into stonework

Bondstone

Mud sill

Foundation bolt

3 ft.

½-in. rebar (#4)

2 ft.

Header section at all doors and windows

³⁄₁₆-in. by 6-in. angle iron welded together

12 in.

12 in.

3 in.

½-in. rebar

8 in. set in stone

³⁄₁₆ in.

8 in.

Rafter bracket

lumber. For another $50 per 1,000 bd. ft. we had the lumber planed to exact dimensions.

The framing included a lot of challenging intersections, among them a simple, yet sturdy and attractive, rafter-to-beam joint in the living room. Here, we marked the location of the 4x6 rafters on the 6x12 beam, and cut a 1-in. deep mortise for each rafter. When the beam was positioned, we secured the rafters with a single 10-in. by ½-in. countersunk lag bolt angled in from above.

On the eastern edge of the living room is a circle of 6-in. diameter poles. These poles have several functions. They support the two levels of roofs, and contain a spiraling, oak-treaded staircase to the upstairs bedroom. At their tops, the poles are joined to rafters and beams by a curved steel bracket similar to the one embedded in the stone cylinder.

After the framing members were installed, we decked the roof with 2x6 tongue-and-groove stock, which we butted to the stone wall. We waterproofed this junction by cutting a 1-in. deep notch into the stone with a masonry blade, a few inches above the roof surface. A copper counterflashing overhangs the step shingles that were woven into the perimeter of each course of cedar roofing.

Refinements—When the house was closed in, we discovered that a fire in the see-through fireplace would smoke disconcertingly at times. The fireplace had a high crown to the damper, an adequate smoke shelf and a straight run of 12-in. flue, 20 ft. long and extended well above the roof level. The three-sided fireplace in the bedroom never had this problem, so we temporarily closed off one side of the downstairs fireplace. This eliminated the smoking, so we blocked up one side permanently. Evidently, a change in air-pressure caused by opening a door was the source of the problem.

After extended periods of rainfall, we noticed some minor seepage through the stone wall. A close inspection of the mortar cap revealed some hairline cracks, so we sealed them with a heavy layer of tar. As an extra precaution, we sprayed the entire stone exterior with a silicone compound. The seepage stopped.

The floor area of the house is quite small, but the way the spaces blend gives the home a feeling of spaciousness. At night, the expanse of glass on the main floor reflects the interior lighting like mirrors, multiplying the sense of volume. Notched into its cedar grove and cradled by its curving stone wall, the home radiates an uncommon sense of shelter. □

Sebastian Eggert, a designer and general contractor, lives in Port Ludlow, Wash.

Welded brackets of ³⁄₁₆-in. steel (above right) connect rafters and beams to staircase poles. The curving pattern in the oak and slate floor, right, traces the circular shapes in the plan. The deep grey slate is set in mastic on a particleboard subfloor. The random-width oak floorboards were chosen for their irregular color and grain patterns. At the far end of the room, the circular staircase carries the 6x12 beam and its let-in rafters.

Converting the Forge

A traditional-looking passive-solar addition makes a home from an outbuilding in the English countryside

by Grahame Collyer

Geographically and economically, Cornwall is far removed from England's metropolitan bustle, and people like us who have moved here are looking for something different from the suburban mainstream. For us and many of our friends, part of the rural existence involves modernizing, restoring or converting an old building to serve as a home. This area of southwestern England is steeped in a past of stone walls and slate roofs, of structures built hundreds of years ago with lots of cheap labor and vast quantities of material dug from the ground close by.

Today, town planners and building officials are continuing this tradition, so the regulations that govern the appearance of houses and the materials used to build them are strict. Space is limited and new construction is frowned upon, so instead, old houses are extended or old barns and mills are converted into residences. This ex-

Grahame Collyer has left teaching to work full-time as a designer/builder. His latest project is the conversion of a large 17th-century barn into two apartments. Photos by the author.

plains the fate of the old forge on our property, which has now, after several years of work, become our new home (photo below).

In 1982, Hilary and I and our two young children returned to Cornwall after a year of teaching in Oregon. We had all been excited to see so many different houses built by people who had few constraints in terms of style, building materials and space. Suddenly our stone cottage, which had always felt cozy with its small rooms, diminutive windows and low ceilings, seemed dark and cramped. The open, light-filled solar designs that we'd seen in Oregon inspired us to make a change.

In the gardens not far from our house there was an old forge, a small stone structure built about 100 years ago. Before we bought the property, the forge had been used by the local blacksmith and carpenter who served this area of small hollow-hugging farms. Ten years ago we had hurriedly converted the space into informal guest rooms. Now it occurred to us that we might continue to renovate the forge and also add onto it. We wanted to build a house that

would look as though it had been standing 100 years and yet have a sense of lightness and airiness. At the same time we were determined to insulate to a high standard and gain much of our heat from the sun.

Gathering material—We spent a year planning the new part of the house. I was still teaching school, and we needed time to collect the materials we wanted. Our work on the initial forge conversion had taught us valuable lessons in masonry and carpentry. It had also given Hilary and me an idea of the look we wanted: old weathered granite combined with fresh plasterwork and time-darkened timbers.

We had little problem salvaging old timbers from derelict buildings. It was more difficult to collect quoins—the hand-squared granite blocks that interlock at the corners of a building. These large stones, which weigh from 100 lb. to 300 lb., add strength where it's needed at the corners of a structure. We eventually collected about 25 from a number of old buildings.

The stones (we call them slates) that make up the rest of the outer walls are easily available, but very irregular in shape. This makes the stonework around doors and windows a real problem, because regular angles are needed for framing the openings. Fortunately, we learned that the city of Plymouth was taking up old granite cobblestones, or *setts,* from streets a few yards from the quay where the Mayflower set sail. The setts were hand-cut and brick-shaped—just right for the window and door openings in our addition. We bought 10 tons. These old stones seemed none the worse for wear after four centuries of traffic. The only clues to their earlier use are the occasional patches of yellow no-parking paint that we were unable to remove.

An L-shape—As we collected stone and salvaged old timbers, the design of the addition became clearer. The new part would be at right angles to the old and roughly parallel to a small stream that runs approximately east-west across our property. From all sides except the south, our house would have traditional small windows and massive stonework under its slate roof. The south side of the addition would be sheltered from public view and have three 8-ft. by 7-ft. windows to bring in winter sun and views of the stream and meadow beyond. In summer, the glass would be shaded by the nearby sycamore, ash and oak trees.

At the west end of the addition, we planned a

Cornwall, the southwestern tip of England, is a region with a rich historical tradition of stone construction. This addition to an old forge continues this masonry tradition, but adds the appeal of insulated, south-facing glass and energy efficiency. Stones around windows and door openings are salvaged cobblestones; the roof is covered with salvaged slate.

From *Fine Homebuilding* magazine (December 1987) 43:58-61

large master bedroom that would be our study and workroom too. With its own bathroom, this section of the house would give Hilary and me our own private wing. The children would also gain some independence with their bedrooms, a bathroom and a loft playroom at the north end of the original forge. Between these two areas, the old living room would lose its corner kitchen and gain an extra window. This room is open to the roof, with its rafters exposed, and has a big Jøtul woodstove that heats most of the house. Copper tubing behind the stove enables us to preheat our water as well.

The new kitchen would occupy the southeast corner of the house, at the intersection of new and old. We planned for it to be at the same floor level as the original structure. The new dining room would be next to the kitchen but separated from it by a 12-ft. long curved elm step. Except for the kitchen, the floor level of the addition is one step down from the old forge.

In our addition we also made space for a sunroom (top photo, p. 185). This room would have a curved granite wall facing a large south window. Our aim was to trap and store solar heat that would gradually be given back to the house through the night. Finally, we squeezed a small utility room into the plan for our washing machine, freezer and various broom-closet items.

Starting work—Having sold our old cottage to finance the addition, we had to move into the old forge as we began to enlarge it. This meant spending a year in the rooms we had hastily converted ten years earlier. With two young children and a very rainy winter, this was probably the hardest part of the project.

The footings for the new wing are poured concrete. In this part of England, frost heaving isn't a problem, so the footing didn't have to be deep in the ground. Along the south side of the addition, I set the top of the footing about 1 in. below grade. On the north side, we had to cut about 4 ft. into sloping ground to foot the bermed wall. The footings are 12 in. deep by 24 in. wide. The substantial size was necessary for stone walls 16 in. thick. Assisted by some hard-working friends, we spent a backbreaking week mixing and pouring to get the footings in. Inside them, we then poured a 4-in. thick concrete slab over 3 in. of rigid insulation that we had set on compacted earth.

The cavity wall—In England, masonry walls (usually concrete blocks or brick) are nearly universal, and so is cavity-wall construction, a type of masonry wall designed to cope with weather that is often cool and damp. Here, with nearly 60 in. of rain a year, cavity walls have been used for centuries to prevent dampness from making its way inside a house. These old hollow walls have stood the test of time well, though their irregular voids make an ideal habitat for mice.

Today, strong mortar and modular concrete block have made cavity-wall construction fairly straightforward. As the name suggests, there's an air space that separates an inner masonry wall from an outer one. Usually, the inner wall is the structural one, but both walls are linked by

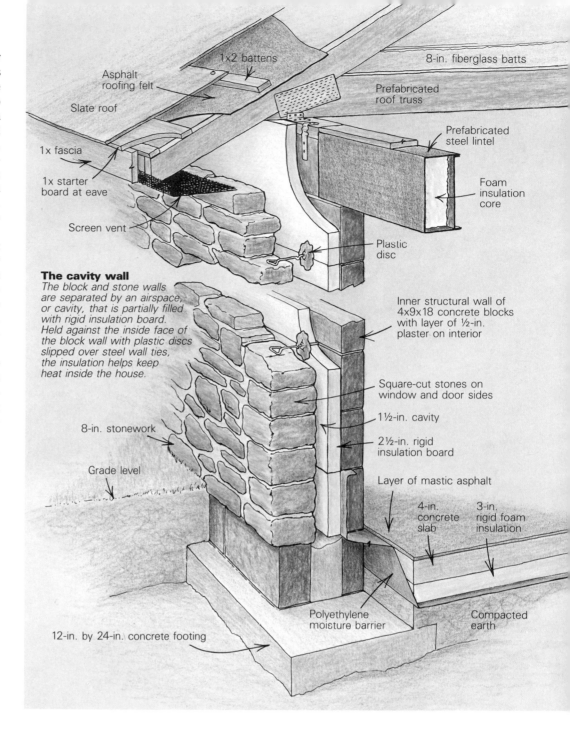

The cavity wall
The block and stone walls are separated by an airspace, or cavity, that is partially filled with rigid insulation board. Held against the inside face of the block wall with plastic discs slipped over steel wall ties, the insulation helps keep heat inside the house.

Labels: 1x2 battens; Asphalt roofing felt; Slate roof; 8-in. fiberglass batts; Prefabricated roof truss; 1x fascia; Prefabricated steel lintel; 1x starter board at eave; Foam insulation core; Screen vent; Plastic disc; Inner structural wall of 4x9x18 concrete blocks with layer of ½-in. plaster on interior; 8-in. stonework; Square-cut stones on window and door sides; 1½-in. cavity; 2½-in. rigid insulation board; Grade level; Layer of mastic asphalt; 4-in. concrete slab; 3-in. rigid foam insulation; Polyethylene moisture barrier; Compacted earth; 12-in. by 24-in. concrete footing

galvanized steel ties that span the cavity, which is otherwise continuous.

During prolonged damp spells, moisture will make its way through the outer wall into the cavity. Cavity moisture can also have warm interior air as its source. Either way, it's important to ventilate the cavity so that moisture can evaporate and be carried outside.

In our walls, cavity ventilation is provided by six small grilles set into the stone about 2 ft. up from the base of the wall and by openings at the top of the wall, as shown in the drawing above. I made the grilles from old cast-iron heat registers. The openings at the top of the wall allow cavity air to diffuse up to the attic. A vented gable and continuous eave vents ensure proper air circulation in this space.

Insulating cavity walls is a recent improvement, and many builders install rigid insulation board inside the house and plaster or panel over it to make the finished wall surface. The problem with this approach is that it makes the house thermally "light." All the mass of the wall

is outside the insulation, and the strong, solid feel of masonry is lost.

We chose to add insulation inside the cavity in the form of 2½-in. thick rigid polystyrene boards. This way the thermal mass inside the insulation smooths out temperature changes. As shown in the drawing, the polystyrene is held against the inner side of the cavity with plastic discs that slide over the galvanized wall ties.

We began by laying up the 4-in. thick inner walls of concrete block. Galvanized wall ties were set in the mortar between alternate courses, spaced about every 2 ft. With these protruding ends bristling from the wall, our construction project had the look of a giant hedgehog. Each wall tie has a crimped section, or elbow, which should point down when the tie is installed. This way, any water droplets that might be conveyed to the inner wall via the tie will instead drip into the cavity off the elbow.

The three large window openings in the south wall were spanned with Catnic lintels (Catnic Components Ltd., Pontygwindy Estate, Caerphilly, Mid Glamorgan CF8 2WJ, United Kingdom). These prefabricated, hollow-steel box sections are strong but lightweight and easy to install in a block wall. The perforated sides of the steel box create a keyed joint for mortar or plaster. We stuffed the hollow centers with foam insulation before mortaring them in place over the window openings (photo below). Later the inside walls were finished with a ½-in. coat of sand/cement plaster followed by a ⅟₁₆-in. thick skin of hard

Lightweight, prefabricated steel lintels filled with rigid foam insulation span the large window openings. Perforations in the lintels act as keys for mortar and plaster.

plaster. We fastened the insulation in place with the plastic discs as the outer stone wall went up (photo bottom left). Working primarily on dry winter days, the stonework went slowly—about 10 sq. ft. on a good day. Given this rate of progress, it made sense to put the roof on before the outer walls were done. The roof rests on the structural inner wall of concrete block. We also positioned the sunroom window frames to span the openings and sealed around them with can-sprayed polyurethane foam. The frames were made for us from Philippine mahogany, a strong, straight-grained, decay-resistant wood. They were sized to hold panels of double-glazed, low-emissivity coated glass.

After some trial and error, we found that it was helpful to use an 18-in. high plywood form behind the outer stone wall, moving it upward as the wall progressed. This temporary backing made it possible simply to throw mortar in to fill voids near the back of the stone wall without having it drop down into the cavity.

Placing the large corner quoins was very satisfying. With the corners up and the setts mortared up around the windows, the wall really began to take shape. We hoisted an old granite mill-step (weighing about 600 lb.) up to form the lintel of a small window in the gable end (photo below right).

The roof—Except for the hip-roofed section above the new kitchen and dining room, the roof structure is factory-made trusses, which are now very common in England. High-priced imported timber has led to computer-designed trusses that minimize timber sizes without sacrificing strength. You specify the span and the loads, and the gang-nailed trusses are delivered

on site, complete. It took a morning to erect and brace the twelve roof trusses, and I spent another week constructing the remaining third of the roof the old-fashioned way, with rafters, purlins and ceiling joists. I found two 9-in. by 6-in. elm beams from a horse-powered grain mill and used them to make a large truss that is the principal structural member of this corner of the roof. Cleaned, sanded and preservative-treated, these old elm timbers remain exposed in the kitchen and dining area.

Both the planners and our preference dictated a slate roof, but because little slate is quarried here today this was another item that had to be bought secondhand. Fortunately, English slate is long-lived. At 80 years, a slate roof here isn't yet middle aged, so salvaged slate is quite an acceptable substitute for new material.

On many older roofs, slates were held in place by oak pegs. One or two pegs were driven into holes in the top of each slate so that they would protrude over horizontal lath strips about 1 in. by ¼ in. in section. Today, the technique is slightly different. Instead of thin lath, we use 1x2 stock, which holds galvanized nails that are driven through holes in the tops of the slates.

Before nailing down the 1x2s, we installed a moisture barrier of asphalt roofing felt across the tops of the rafters. Less expensive polyethylene sheeting has been used for this purpose, but its greater impermeability has often led to serious condensation problems in the roof, especially in well-insulated houses where the roof space above the insulation is cooler. Because of this problem, building regulations now specify sound ceiling vapor barriers and a high standard of roof ventilation at eaves and ridges. We played it safe, using the more expensive bitumen felt.

Technology hides beneath the traditional exterior. Before the outer stone wall was built, an inner, load-bearing wall of concrete block was laid up and insulated on the inside wall of the cavity with 2½-in. thick rigid polystyrene. At right, an old granite millstone turned lintel is lifted into place.

Inside—After a year of work, the shell of our addition was complete. We relaxed, thinking that the end was in sight. Not by a long shot! Finishing the interior has been rewarding, but painfully slow, chiefly because I've had to stop full-time work on our house in order to begin a major barn-to-residence conversion down the road. Over the last few years, Dutch Elm disease has killed off all the elms in Britain, ravaging the landscape but providing woodworkers with lovely stock. Gradually, my supply of beautifully figured elm is being made into interior doors, cabinets and trim for our new home.

The floor was one of the few jobs we contracted out. We chose a brown-colored mastic-asphalt, laid down over the concrete slab at 200°C and troweled smooth while hot. There were several advantages to this floor choice. First, it's a simple one-step, monolithic finish surface, faster to install and less expensive than tile. Second, the brown color is very effective in absorbing the sun's heat, and conductivity to the concrete slab below is excellent. Finally, the asphalt is more resilient than concrete, and should prove to be quieter and easier on our feet.

Our first impressions of the new space were just as we had hoped. It was nice to experience the open, light-filled interior that our old stone house had lacked. The large windows give us a strong connection with the outdoors. The insulated walls and ceiling (8-in. fiberglass batts between the roof trusses) keep the legendary English dampness and chill out, and the mass of the floor and sunroom wall (photo right) effectively store solar heat, so that the house stays warm long after the sun goes down. So far, we haven't felt the need to install a woodstove in the new part of the house. □

The curved wall of the sunroom (above), shared by the master bedroom, stores solar heat during the day and radiates it back to the living space at night. The resilient asphalt mastic floor was troweled on over the concrete slab. Below, the original forge is on the right, and meets the addition at 90°. The small hip-roofed section covers the kitchen and part of the dining room.

Laying Flagstone Walks
Tips from a professional mason

by Joe Kenlan

Flagstone walks enhance the appearance of a yard or an entry. Above, a pathway with stones in place but not grouted; above left, the finished walk. A dry-laid path (above right) is less formal and rougher to walk on than a mortared walkway.

Flagstone is not a particular kind of stone, but a generic term. It designates any hard stone that fractures into broad, flat pieces suitable for paving walks, patios and floors—most commonly slate, limestone or sandstone (marble is often used for expensive installations). Most large masonry-supply stores carry flagging material; the particular type will depend largely on what is available locally. Sandstone and limestone can be found throughout the country, but the metamorphics—such as marble and slate—are limited for the most part to mountainous areas. Shipping stone is expensive, and if you order it from far away you can expect the cost of shipping to exceed the cost of the stone itself. For example, I can buy Tennessee crab orchard stone for $50 to $60 a ton where it is quarried. But it sells for $150 to $165 a ton here in North Carolina, just 300 miles away.

The most important characteristic of good flagstone is hardness. If you can crumble the edges of the stone with your hands it definitely won't stand up to traffic. Good flagstone also has to be flat and smooth so that tripping will not be a problem.

Flagstone can be laid over a bed of either concrete, sand or gravel, or directly in the earth. It lends itself to everything from formal entrances to meandering garden paths. Laying a stone walkway requires little expertise (skill will come with practice) and no exotic equipment. You'll need a rule and level to lay out and set the screeds (the boards that define the contours and level of the walk), a circular saw for cutting screeds and stakes, and a trowel and a brick hammer for setting and trimming the stones. Concrete can be mixed in a wheelbarrow with a hoe. An ordinary steel garden rake is handy for spreading the bed and for pulling the reinforcing wire up into the concrete. Another useful item is a pair of cutting pliers for trimming the wire. Grouting requires a stiff brush, large sponges, buckets and a piece of burlap for final cleaning.

Laying out a mortared walk—Most flagstone walks are functional rather than decorative, and should go from point A to point B as directly as possible because pedestrians are likely to do that whether you put the walkway there or not. Moderate curves for appearance are okay, but any large deviations should be justified by taking the walk around something, such as a tree, a bush or a boulder.

The walk should be wide enough for its expected traffic. A path around a little-used part of a yard might be only 16 in. wide, but a main entry requires a minimum width of 32 in. The busier the walk, the wider it should be.

Once the layout is decided on, mark it out on the ground using string for the straight sections and marking powder (mortar, flour or lime all work well) for the curves. Alternatively, you can cut its outline into the ground with a shovel. The layout width includes the finished width of the walk plus a 3-in. allowance for screed boards.

Digging the trench—Once the walk has been marked out, dig a 5-in. to 6-in. deep trench. This depth allows for a 3-in. or 4-in. reinforced concrete slab and the inch or two the stone top requires. Cut the bottom of the trench flat (a

Photos: Michael Yarborough

square shovel is useful here) and square up the sides. Unless you are working in solidly compacted earth (for example, an old existing walkway), thoroughly tamp the bottom of the trench by hand or with a power tamper.

In wet or filled areas or when working on uncertain subsoils, my helpers and I usually add at least 2 in. of crushed stone to the bed of the walkway to provide a firm base and to allow some drainage. If crushed stone is used, the trench must be deepened accordingly. Some areas may require more extensive measures, such as culverts, to provide drainage.

Next the screeds are set. These boards act as forms for the concrete pour, defining the shape and the finished height of the walk. We use 2x6 lumber for the straight sections because it is stiff enough to resist bowing under the pressure of the poured concrete. We use strips of plywood kerfed vertically for the curved sections (inset photo at right). It's faster and easier to kerf an entire sheet of plywood and then rip it into strips for curved screeds.

Unlike brick walks, stone walks cannot easily be crowned since the stones don't follow a regular pattern across the walkway. We usually try to set the screed boards even with or just above the adjoining grade with a slight slope (about ¼ in. per foot) to one side, so water will drain. This keeps the walk dry but doesn't interfere with lawnmowers or with the occasional traffic across the walk.

We prefer to work with stone that is between 1 in. and 1½ in. thick for mortared work. Anything thinner than ¾ in. may be knocked loose. Thicker stone costs more (stone is sold by weight) but is perfectly suitable otherwise. Using thicker stones here and there in a walkway makes screeding more of a problem because the mortar bed must be adjusted to accommodate the additional thickness.

Sometimes there will already be a set of steps at one end of the walk. In laying out the walk, remember that steps at either end of a walkway will determine its height since the surface of the walk becomes, in effect, another tread in the step and must be set the same distance down (or up) as the rest of the risers. Add another step if you have to to even things out. Since steps are virtually always set level side to side, set the walkway level at this point, and begin to slope it gradually about one tread width from the first step.

Stakes, either of wood or rebar, are used as needed to hold the screed boards in position, and earth is backfilled against the outsides of the boards to help hold them and resist bowing. To set the boards at the proper height, we either tack them to the stakes or tie them to the rebar in such a way that they can be easily removed later. They can be shimmed up using small stone wedges.

The screed boards are checked with a level to ensure sufficient drainage and with a rule to make sure the two sides are equidistant. This is particularly important when laying out curved sections. We lay out one side completely and then use the rule and level to set the other.

Once the screed boards are set and checked, we cover the bottom of the trench with polyeth-

After the walkway has been laid out and excavated, it is edged with screed boards to contain the pour and mark the finished level of the stones. Straight sections are done with 2x6s; curves (inset) are edged with strips of plywood kerfed vertically. Rebar braces and a polyethylene moisture barrier complete the preparations. Above, concrete is placed over the reinforcing wire. On this walkway, flagstones are being set directly into the wet concrete.

ylene. This limits moisture passage up into the walkway, which can cause unsightly efflorescence in the flagstone.

Pouring the concrete—On top of the plastic we roll out standard 4-in. by 4-in. slab reinforcing wire (or "hog-wire" fencing), cut it and fit it.

Now we go one of two ways. If the project is large we generally pour the whole walk in one pass from a truck, let it set up and then place the stone in a mortar bed on top. If the project is small enough to do in one day, we pour it and set the stone in one operation (photo above). The concrete mix (1 part cement to 3 parts sand, or 1 part cement to 2 parts sand to 3 parts gravel) should be fairly stiff, as for normal slab work.

Using the first method we essentially pour a concrete sidewalk, except that instead of

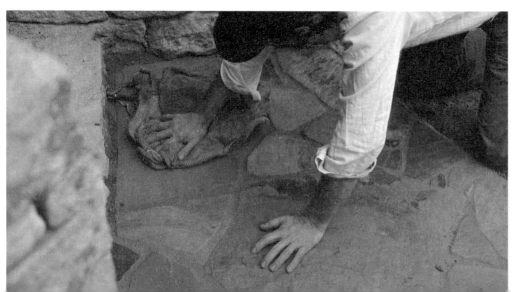

screeding off at the top of the boards as would normally be done, we use a notched 2x4 to screed off so that the finished top of the pour is 2 in. below the forms.

To make a screed guide we take a 2x4 that is 6 in. or so wider than the walk and screed boards and notch it 2 in. at both ends. The notched-out part of the board should be an inch or so narrower than the finished walk. When this board is pulled along the tops of the screed boards it smoothes off the concrete 2 in. below the finish height of the walk.

Next we pour the concrete to the approximate depth, pulling the reinforcing wire up into the concrete as we go, and screed it off. The surface is left rough so that the mortar bed for the stone will bond to it. As we pour, we remove any stakes that are inside the screed boards. The concrete should hold the screed boards in place at this point. If the screed boards are bulging anywhere, we add stakes as necessary to hold back the pressure of the pour.

Setting the stones—Once the slab has hardened (a day is usually sufficient), the stones can be set in a shallow mortar bed. Before starting to set them, it's helpful to look at the type of stone you'll be using and to decide if you want to lay it in a particular pattern. In our work, we usually let the stones decide the finished pattern. What we do emphasize is the relationship of the stones to each other so that we have evenly spaced, parallel joints. Using a brick hammer, we shape and fit each stone so it interlocks with its neighbors (sidebar, facing page). There is no structural reason for this, so many masons lay the stone randomly to save time. But we feel the result is well worth the extra effort. After all, the walk will be there a long time.

As in most of the other stonework we do, we work from the outside in, laying the two edges and then filling in the space between them. Edge pieces need to be large enough to keep them from being accidentally dislodged over time. In laying out the stone we first take stock of its natural shapes. Some stones break at sharp angles—others more gently—so we try to leave spaces that will conform to the stone we have.

The stones should be well washed before using them to rid them of quarry dirt and dust, which will interfere with the bonding. We like to keep them damp so that they don't suck moisture from the cement while they are being set.

The stone is set using a rich mortar (1 part cement to 2 parts sand) that is mixed to about the consistency of mortar used for laying block. (If you find that the stones "float" while you are setting them, use a slightly drier mix.) With this mix, pour a bed thick enough to bring the top of the stone slightly above the screed boards, and

Grouting. After the stones have been allowed to cure for a day, a slightly damp grout mix is worked into the joints with a stiff brush (top left). Then more mortar is worked over the entire surface of the walk with a wet sponge, forcing it into the joints (second from top). Subsequent spongings remove most of the excess mortar (third from top); a final rub with burlap (bottom left) polishes the stone.

begin to set the stone. Using a 2x set across the width of the walk as a guide, gently tap the stones down to finish height using the handle of a hammer or trowel.

As we work along, we also use the board to make sure that all the stones are in the same plane by laying it diagonally across the walk in several directions. This is important to prevent toe-catching irregularities.

Grouting—Once the stone has been set and allowed to cure for a day, we grout the joints with a mix of 1 part sand to 1 part portland cement. The grout is mixed slightly damp and is swept and worked into the joints with a stiff brush (photo facing page, top left). When the joints are well filled, additional mortar is placed on the surface of the walk and is worked into the joints using a large soft sponge. We use a generous amount of water on the sponge to make sure that the grout is wetted all the way down into the joint (photo second from top, facing page).

The sponge is gently worked over the tops of the stones to smooth the joints. The first pass leaves a slurry of cement over the surface of the walk that is gradually removed by subsequent spongings. We use plenty of clean water until all the visible mortar is washed from the surface of the stone and the joints are filled smoothly and uniformly (photo third from top, facing page).

Generally, three or four washings with the sponge and water are needed to clean the walk. One final cleaning, using a piece of damp burlap, polishes the stone and removes any remaining mortar film (bottom photo, facing page). Some masons use straw instead of burlap, but we don't like to use straw because it leaves debris in the joints.

If, after the walk has been cleaned and dried, there is still some film remaining on the surface of the stone, we let the walk set for several days (until the mortar changes color) and wash it down with a non-acid masonry cleaner. We test the wash on a scrap piece of stone to see what effect it will have. The wash and a rinse will take care of any remaining film. Rubber gloves and eye protection are essential.

In exterior work we generally do not apply a sealer, but if one is required we test it on several unlaid stones to see how it affects their color and surface. Sealers can cause some stones to become slippery when wet.

After the walk has been allowed to set thoroughly we remove the screed boards and backfill the edges with dirt.

Dry-laid walks—Stone paths without mortar are usually reserved for less formal entrances than the grouted type since they are less even and smooth (photo at right, p. 186). For dry-laid walks, we select stones at least 2 in. thick since the mass of the stone is what will anchor it in place. The stones can either be set individually with a moderate space between them (this method is useful if the stones vary a great deal in thickness), or by using a modified version of the concrete-set method described above. In the latter case we dig the trench as before, except that the depth is determined by the thickness of

Cutting flagstone

The surface of a stone can be smoothed out with a mallet and chisel.

Flagstones—whether slate, limestone or sandstone—are generally strongly layered, a characteristic that produces flat, thin pieces. But such stones resist attempts to cut them across their natural rifts. Consequently, in our work my helpers and I first look for a stone that nearly fills the space we are dealing with and then trim it by nibbling in from the edges with a brick hammer or similar tool. Attempts to take off large sections at once usually result in unpredictable breaks.

Undercutting the backside of the stone before working the face is helpful, as this provides a weak point to break against. Smoothing can be done by working the stone at an angle from the edge. Don't ever work toward the face of the stone because more often than not, you will break out a larger chip than you intended. For more precise cuts, use a skillsaw equipped with a masonry cutting blade. Score the stone about halfway through and then snap off the waste piece with a sharp hammer blow. Again, always work away from the face of the stone. If you have access to either a stone saw or a splitter you can of course use them. But a sharp brick hammer or hammer and chisel will usually suffice.

When sawing stone it is important to use plenty of water, even when using dry-cutting blades, to reduce dust. Ear protection and a dust mask should be worn, and eye protection is a must at all times.

On most flagstones, surface imperfections can be removed by chipping them off with a sharp thin-bladed chisel held at an angle of about 30° to the face of the stone (photo above). You simply flake off the undesired layer.   —J. K.

the stone to be used plus a 2-in. to 3-in. leveling bed of crushed rock and sand.

Next, the trench is tamped and crushed stone is added as required. Then we roll out polyethylene, tar paper or polyester cloth (which inhibits the growth of weeds and keeps the sand leveling bed from washing into the stone below), and lay a bed of sand on top of that. The sand is screeded, and the stones are set using more sand where needed to bring them to ground level. Stones should be laid tightly together to help them bind on each other. Then sand or fine dirt is spread on top and allowed to work down into the joints. We leave the walk covered with sand this way for as long as is practical so as much as possible works down between the stones.

Walks laid this way have a tendency to spread a little over time but we feel that this adds to their simple, rustic character. It does, however, make them less sure footing for spiked heels.

Stepping-stone paths—The most informal type of walk is the stepping-stone path, with the stones spaced just closely enough to accommodate footsteps. For this, thicker stones are preferred, and each stone is set individually. Stepping stones are usually set flush with grade and follow the contours of the ground.

For a typical path, we begin by laying out the entire walkway, testing occasionally to be sure that it will be comfortable to use. We try to emphasize parallel lines between adjacent stones. Then we scribe the outline of each stone on the ground with a trowel, remove the stone, dig out and smooth the bed and reset the stone, making sure that it sits firmly and does not rock. In muddy or soft soils, we excavate deeper and add a 1-in. to 2-in. thick layer of crushed stone beneath each stepping stone to stabilize it.   □

Stonemason Joe Kenlan lives in Pittsboro, N. C.

Fine Homebuilding
Editorial Staff, 1981-1988

Mark Alvarez
Fran Arminio
Joanne Kellar Bouknight
Ruth Dobsevage
Mark Feirer
Bruce Greenlaw
Kevin Ireton
Linda Kirk
Kenneth Lelen
Betsy Levine
Michael Litchfield
John Lively
Lori Marden
Lynn Meffert
Charles Miller
Debra Polakoff
Don Raney
Paul Roman
Tim Snyder
Paul Spring

Fine Homebuilding
Art Staff, 1981-1988

Frances Ashforth
Elizabeth Eaton
Deborah Fillion
Lee Hov
Betsy Levine
Chuck Lockhart
Michael Mandarano

Fine Homebuilding
Production Staff, 1981-1988

Claudia Applegate
Barbara Bahr
Lisa Carlson
Mark Coleman
Deborah Cooper
Kathleen Davis
David DeFeo
Dinah George
Barbara Hannah
Annette Hilty
Nancy Knapp
Margot Knorr
Gary Mancini
Robert Marsala
JoAnn Muir
Swapan Nandy
Cynthia Lee Nyitray
Ellen Olmsted
Priscilla Rollins
Thomas Sparano
Austin E. Starbird

Book jacket design:
Jeanne Criscola

Book copy editor:
Kathryn A. de Koster

Manufacturing coordinator:
Peggy Dutton

Production coordinator:
Deborah Fillion

Design director:
Roger Barnes

**Editorial director,
books & videos:**
John Kelsey